AN INTRODUCTION TO CANADIAN CRIMINAL PROCEDURE AND EVIDENCE

FOURTH EDITION

Joan Brockman
Simon Fraser University

V. Gordon Rose
Simon Fraser University

NELSON / EDUCATION

NELSON / EDUCATION

ISBN-13: 978-0-17-638055-7
ISBN-10: 0-17-638055-8

Contents

Preface xiii

INTRODUCTION: Perspectives on Criminal Procedure and Evidence, and the Constitutional Framework 1

Our Criminal Justice System–One of Many Forms of Social Control 1
Perspectives for Studying Criminal Procedure and Evidence 2
    Social Science of Law 4
    Social Science in Law 4
    The Insiders' Perspectives 5
Tensions in the System 6
    The Nature of the System–Conflict Versus Consensus 6
    Crime Control Versus Due Process 6
    Factual Guilt, Legal Guilt, and the Qualified Search for Truth 6
    Codification or the Common Law 7
    Standards/Principles or Rules 8
    The Adversarial System 8
Players in the System 9
    The Role of the Complainant/Victim 9
    The Police 10
    The Accused 10
    Witnesses, Front-Line Workers 10
    The Role of Crown Counsel and Defence Counsel 10
    The Role of *Amici Curiae* 10
    The Role of the Judges 11
    The Role of Jurors 11
    The Role of Interveners 11
Division of Powers under the Canadian Constitution 12
    Federal, Provincial and Municipal Levels 12
    Employment Relationships Between Governments and Prosecutors 13
The Federal Court of Canada 14
The *Canadian Charter of Rights and Freedoms* 14
    Section 1 15
    Section 24 16
Questions to Consider 16
Bibliography 17
Cases 20

PART I: THE PROCEDURES FOR BRINGING AN ACCUSED TO COURT, TRIALS, DISPOSITIONS, AND APPEALS

Chapter 1: Classification of Offences, Elections, and Jurisdiction of the Court 23

Chapter Objectives 23
Classification of Offences 23
    Indictable Offences 24
    Summary Conviction Offences 24
    Hybrid or Dual Offences 24
Reforming the Classification of Offences 24
Elections 26
    Overriding Elections and Re-Elections 28

Contents

Jurisdiction   28
   Time Limitations   29
   Territorial Limitations   31
      Waiver of Charges   31
      Change of Venue   31
The Effect of Delay on *Charter* Rights   33
   Pre-Charge Delay   33
   Post-Charge Delay   34
   Delay Between Verdict and Written Reasons   36
   Delay Between Conviction and Sentence   36
   Delay Caused by Appeals   37
Summary   37
Questions to Consider   38
Bibliography   38
Cases   40

Chapter 2: Compelling the Appearance of the Accused and Judicial Interim Release   43

Chapter Objectives   43
Compelling the Appearance of the Accused   43
   Appearance Notice   44
   Summons   45
   Arrest Without Warrant by a Private Citizen or a Peace Officer   45
   Arrest Without Warrant by a Peace Officer Only   46
      The Duty Not to Arrest   47
      Reasonable Grounds to Make an Arrest   47
      Entering a Private Dwelling to Make an Arrest   48
   Arrest With Warrant   49
   Arrest for Breach of Judicial Interim Release, With or Without Warrant   50
Charter Considerations   50
   The Meaning of Detention   50
      Investigative Detention   51
      Investigative Detention and Racial Profiling   52
   Arbitrary Detention–Section 9 of the *Charter*   52
   Right to Counsel—Section 10 of the *Charter*   53
Protection and Liability of Persons Enforcing the Law   54
Releasing a Suspect   54
Failure to Appear   56
Judicial Interim Release   56
   Section 469 Offences   57
   Offences Other than Section 469 Offences   57
   Forms of Judicial Interim Release   57
   Grounds for Judicial Interim Release   58
   Reverse Onus Provisions at the Show Cause Hearing   59
   Release Pending Appeal   60
Bail Variation   60
Bail Review   60
Section 525 Review   61
Estreatment and Forfeiture   61
Summary   62
Questions to Consider   63
Bibliography   63
Cases   65

Contents

Chapter 3: Informations and Indictments, Arraignment and Plea   69

Chapter Objectives   69
Information and Indictments   69
    The Historical Development of Prosecutions   69
    Informations   69
    Indictments   72
    Amending an Information or Indictment   74
Private Prosecutions Today   74
Discretion to Lay Charges and Charge Screening   75
Attorney General's Control Over Prosecutions   76
Direct Indictments   76
Control over Prosecutorial Discretion and Misconduct   77
Arraignment   78
Pleas   79
    Plea Bargaining   79
Summary   80
Questions to Consider   81
Bibliography   81
Cases   84

Chapter 4: Crown Disclosure and Preliminary Inquiry   85

Chapter Objectives   85
Crown Disclosure   85
    Particulars   88
    Pre-Trial Conference   88
    Disclosure Court   89
Defence Disclosure   89
Evidence in Possession of Defence Counsel   91
Procedure Before a Preliminary Inquiry   92
The Preliminary Inquiry   92
    Exclusion of the Public and Publication Bans   94
    An Absconding Accused   94
    Uses of Preliminary Inquiry   94
    The Abolition of the Preliminary Inquiry   95
Summary   96
Questions to Consider   97
Bibliography   97
Cases   99

Chapter 5: Juries and Procedure at Trial   101

Chapter Objectives   101
Jury Trials   101
    The Right to Trial by Judge and Jury   101
    Qualification and Selection of Jurors: The Out-of-Court Process   102
    Challenging the Array   103
    Selecting the Jurors: The In-Court Process   104
        The Effect of Pretrial Publicity and Notoriety on the Selection of Jurors   106
        The Impact of Systemic Racism on the Selection of Jurors   107
        Broadening the Challenge for Cause   109
    Talesmen–Running out of Potential Jurors   109
    Discharging a Juror   109

Contents

Defects in Selecting a Jury   110
Jury Unanimity   110
Directed Verdicts   110
Secrecy of Jury Proceedings   111
Social Science and Juries   111
Jurors' Comprehension of Evidence and Instructions   113
The Role of the Jurors   113
Jury Nullification   113
Should the Jury System be Abolished?   115
Voir Dires   115
Exclusion of the Public and Publication Bans   116
Procedure at Trial   118
Reasons for Judgment   118
Summary   119
Questions to Consider   120
Bibliography   121
Cases   123

Chapter 6: Sentencing and Appeals   127

Chapter Objectives   127
Alternative Measures   127
Sentencing Purpose, Objectives, and Principles   127
The Sentencing Hearing   128
Probation Reports   129
Victim Impact Statements   130
Previous Criminal Record   130
Time in Custody   131
Sentences and Other Orders   131
Absolute and Conditional Discharges   131
Suspended Sentence and Probation   132
Fines   133
Conditional Sentences of Imprisonment   134
Incarceration   134
Intermittent Sentences   135
Dangerous Offender Declaration   136
Restitution   136
Victim Fine Surcharge   136
Prohibitions   137
Sex Offender Registration   137
Proceeds of Crime and Forfeiture   137
Proceeds From Publications About Crimes   138
Appeals   138
Appeals–Procedure by Indictment   140
Appeals–Summary Conviction Procedure   142
Fresh Evidence on Appeal   143
The "Faint-Hope" Clause   143
Applications for Ministerial Review—Miscarriages of Justice   144
Summary   145
Questions to Consider   146
Bibliography   147
Cases   151

Contents

PART II: GATHERING EVIDENCE AND ITS ADMISSIBILITY

Chapter 7: Evidence That is Illegally or Improperly Obtained   157

Chapter Objectives   157
The Common Law: Illegally Obtained Evidence   157
    The Fruit of the Poisonous Tree in the United States   157
    The Canadian *Bill of Rights*   158
    Suggestions for Law Reform Prior to the *Charter*   158
Section 24 of the *Charter*   159
    Historical Development of Section 24   159
    Section 24(1) Remedies   159
        Who Qualifies for Remedies, and Under What Circumstances?   159
        What is a Court of Competent Jurisdiction?   160
        What Remedies Can the Court Order?   160
    Section 24(2)   161
        The Rationale for Applying Section 24(2)   161
        Who Has What Onus?   162
        "Obtained in a Manner"   162
        Bringing "the Administration of Justice Into Disrepute"   163
        The *Grant* Framework   164
Summary   166
Questions to Consider   166
Bibliography   166
Cases   168

Chapter 8: Search and Seizure   171

Chapter Objectives   171
Section 8 of the *Charter*   171
    Excluding Evidence Under Section 24(2)   175
Search by Warrant–*Criminal Code*   175
    Section 487 Search Warrant   175
    DNA and Bodily Impression Warrants: Sections 487.04 to 487.091   177
    Other Warrants under the *Criminal Code*   178
    Provisions Regarding Search Warrants   180
    Production Orders   181
Searching a Lawyer's Office   181
Search by Warrant–Non-Code Provisions   181
Warrantless Searches to Obtain Information to Support an Application for a Search Warrant   182
Searching Motor Vehicles   184
Search Without Warrant–Legislation   184
Search Without Warrant–Common Law   185
    Search by Sniffer Police Dogs   185
    Search During Investigative Detention   185
    Search as an Incident to Arrest   185
    Plain View   187
    By Consent   188
Summary   189
Questions to Consider   190
Bibliography   190
Cases   193

Contents

Chapter 9: Electronic Surveillance and the Interception of Private Communications   197

    Chapter Objectives   197
    History of Electronic Surveillance   197
    The Legislative Scheme: What it Covers   198
        Offences for Which Authorization is Available   199
        The Meaning of "Intercept" and "Means of Interception"   199
        Private Communication   200
    The Legislative Scheme: Protecting Privacy   201
        Offences Designated to Protect Privacy   201
        Damages   203
        Notification to Target   203
        Reporting   203
    Judicial Authorization to Intercept   203
        Investigative Necessity   204
        Interception of Communications by Lawyers   205
        Renewals   205
        Emergency Authorizations, Section 188   205
    Consent Interceptions   206
        Judicially Authorized Interceptions by Consent, Section 184.2   206
        Unauthorized Interception to Prevent Bodily Harm, Section 184.1   207
        Unauthorized Emergency Interceptions, Section 184.4   207
    Camera Surveillance, Section 487.01 and Part VI   207
    The Admission of Electronic Surveillance as Evidence   208
        Notice, Section 189(5)   208
    Review of Authorizations   209
    Summary   210
    Questions to Consider   211
    Bibliography   212
    Cases   214

Chapter 10: Admissions and Confessions   217

    Chapter Objectives   217
    The Principle Against Self-Incrimination   217
    Admissions   217
        Formal Admissions   217
        Informal Admissions   218
    Confessions   218
        A Historical Note and Rationale   219
        The Confession Rule—Voluntariness   219
        Operating Mind   220
        Persons In Authority   221
        The Relationship Between the Confession Rule and Section 7   222
        Derived Confessions Rule   223
        Other Derivative Evidence   223
    The Right to Retain and Instruct Counsel, Section 10(b)   224
        Duties on Police Officers Regarding the Right to Counsel   225
        Informational Rights Under Section 10(b) of the *Charter*   225
        The Right to Counsel When There is a Change in Jeopardy   226
        Waiver   227
        The *Grant* Analysis for Violation of the Accused's Right to Counsel   227

Contents

The Right to Remain Silent, Section 7   227
  Jail-House Informants (In-Custody Informers)   228
  In-Custody Admissions to Undercover Police Officers   229
  Out-of-Custody Admissions to Undercover Police Officers   230
The Privilege Against Testimonial Self-Incrimination   232
  Common Law   232
  Canada Evidence Act, Section 5(2)   232
  Section 13 of the *Charter*   232
  Limited Derivative-Use Immunity   234
  Can Witnesses be Exempted from Testifying?   235
  Anti-Terrorism Investigative Hearings   236
  Testifying in Non-Criminal Matters   236
  Non-Testimonial Statements Compelled by Statute   236
Summary   238
Questions to Consider   239
Bibliography   240
Cases   243

Chapter 11: Types of Evidence   247

Chapter Objectives   247
Introduction to Evidence   247
  The Burden (or Onus) of Proof and Standard of Proof   247
  What is Evidence?   248
  A Question of Relevance   249
  Why Have Rules of Evidence?   249
  Where are the Rules of Evidence?   250
The Law Reform Commission of Canada   251
Evidence: Getting it into Court   251
  Testimonial or *Viva Voce* evidence   251
  Real Evidence   252
  Commissioned Evidence   252
Evidence: What is it used for?   252
  Demonstrative and Illustrative Evidence   252
  Direct Evidence   253
  Circumstantial Evidence   253
Witnesses   254
  The Oath or Affirmation   254
  Compellability   255
  Competence   256
    Spouses   256
  Corroboration   257
  Shielding Rape Victims   258
  Assistance/ Protection of Child Witnesses, Witnesses with Disabilities and Other Witnesses   260
    Support Person for Witnesses–Section 486.1   261
    Testimony Outside the Courtroom or Behind a Screen–Section 486.2   261
    Accused Not Allowed to Cross-Examine some Witnesses–Section 486.3   262
    Videotaped Evidence of Victim or Witness under 18—Section 715.1   262
Summary   263
Questions to Consider   264
Bibliography   265
Cases   267

Contents

Chapter 12: Exclusionary Rules   269

Chapter Objectives   269
The Exclusion of Evidence by Privilege or Immunity   269
     Solicitor-Client Privilege   269
     Spousal Communications   271
     Extending Privilege With Wigmore's Criteria   272
     Public Interest Immunity   272
       Police Informers' Privilege   274
The Privacy of Complainants   274
     Complainants' Records–the Mills Regime   274
     Examination of Complainant by Defence Experts   275
The Exclusion of Hearsay   276
     The Principled Exception to the Hearsay Rule   277
Character Evidence   280
Credibility   282
     Reasonable Doubt and Credibility   283
     Prior Consistent Statements   283
     Prior Inconsistent Statements   284
Similar Fact Evidence   284
Summary   285
Questions to Consider   287
Bibliography   287
Cases   290

Chapter 13: Judicial Notice, Secondary Sources, and Opinion Evidence   293

Chapter Objectives   293
Introduction   293
     Tacit Assumptions--Bias in Adjudication and Interpretation of the Law   294
Judicial Notice of Law   295
Judicial Notice of Facts   295
     Rationale and Implications   296
     Adjudicative (or Social) Facts   297
     Social Authority (Legislative Facts)   298
     Social Framework   299
     Can a Court Rely on Secondary Sources and Expert Evidence to Assist in Taking Judicial Notice?   299
Secondary Sources   299
The Use of *Facta* to introduce Social Science Research   300
Opinion Evidence   301
     Expert Opinion Evidence   302
     Opinion Based on Hearsay   306
     Oath Helping   307
Summary   308
Questions to Consider   308
Bibliography   309
Cases   310

Appendix A: Commissions of Inquiry and Studies into the Criminal Justice System in Canada   313
Appendix B: Reading a Case and Legal Research   317
Appendix C: True Canadian Crime and Other Misconduct   325
Glossary   335
Subject Index   347
Index of Cases   359

Preface

Since the second and third editions of this text, we have discovered that Criminology students are carting our book off to law schools, sharing it with their classmates, and then using it as a research tool during articles and practice. Lawyers, returning to criminal law after years away from it, are also finding it to be a good refresher course on criminal procedure and evidence in Canada. Although our intentions were, and still are, to provide a textbook for students in the social sciences, we welcome this additional audience and appreciate feedback from them.

The book is laid out such that Part I can be used for a course on criminal procedure, covering how an accused is compelled to attend court, judicial interim release, informations and indictments, Crown disclosure, preliminary inquiries, the conduct of a trial, jury trials, sentencing, and possible avenues of appeal following a trial. Part II can be used for a course on evidence in criminal trials. It examines the gathering of evidence and rules of evidence that determine what the trier of fact (the jury, or the judge if there is no jury) can consider in coming to a decision in a criminal prosecution. There are separate chapters on search and seizure, electronic surveillance, admissions and confessions, and types of evidence and exclusionary rules. Case examples of the effect of the *Canadian Charter of Rights and Freedoms* on the rules of procedure and evidence are covered throughout the text.

The Introduction, which is relevant to both Parts, identifies a number of tensions in the criminal justice system, and provides a basis for a critical assessment the system. Social science research is increasingly exposing problems and fallacies in our criminal justice system, and this text provides a framework within which social science research can be discussed, and in some cases used in the courtroom by lawyers and experts.

There are many studies and commentaries concerning what is wrong with the criminal justice system and how it might be improved. The Law Reform Commission of Canada was formed in 1972, in part to study and reform the system. The Commission was abolished in 1992, and reintroduced as the Law Commission of Canada in 1997 which was in turn again abolished in 2007. Criticisms of the criminal justice system have also come from other commissions, committees, inquiries, and *ad hoc* studies of sexist and racial bias within our legal system and more specifically, within our criminal justice system. Appendix A contains a list of commissions of inquiry and studies into the criminal justice system in Canada. An increasing number of these critiques have to do with wrongful convictions and miscarriages of justice. These works, as well as Law Reform Commission of Canada reports referred to throughout this text, are useful resources when considering reforms to our present system.

This text contains several other features to assist both instructors and students in their study of criminal procedure and evidence. Each chapter contains a series of "Questions to Consider," which are useful for guiding the reader through a review of the chapter. A bibliography and list of cases at the end of each chapter provide instructors with references for greater detail on a specific topic, and provide students with resources if they are writing a paper on one of the topics raised in the chapter. The bibliography contains a combination of both legal and social science sources so that the reader can consult more detailed examinations of the topics discussed in the text. Appendix B contains a brief discussion on how to read a case and conduct legal research. Many students will have already been exposed to these techniques through other courses; for them the Appendix should provide a useful summary. Appendix C provides a brief description of true Canadian crime and other misconduct. For some books we have also provided case citations so that the reader can look up the court's judgment in the case. Instructor or students can use these books and cases to illustrate that there is "more than one side" to any criminal event.

New to This Edition

As anyone who works in the criminal justice system or who follows news from the Supreme Court of Canada knows, criminal procedure and evidence is an area of the law that can change every week, when Supreme Court of Canada releases its judgments. In addition, Parliament has been quite active since the third edition of this book in making changes to the *Criminal Code*. This fourth edition reflects these changes, including:

Legislative changes

- changes to a number of offences to make them hybrid offences
- fines for summary conviction offences increase from two to five thousand dollars
- increase in maximum terms of imprisonment to eighteen months for some summary conviction offences; others remain at six months
- an increase in the number of offences that stipulate a reverse onus for bail
- a reworking of sobriety tests for driving while intoxicated by drugs or alcohol
- a reduction in the number of offences that qualify for conditional sentences

Supreme Court of Canada decisions:

- confirmation of a flexible approach under section 11(d) of the *Charter* (*Godin*)
- clarification on the Crown proceeding by indictment in summary conviction offences when the time limitation to proceed by summary conviction has expired (*Dudley*)
- clarification on the concept of detention (*Grant*)
- a narrowing of the concept of "malice" for malicious prosecution law suits (*Miazga v. Kvello Estate*)
- further clarification of Crown disclosure requirements (*McNeil*)
- jury nullification (*Krieger*)
- remedies under section 1 of the *Charter* (*Bjelland*)
- a rejection of the *Collins/Stillman* test in favour of a new framework under section 24(2) of the *Charter* (*Grant*)
- confirmation that the approach to privacy under section 8 of the *Charter* is normative, not descriptive or societal (*Patrick*)
- there is a common law power for police to use sniffer dogs provided they have reasonable suspicion that there is evidence of a crime (*Kang-Brown*)
- the police cannot do full scale searches of schools to look for drugs; they require reasonable suspicion for specific searches (*A.M.*)
- the impact of promises on the voluntariness of and accused's statement (*Spencer*)
- the right of a detainee to remain silent (*Singh*)

We have included Supreme Court of Canada decisions up until January 1, 2010.

PREFACE

Background of Authors

We arrived at writing this text from two similar backgrounds, but by two different paths. As a graduate student in sociology at the University of Alberta in the mid-1970s, Joan wanted to conduct research on plea bargaining for her doctoral dissertation. Since the Department of Sociology did not teach law courses, she sought enrolment in a course on Criminal Law and Procedure at the Faculty of Law. The exclusivity of the educational process convinced her that if she were to ever be able to conduct research on the justice system she would have to become a lawyer. So she did.

Joan attended the Faculty of Law at the University of Calgary, graduated with an LL.B., and articled at a firm located in a small shopping mall. The day after her call to the Alberta Bar in 1981, she moved to Vancouver and earned her LL.M. at the University of British Columbia. She articled with the federal Department of Justice, gaining experience in criminal prosecutions, civil litigation, tax litigation, and property law. Following her call to the Bar in British Columbia in 1983, she spent a number of years moving back and forth between teaching at the School of Criminology at Simon Fraser University and working as legal counsel for the Department of Justice. She became a faculty member at the School of Criminology in 1987 and is presently a full professor at the School. From 1998-2006, she was also a part-time Commissioner with the British Columbia Securities Commission. As a Commissioner she sat on three-member panels that heard cases alleging violations of securities law and appeals from decisions by the self-regulating organizations that discipline their members who work in the securities area. As a Commissioner, she was also a member of the Board of Directors for the Securities Commission. The Board establishes regulatory policy and rules that govern the securities market and oversees the employees of the Commission.

Gordon followed a different route. After completing his undergraduate degree in psychology at the University of Calgary, he applied to law school. Although having no preconceptions about the study of law, it quickly became apparent to Gordon that the only area of law that interested him was criminal. Following graduation with an LL.B. from the University of Victoria in 1980, Gordon attended the University of Toronto, obtaining an LL.M. in 1981. His graduate studies focused on criminal law, and his thesis topic was "Parties to an Offence," which was later published (1982) by Carswell in their Criminal Law Series.

Gordon then articled with the Department of Justice in Vancouver and subsequently joined the Criminal Prosecutions section of that office. Following approximately 10 years in federal criminal prosecutions, Gordon returned to university, initially studying mathematical statistics. In 1995, he entered the graduate program in law and psychology at Simon Fraser University, where he received his M.A. and Ph.D., studying criminal juries, and devising empirical tests of all the legal folklore and "conventional wisdom" he encountered in 15 years of legal education and practice. Gordon was awarded a post-doctoral fellowship by the Social Sciences and Humanities Research Council of Canada to conduct a course of jury research. He ultimately declined that award to accept a teaching position in the Department of Psychology at Simon Fraser University. Gordon has written about the psychology of various aspects of criminal procedure and evidence, and has given expert opinion evidence in court regarding juries.

In 1983, we teamed up to teach CRIM 330 (Criminal Procedure and Evidence) at Simon Fraser University. Then, we were focusing primarily on the mechanics of the legal system. Since then, our respective experiences have led us to an appreciation of some problems and grey areas in the legal system. As well, the impact of the revised Canadian constitution has greatly changed the criminal trial process since we first taught the course, introducing new levels of complexity. This text attempts to cover the mechanics of a technical and essential area of the law, but also to reflect some of the concerns and qualifications suggested or supported by the social sciences.

NOTE TO INSTRUCTORS

When Joan uses this text she likes to have students read relevant cases along with the text. Here are some suggestions for required readings. All of the cases are available through the Supreme Court of Canada website and other data bases such as Quicklaw, Criminal Source, and Criminal Spectrum:

Chapter 1 Classification of Offences, Elections, and Jurisdiction of the Court
R. v. MacDougall, [1998] 3 S.C.R. 45.

Chapter 2 Compelling the Appearance of the Accused and Judicial Interim Release
R. v. Buhay, [2003] 1 S.C.R. 631.
R. v. Mann, [2004] 3 S.C.R. 59.
R. v. Grant, 2009 SCC 32 (on concept of "detention").

Chapter 3 Informations and Indictments, Arraignment and Plea
R. v. Regan, [2002] 1 S.C.R. 297.

Chapter 4 Crown Disclosure and the Preliminary Inquiry
R. v. Bjelland, 2009 SCC 38.

Chapter 5 Juries and Procedure at Trial
R. v. Find, [2001] 1 S.C.R. 863.

Chapter 6 Sentencing and Appeals
R. v. Gladue, [1999] 1 S.C.R. 688.

Chapter 7 Evidence That is Illegally or Improperly Obtained
R. v. Goldhart, [1996] 2 S.C.R. 463.
R. v. Grant, 2009 SCC 32 (on the framework under section 24(2) of the *Charter*)

Chapter 8 Search and Seizure
R. v. Tessling, [2004] 3 S.C.R. 432.
R. v. Mann, [2004] 3 S.C.R. 59.
R. v. Patrick, [2009] 1 S.C.R. 579.

Chapter 9 Electronic Surveillance and Interception of Private Communications
R. v. Araujo, [2000] 2 S.C.R. 992.
R. v. Pires; R. v. Lising, 2005 SCC 66.

Chapter 10 Admissions and Confessions
R. v. Oickle, [2000] 2 S.C.R. 3.
R. v. Spencer, [2007] 1 S.C.R. 500.

Chapter 11 Types of Evidence
R. v. Turcotte, [2005] 2 S.C.R. 519.
R. v. Nikolovski, [1996] 3 S.C.R. 1197.

Chapter 12 Exclusionary Rules
Smith v. Jones, [1999] 1 S.C.R. 455.
R. v. Brown, [2002] 2 S.C.R. 185.

Chapter 13 Judicial Notice, Opinion Evidence, and Secondary Sources
R. v. J.L.J., [2000] 2 S.C.R. 600.

PREFACE

February 7, 2010.

Joan Brockman
School of Criminology
Simon Fraser University
Burnaby, BC V5A 1S6
brockman@sfu.ca

V. Gordon Rose
Department of Psychology
Simon Fraser University
Burnaby, BC V5A 1S6
vgrose@sfu.ca

INTRODUCTION: Perspectives on Criminal Procedure and Evidence and the Constitutional Framework

This Introduction provides a brief overview of the criminal justice system and other legal systems that might be engaged to react to unwanted behaviour. It examines how social science research can be used to answer questions about criminal procedure and evidence, and discusses different perspectives which can be employed to assess issues raised about the criminal justice system. It then examines the constitutional framework of our criminal justice system, including the *Canadian Charter of Rights and Freedoms*. The material in the Introduction is relevant to the discussions in both Parts I and II of this text.

OUR CRIMINAL JUSTICE SYSTEM: ONE OF MANY FORMS OF SOCIAL CONTROL

The criminal justice system is part of a larger legal system. Here is an example that illustrates how the different components of the legal system might characterize the same behaviour in different ways. A professor, a member of a provincial College of Physicians and Surgeons, teaches at a university. The professor sexually assaults a student. The student could decide to do nothing, or take one or more of the following actions:

> 1) file a complaint with the police, in order to commence criminal proceedings against the professor (thereby initiating a process which may result in the behaviour being characterized as a crime);
> 2) sue the professor civilly, in order to recover damages for harm suffered as a result of the assault (this civil suit could result in the behaviour being treated as a tort, requiring compensation);
> 3) file a sexual harassment complaint against the professor at the university, under the university's internal sexual harassment or human rights policy, resulting in the professor being disciplined through the university's internal policies;
> 4) make a complaint against the professor under provincial human rights legislation (sexual harassment is a violation of human rights);
> 5) file a complaint with the College of Physicians and Surgeons, which might result in the professor being expelled from the College, thus precluding the professor from practising medicine in the province; and
> 6) seek compensation through a provincial criminal injuries compensation scheme.

Other actions could be commenced because of the incident:

> 1) The College of Physicians and Surgeons might set up a Committee to investigate, more generally, the sexual misconduct of physicians.
> 2) The president of the university might decide that the department within which the professor teaches is so corrupt that a more widespread investigation or inquiry is needed.
> 3) The government might set up a public inquiry to evaluate the College of Physicians and Surgeons self-regulating powers.

Introduction

A similar set of agencies may be involved if, for example, a lawyer misappropriates funds held in trust for a client. The client might try to do one or more of the following: 1) complain to the police, who may start a criminal investigation; 2) sue the lawyer civilly to recover the money; or 3) complain to the provincial Law Society. These are formal legal means of resolving disputes arising out of the misconduct. Again, more general investigatory committees could be established. The client could also try to negotiate the return of the money without involving any of the formal agencies. Another example, securities violations, are more complicated because they can involve not only self-regulatory bodies (such as the Investment Industry Regulatory Organization of Canada and the Mutual Funds Dealers Association of Canada), but also provincial Securities Commissions (government agencies set up to oversee the securities industry, including the self-regulatory bodies). While the various agencies often coordinate investigations, it is possible for someone to be investigated by multiple agencies in one province, and by agencies in more than one province. Governments might also establish inquiries where there are allegations of widespread problems. For example, numerous federal and provincial committees have examined the merits of having a National Securities Commission.

One might contrast these examples with the situation involving two individuals who pass each other in the street one evening. Suppose one assaults the other and steals her leather jacket. If issues surrounding the assault and theft are not resolved between the parties, and if the victim can identify her assailant, she can complain to the police and criminal proceedings may be commenced. She also has the option of suing civilly for damages. This usually occurs only if the assailant has some money or assets, such that the plaintiff can collect on a judgment, although on occasion civil suits are commenced in order to make a point, with little or no expectation of collecting on a judgment. The main difference between this dispute and the earlier examples is that there are no administrative or quasi-judicial bodies (like the College of Physicians and Surgeons or the Law Society) having authority over this assailant and how he conducts himself on the street.

An investigation may be directed at aspects of the criminal justice system itself (see Appendix A for some examples). Other investigations might be conducted by a Coroner if there is a death, or the Fire Marshall in the case of a fire. The criminal justice system is only one of a number of different avenues of redress, and any of the above non-criminal proceedings can occur when "criminal" conduct takes place. Often the actions of the victim, the police, or some other investigating person or body will determine whether the perpetrator enters the criminal justice system. It is important to remember that the same behaviour may be characterized differently, depending on a number of factors.

These various systems are not mutually exclusive, and often the commencement of an action in one system causes problems within another. For example, a person might be compelled to attend and testify at an inquiry, and may at some later date be charged with an offence related to the same subject matter. Some of the problems surrounding these overlapping systems are dealt within Part II of this text (see also Murdoch and Brockman 2001).

PERSPECTIVES FOR STUDYING CRIMINAL PROCEDURE AND EVIDENCE

There are a number of overlapping theoretical perspectives or frameworks that can be used in studying criminal procedure and evidence, both within the context of criminology specifically, and the social sciences in general. A perspective, framework, or theoretical position has the effect of focussing one's thoughts in a certain direction, and in doing so tends to filter out other perspectives. The law itself is one of these filtering mechanisms, used largely by lawyers and judges. The police have their own filtering mechanisms. Everyone who examines any aspect of the criminal justice

2

system does it from a particular perspective. It is often the case that those who work within the system acquire a particularly narrow, or perhaps even jaded, view of how the system works. One of the arguments in favour of jury trials has been expressed this way:

> The horrible thing about all legal officials, even the best, about all judges, magistrates, barristers, detectives and policemen [*sic.*], is not that they are wicked (some of them are good), not that they are stupid, several of them are quite intelligent, it is simply that they have gotten used to it. Strictly, they do not see the prisoner in the dock; all they see is the usual man in the usual place. They don't see the awful court of judgement; they only see their own workshop (Chesterton 1915, 50).

Convinced that they have arrested the right person, police officers may consciously or unconsciously construct a case against the accused, despite evidence which would point to another perpetrator, or to the innocence of the accused. Such **tunnel vision** has been identified as one of the causes of wrongful conviction (Federal-Provincial-Territorial Heads, 2004). This is clearly demonstrated in a number of the true crime books referred to in Appendix C to this text, such as, Michael Harris, *Justice Denied: The Law Versus Donald Marshall*, Carl Karp and Cecil Rosner, *When Justice Fails: The David Milgaard Story,* Kirk Makin, *Redrum the Innocent*, and Julian Scher, *'Until You are Dead.'* For example, Guy Paul Morin, whose conviction was questioned by Makin, was exonerated in January of 1995 after DNA testing demonstrated that he was not the person who sexually assaulted Christine Jessop in Queensville, Ontario in 1984.

The public, police, lawyers, and other authorities may also work the opposite injustice, allowing crimes to continue in spite of the widespread knowledge that they are occurring; see Michael Harris, *Unholy Orders: Tragedy at Mount Cashel*. Other crimes may remain unprosecuted for other reasons; see Lisa Priest, *Conspiracy of Silence*, Bridget Moran, *Judgement at Stoney Creek*, and Warren Goulding, *Just Another Indian: A Serial Killer and Canada's Indifference*, described in Appendix C. Sometimes victims of crime do not want to bother with invoking the criminal justice system. For example, a corporation, or the police if they are contacted, may decide that it's not worth the time and effort to investigate internal theft, even if it is in the tens or hundreds of thousands of dollars.

Perspectives have a great deal to do with what we discover about the criminal justice system, and with court decisions. Both social science and law are ways of interpreting the world and arriving at some form of "truth" about what has occurred. The questions we ask and the perspectives we take are crucial in determining what our results will be. Questions which remain unasked are as important as questions that are asked.

The same perspective does not necessarily yield the same result, when different people re-examine the same event. Rui-Wen Pan, accused of killing his former girlfriend Selina Shen in Ontario, went through three jury trials for murder. At his first trial, eleven out of the twelve jurors were prepared to convict, but since the decision had to be unanimous, a new trial was ordered. At his second trial, all but one juror wanted to acquit him, and again a new trial was ordered. At his third trial, he was convicted; see Doug Clark, *Unkindest Cut: The Torso Murder of Selina Shen,* and Nick Pron and Kevin Donovan, *Crime Story: The Hunt for the "Body Parts" Killer* (Appendix C). Further, we can never underestimate the effect of the individual actors who play a role in the system. At Pan's first two trials there were male prosecutors; two women prosecuted at the third trial. Pan testified at his second and third trial, and Clark is of the view that Pan was pushed over the edge by the female prosecutor, and probably convicted himself as a result. Obviously, there were many other factors, and each trial had a different combination of twelve jurors. Both the Ontario Court of Appeal and the Supreme Court of Canada dismissed Pan's appeals.

The two books on Pan's trials are also interesting because they give two different perspectives of the same case. Clark seems to identify more with how a police officer might view the case. As a reporter, Clark had covered many police investigations and was subsequently awarded an Ontario Provincial Police Commissioner's commendation in 1984. Pron and Donovan are two investigative reporters who tried to capture Pan's side of the story, and they became as much a part of the story as were Shen and Pan. They comment, "Thirty-six jurors and only two-thirds voted guilty. Maybe Pan didn't do it" (350).

Social Science of Law

The social science of law (in this case, the law of procedure and evidence) can be studied from a number of different criminological or social science perspectives–for example, the sociology of law, or the psychology of law. It is the study of how law works from the social scientist's (or, as some would say, the outsider's) perspective. The analysis might be descriptive, empirical, analytical, experimental, critical, feminist, and so on. It can take place at the micro- or macro-level. It may be functional, conflict-oriented, critical, feminist, liberal, Marxist, or post-modern (see for example, Boyd 1986; Comack and Brickey 1991; Burtch 2003; Cotterrell 1992; Hinch 1994; Ratner and McMullan 1987).

Such research is often used in policy planning and law reform. So, social scientists might determine through experimentation that there is no significant difference between the way twelve jurors deliberate and the way six jurors deliberate. This finding could be used as an argument in Parliament to change the twelve-person jury to a six-person jury.

Social Science in Law

Law is, to a large extent, based on assumptions of fact. Such "facts" may be based on myths or stereotypes held by judges. They may be true or false. Students of human behaviour often try to uncover the assumptions of fact that underlie the law. There is nothing new in this approach to the law. In 1881, Oliver Wendell Holmes touched on this aspect of judicial reasoning:

> The life of law has not been logic: it has been experience. The felt necessities of the time, the prevalent moral and political theories, intuitions of public policy, avowed or unconscious, even the prejudices which judges share with their fellow-men, have had a good deal more to do than the syllogism in determining the rules by which men should be governed (1881, 1; quoted by L'Heureux-Dube, J. in *Symes* 807).

This perspective recognizes that the law is not a self-contained system; rather, the sources that judges use to interpret or "find" the law go far beyond the traditional legal sources, to include what might be considered facts from the realm of common sense or social science.

The "social science in law" perspective looks at the use of social science research as evidence in court. In 1942, Kenneth Davis distinguished between the use of social science research as evidence to assist a judge in determining questions of law (legislative facts), and social science research used to assist the judge or jury in determining the facts in dispute between the parties (adjudicative facts). In the mid-1980s, John Monahan and Laurens Walker came to the conclusion that Davis's model had outlived its usefulness, and that it made more sense to examine social science research used as evidence in court as falling into three categories:

(1) social authority,
(2) social fact, and
(3) social framework.

According to Monahan and Walker, social authority is social science research used to determine (create or interpret) the law. As Walker and Monahan paraphrase Holmes, "what begin as questions of fact often become, in time, questions of law." For example, the definition of discrimination is interpreted differently when judges have a better understanding of the perspective of someone who has been sexually harassed (see Brockman 1992). Social authority is similar to legislative facts.

Social facts (or adjudicative facts) are those facts which might assist the trier of fact in deciding a specific issue before the court. For example, the accused might try to introduce evidence of the results of a social science survey showing that a certain performance was not "immoral, indecent or obscene" under section 167 of the *Criminal Code*.

According to Monahan and Walker, social science research that assists the judge or jury in understanding the broader social or psychological context, in which a complainant or an accused find themselves, is social framework evidence. For example, if the trier of fact does not understand how battered women or abused children generally respond to crimes against them, social science research may be necessary for the trier of fact to better understand how victims might react.

In some respects, the use of social science research leaves one in a somewhat conservative position, as one has to "buy into the system" in order to put social science research to use. There is much debate as to whether this is a successful way of furthering the rights of those who are disadvantaged (see for example, Mandel, 1994). We will return to social science in law in Chapter 13.

The Insiders' Perspectives

Another approach is to examine the system from insiders' perspectives. How do each of the various groups of actors see the system in operation? One of the perspectives, which some insiders believe to be the only perspective, is that of the legal actors in the system. Lawyers and judges take over the case and determine its outcome, sometimes with little regard to what the accused, or perhaps the victim, might think or want to happen. This is discussed further in the section "Players in the System."

Much of this text is devoted to law from the legal perspective of lawyers and judges. As you read about the cases in this text, think about some of the other perspectives from which one could analyze the cases. There are many perspectives other than the legal one: that of the accused, the victim, family, friends, front line workers, the police, expert witnesses, other witnesses, jurors, and so on. Observers of the process–journalists, friends, or other members of the public–have yet another perspective. The lawyers' perspective–what the rules are and how they work in practice–may result in the construction of a case which bears little resemblance to what either the accused or the victim think happened. The legal perspective is so dominant in our criminal justice system that a judge can refuse to accept a plea of guilty and force the accused to go through a trial. To what extent do the rules of procedure and evidence ignore the stories that victims and accused persons have to tell? To what extent are their versions of reality distorted in the legal system? How might this distortion be reduced?

TENSIONS IN THE SYSTEM

There are persistent tensions, both within the criminal justice system, and among those who study it or try to reform it. This section identifies a number of those tensions, and you should consider them as you work your way through this text.

The Nature of the System–Conflict Versus Consensus

Why do we have criminal law? What purposes does it serve? Who does it really apply to? Are our criminal laws the result of a consensus in society, or do they serve the purposes of one class or group of people at the expense of another? Are physicians who commit medicare (or OHIP) fraud treated differently than people who commit welfare fraud (for example, see Brockman 2010b, and Mosher and Hermer 2010). Some of the reports in Appendix A conclude that the law and its application have a disproportionate impact on the working poor and minority groups. In a research study on how we criminalize some violence, Comack and Balfour examine two contradictory versions of law–"law as a fair and impartial arbiter of social conflicts; and law as one of the sites in society that reproduces gender, race and class inequalities" (2004, 10). These issues are more directly the subjects of other texts, but you should not forget their importance as you study criminal procedure and evidence.

Crime Control Versus Due Process

Discussions about whether the purpose of the criminal justice system is to convict the guilty or to ensure that the innocent are not convicted are often couched in terms of whether the system was designed with crime control or due process in mind. Obviously, the answer is both. We are constantly hearing criticisms, however, that one goal is dominating at the expense of the other. The Law Reform Commission of Canada identified justice (or fairness) as the primary concern of the criminal justice system, but also recognized that this raises the question of "fairness for whom?" (1988, 15).

Factual Guilt, Legal Guilt and the Qualified Search for Truth

The criminal justice system is concerned with legal guilt, not factual guilt. It requires that the proper procedures be followed, and that only evidence which is admissible under strict rules be used to decide whether a person is to be found guilty (legally speaking). This may lead to a decision which may not coincide with factual guilt, that is, whether the person actually engaged in the alleged behaviour. There are undoubtedly many cases in which those who are factually guilty are acquitted and escape criminal sanction. This becomes obvious to the public when members of the press report that judges have excluded evidence that would otherwise appear to point to the guilt of the accused.

Mr. Justice Samuel Freedman, a former Chief Justice of the Manitoba Court of Appeal, wrote:

> The objective of a criminal trial is justice. Is the quest of justice synonymous with the search for truth? In most cases, yes. Truth and justice will emerge in a happy coincidence. But not always. Nor should it be thought that the judicial process has necessarily failed if justice and truth do no end up in perfect harmony. Such a result may follow from law's deliberate policy. . . . The law makes its choice between competing values and declares that it is better to close the case without all the available evidence being put on the record. We place a ceiling price on truth. It is glorious to possess, but not at an unlimited cost. "Truth, like all other good things, may be loved unwisely–may be pursed too keenly–may cost too much" (1972, 99)

Introduction

The above statement was partially quoted recently by the Supreme Court of Canada in *Bjelland*, 2009, para. 65 in emphasizing the importance of justice in criminal trials. The Law Reform Commission of Canada (1988, 10) has stated that the Canadian criminal justice system engages in a "qualified search for truth." The Supreme Court of Canada endorsed this position in *Noël*, in which Arbour, J., for the majority, stated "it has never been the case in our criminal justice system that the search for truth could be pursued at all costs, by all means" (para. 57). In dissent, L'Heureux-Dubé, after referring to articles by Paciocco (2001) and Peck (2001), wrote:

> In "The Adversarial System: A Qualified Search for the Truth", supra, [Peck] forcefully defends the notion that the search for truth must be qualified in appropriate circumstances where other more valuable principles apply. Ensuring that an accused receives a fair trial, deterring police misconduct, and preserving the integrity of the administration of justice are all laudable goals to which this Court must strive in its rules of evidence, at times to the detriment of full access to the truth. Where these goals are met, however, the search for the truth must, in my view, be the preponderant consideration (para. 85).

Although in a number of cases across Canada, people have been found legally guilty of a crime they did not actually commit (see Braiden and Brockman 1999) and have even pleaded guilty to crimes they did not commit (Brockman 2010a), it is an important value in our society that the innocent not be convicted (see Chapter 6). We are prepared to attempt to ensure this, even at the expense of allowing "guilty" people to go free. We have elaborate rules of procedure and evidence, theoretically designed to guarantee this. What these rules are, and whether they should be codified, are topics of debate.

Codification Or the Common Law

From the earliest times people have argued over whether it is better to codify the law or to rely on the common law. Much of what is considered criminal law and the law of criminal procedure today is contained in the *Criminal Code*. This, of course, does not mean that criminal law and procedure are completely codified, because every section of the *Code* can be interpreted and reinterpreted by the courts. Mr. Justice George L. Murray, in 1980, gave the example that it took "nearly ninety years to determine if there is a difference of meaning between the words 'know' and 'appreciate' in section 16 of the *Criminal Code*" (Uniform Law Conference 1982, 499).

Although the federal government has enacted the *Canada Evidence Act*, most of what we consider to be the law of evidence is not found there, but rather is in the common law, developed by judges on a case-by-case basis. Thus, for a question such as what constitutes a criminal offence, judges start with the words of a statute (typically the *Criminal Code*), and then rely upon other cases and their own initiative to interpret that legislation. With questions of evidence, however, judges find the law as developed, maintained, and interpreted in judicial decisions. (Consider, for example, the law with regard to similar fact evidence, discussed in Chapter 12).

A major effort to develop a uniform Code of Evidence was made between 1975 and 1980; however, in 2010, we still do not have anything like a complete code of evidence in Canada. Of course, even if we did have a code of evidence, we would still have to depend on judicial interpretations of what the legislation actually meant. The effort to codify the law of evidence in Canada is discussed in more detail in Chapter 11.

Standards/Principles Or Rules

Related to the concerns about codification is the concern over whether legislation (or the law developed by judges) should be in the form of clearly defined and very precise rules, leaving little discretion to judges, or in the form of more general standards or principles, giving judges more flexibility and discretion in refining and applying the law (see, for example, the Law Reform Commission of Canada 1988, 19-20). To illustrate, a rule might require that an accused be tried within eight months of a charge being laid, while a standard or principle might say that someone charged with an offence has a right to be tried "within a reasonable time." This tension is illustrated in many of the cases referred to in the text, although the terminology used in describing it varies.

The Adversarial System

Another series of questions revolves around the nature of the adversarial system, its effectiveness, and whether we actually *use* the adversarial system in the way we *define* it. For example, a large number of offences are disposed of by way of guilty pleas, many of which are the result of plea bargaining (Di Luca 2005; Fitzgerald 1990; Klein 1976; Law Reform Commission of Canada 1989). Is this compatible with an adversarial system? We return to plea bargaining in Chapter 3.

One characteristic of the adversarial system is that a case is heard by an impartial and essentially passive judge who decides it on the basis of two competing presentations. Is this the best means of arriving at the truth of what occurred in the past? The use of the adversarial system as a method of finding the truth has been criticized by American Realists such as Jerome Frank (1949), who outlined the difficulties of establishing "what happened," such as, faulty comprehension and recollection of what occurred, honesty of witnesses, reaction of a witness to cross-examination even if they are telling the truth, and so on.

Criticisms have also come from social scientists, who have studied the legal system from the outside and found that witnesses have faulty recall, both in terms of eyewitness identification of persons and of what actually occurred in their presence. In a critique of a case in the United Kingdom which resulted in a review of criminal procedure, McBarnet (1976, 456) wrote that the conviction was "based on an inaccurate version of events [that] was not an accident or freak occurrence but exactly what was aimed at . . . in a system based on adversary investigation and advocacy." Some of the books listed in Appendix C illustrate the same point (see, for example, Harris 1986; Karp and Rosner 1991; and Makin 1998).

The criminal justice system also addresses questions of law and policy. For instance, what is the meaning of "in the interest of public morals" under section 486(1) of the *Code*, when a judge makes a decision whether to exclude the public to protect a witness who is under the age of eighteen years? Would the admission of certain evidence bring the administration of justice into disrepute under section 24(2) of the *Charter*? Are questions such as these best resolved through the adversarial method? Young (1997a, 1997b) suggests that the only way to ensure constitutional rights is to have the police, Crown, defence counsel and the judge all responsible for them, because "the logic of adversarial justice and the demands of Constitutional rights cannot comfortably co-exist with respect to many issues" (1997b, 434).

PLAYERS IN THE SYSTEM

Many people are involved as cases are processed through the criminal justice system; some are marginalized while others are brought to the forefront.

The Role of the Complainant/Victim

At one time, criminal prosecutions were private matters, brought by individuals against individuals. The system evolved to the point where a criminal prosecution can now be processed by state agents without the victim's consent, and even against the victim's wishes. In *X*, the Ontario High Court dismissed an application to quash a subpoena served on a rape victim who did not want to testify at the preliminary hearing of one of the accused who had raped her (having already testified at the earlier preliminary hearing of another accused). According to her doctor, "she was not emotionally capable of withstanding the rigors of yet another court appearance, and that to do so would be to her emotional detriment" (88). She also felt threatened by the accused. In dismissing her application, Linden, J. stated:

> Although it is clearly a stressful situation for her to testify, and it would certainly be to her emotional detriment, the evidence is not strong enough for me to conclude that her security of the person would be interfered with. Anxiety and stress, as real and as unpleasant as they may be, are not enough to qualify as infringements of the security of the person [under section 7 of the *Charter*]. It is hard to differentiate the applicant's distress from that of many other rape victims, who often suffer emotional trauma in giving evidence, yet still proceed to do so as their public duty. . . .

> If we were to permit anyone who is frightened or apprehensive of giving evidence to refuse to do so, desperate and dangerous accused persons would be encouraged to make these threats in the hope of discouraging witnesses from testifying. It could produce a situation where the very worst offenders could avoid conviction by threatening the witnesses who have the evidence to convict them. The legal system cannot tolerate that. Hence, the courts are unable to excuse witnesses from giving their testimony, even when it is fraught with danger and emotional trauma for the individual (89-92).

Some provincial Attorneys General have made prosecution in spousal assault cases mandatory, in order to force victims to testify against their abusers. Should victims have a say in whether the Crown proceeds with charges? (see, for example, Bonnycastle and Rigakos 1998; Comack and Balfour, 2004; Faubert and Hinch, 1996; and other articles annotated in Bouchard, Boyd and Sheehy 1999, 150-181). Ross (2002) discusses the apparent disconnect between what victims expect of the criminal justice system and what they get. He suggests that prosecutors take a more "relational" approach, and recognize the social context of each crime.

Although victims are sometimes compelled to testify against their wishes, they are often excluded from other aspects of the criminal trial process (see Barrett 2001). In *O'Connor*, a priest (and principal) of a residential school in Williams Lake, British Columbia was accused of rape and indecent assault (the terms for the offences when they occurred in the mid-1960s). At the trial in 1992, the trial judge stayed the proceedings because the Crown had failed to make full disclosure to the defence. When the Crown appealed the trial judge's decision to the British Columbia Court of Appeal, the complainants and five organizations (Aboriginal Women's Council, Canadian Association of Sexual Assault Centres, Disabled Women's Network of Canada, the Women's Legal and Education and Action Fund, and the Canadian Mental Health Association) applied to the Court to intervene (discussed below under "The Role of Interveners") in the proceedings. The Court granted intervener status to the organizations, but denied the complainants' application, on the grounds that a crime is "not a wrong against the actual person harmed . . . but a wrong against the community as a whole" (503). The ruling provides that while the victims have a personal interest in a prosecution, they do not have the type of interest required to be permitted to intervene in the case. What might be some of the arguments for and against allowing complainants to intervene in a case? Paciocco (2005) suggests that entrenching victims' rights in the *Charter* would inappropriately weaken the presumption of innocence, and would extend state power when the purpose of the *Charter* is to restrict state power (see also Young 2005 and Roach 2005).

The Police

The police are major actors in our criminal justice system. In the earlier examples of the professor, the lawyer, and assailant, the police would likely play a reactive role–that is, they would respond to a complaint. In other situations, the police play a proactive role, using undercover agents, for example, to catch people committing criminal offences such as soliciting for the purposes of prostitution or drug dealing.

The Accused

Richard Ericson and Patricia Baranek (1982) studied the criminal justice system from the perspective of the accused. They found that the accused has little say in the process, and is very dependent on the police, lawyers, and other actors in the system. The dominant forces are the legal actors, who largely determine what will happen to the accused. The element of coercion in guilty pleas is discussed by Fitzgerald (1990) and Brockman (2010a).

Witnesses, Front-Line Workers

Very little research attention has focussed on those who witness crime, or who deal with victims of crime shortly after the crime occurs. Some of these people's concerns are similar to those of the victims of crime. Currently, records kept by front line workers and therapists are frequently being demanded by defence counsel in sexual assault cases, exposing those workers to serious dilemmas. This issue, which arose in the *O'Connor* case, and which resulted in changes to the *Criminal Code*, is dealt with in Chapter 12.

The Role of Crown Counsel and Defence Counsel

The prosecutor, or Crown counsel, is theoretically an objective non-partisan, who presents the available evidence fairly (for a discussion of inappropriate cross-examination, see Akhtar, 2004). Defence counsel is an advocate, representing the interests of the accused. The roles of Crown counsel and defence counsel are discussed in greater detail in Chapter 4, under "Disclosure."

The Role of *Amici Curiae*

The role of *amici curiae* ("Friends of the Court") dates back to the 17th Century, but their use has recently increased (Carter 2008, 89). An *amicus curiae* is someone appointed to assist the court by making representations in matters of law or fact that might not otherwise be addressed. A demand for an *amicus curiae* might arise when accused are unrepresented, whether it be through a denial of legal aid or the sequential firing of their defence counsel until they are left without counsel (for example, see *Benji* 2006). However, on at least on one occasion, an *amicus curiae* was appointed to assist defence counsel in a jury case where the trial was being constantly postponed because of defence counsel's health issues (Victoria Times Colonist 2009). Recently, the Supreme Court of Canada appointed an *amicus curiae* where proceedings against an accused had been withdrawn, and the Court wanted to proceed with the issue despite the fact that it was moot (*McNeil* 2009 para. 1-2).

The Role of the Judges

The role of the judge in our adversarial system is supposed to be that of a dispassionate listener who does not generally get involved in questioning witnesses. Rather than being an investigator, as in some civil law countries, the judge in the Canadian system is an adjudicator, making decisions based solely on the evidence presented to the court by the opposing sides.

The introduction of specialized trial courts or problem-solving courts, such as Drug Treatment Courts (in Vancouver, Toronto and St. John), Domestic Violence Courts (Winnipeg, Calgary), and Mental Health Courts (Toronto), have to some extent changed the role of the judge. Judges is such courts are expected to be more actively engaged in finding creative solutions to underlying problems, rather than to sit as dispassionate listeners (see, for example, Bakht 2005; Chiodo 2001; Heerema 2005; Van de Veen 2004). More recently, Vancouver's Downtown Community Court has opened, and "operates on the principle that collaborative case management can help offenders make long-term changes to their behaviour" (BC Criminal Justice Reform).

The Role of Jurors

Jurors are often ignored in Canadian legal research because the *Criminal Code* makes it an offence for them to talk about their deliberations (discussed further in Chapter 5). Studies that have been conducted find that jurors may experience a great deal of stress (see, for example, Chopra and Ogloff 2001); however the courts and Parliament are reluctant to alter the secrecy provisions of the *Criminal Code*.

The Role of Interveners

The court grants individuals or organizations "intervener" status to assist the court in interpreting the law; however, interveners are not allowed to make arguments based on the merits of the case. Although some argue that it is important for judges to hear public interest groups (see Bryden 1987 for an argument for expanding such participation), others argue that it politicizes the hearing and increases the time that it takes to hear a case. For example, in *Mills* (see Chapter 12) there were 18 interveners. Given the increased political nature of judging following the introduction of the *Charter* in 1982, it makes sense that judges would listen to a wider variety of social, economic, and political arguments when making such decisions (see for example, Manfredi 2004; Mandel 1994).

DIVISION OF POWERS UNDER THE CANADIAN CONSTITUTION

Under the *Constitution Act, 1867* (called the *British North America Act* at the time of Confederation in 1867), the federal government has jurisdiction (amongst other things) over:

> 91(27) The Criminal Law, except the Constitution of the Courts of Criminal Jurisdiction, but including the Procedures in Criminal Matters.
> (28) The Establishment, Maintenance, and Management of Penitentiaries.

The provincial governments have jurisdiction over subjects including:

> 92(6) The Establishment, Maintenance and Management of Public and Reformatory Prisons for the Province.
> . . .
> (13) Property and Civil Rights in the Province.
> (14) The Administration of Justice in the Province, including the Constitution, Maintenance and Organization of Provincial Courts, both of Civil and of Criminal Jurisdiction, and including Procedure in Civil Matters in those Courts.
> (15) The Imposition of Punishment by Fine, Penalty, or Imprisonment for enforcing any Law of the Province made in relation to any Matter coming within any of the Classes of Subjects enumerated in this Section.

In addition, the provincial government can delegate to cities and municipalities the power to make and enforce by-laws. So, in a sense, there are three levels of government that make and enforce laws through the criminal justice system–the federal, the provincial and the municipal.

Federal, Provincial, and Municipal Levels

Offences created by the federal government appear in statutes such as the *Criminal Code, Controlled Drugs and Substances Act* (which in 1997 replaced the *Narcotic Control Act* and parts of the *Food and Drugs Act), Fisheries Act, Competition Act, Harbours Board Act, Income Tax Act,* and many other federal statutes. Offences under these statutes are enforced and prosecuted through the procedures set out in the *Criminal Code*, and by the rules of evidence found in the *Canada Evidence Act* and the common law. The *Contraventions Act*, S.C. 1992, C. 47, provides some variation from the procedures set out in the *Code* for less serious offences, designated as "contraventions" under the legislation.

In *Hauser* and *Kripps Pharmacy*, the Supreme Court of Canada decided that the federal government has the authority under the *Constitution Act* to prosecute offences under all federal statutes. Before these decisions, many constitutional experts thought that the provinces had the sole constitutional authority to prosecute *Criminal Code* offences, and that the federal government had the constitutional authority only to prosecute non-*Code* federal offences. This would have left the federal government at the mercy of provincial Attorneys General in having its own laws enforced (discussed by the Law Reform Commission of Canada 1986, 24-25).

Even though the federal government has the authority to prosecute offences under the *Criminal Code*, if you attend any of the courts which deal with these offences in any province (but not the Northwest Territories, the Yukon Territory, or Nunavut), you will find that provincial employees or agents (private lawyers retained by the government) actually prosecute most of the offences under the *Code*. This is the result of the federal government delegating its prosecutorial powers to the provinces under section 2 of the *Code* (see section 2, definition of "Attorney General"). However, the federal government has exercised its right to retain joint prosecutorial powers over some *Code* offences, such as terrorism and securities and other fraud. Employees of the Public Prosecution Service of Canada (PPSC) (an independent federal government agency created in 2006) or its agents usually prosecute offences under federal legislation other than the *Code*.

In most provinces, federal and provincial prosecutions of federal offences take place in the provincial court, and in the superior court(s) of the province (for example, the Supreme Court of British Columbia, Court of Queen's Bench in Alberta, and the Superior Court of Justice in Ontario). Chapter 3 discusses which level of court hears cases involving which offences.

Offences created by the provincial governments are found in statutes such as the British Columbia *Medicare Protection,* the Ontario *Highway Traffic Act*, the Alberta *Livestock Diseases Act*, and the Saskatchewan *Hotel Keepers Act*. The provincial governments are responsible for enforcing their own laws. Provincial legislation, such as the B.C. *Offence Act* (formerly the *Summary Convictions Act*), specifies prosecution procedures. Typically, such legislation adopts the summary conviction procedures of the *Criminal Code*. All provincial offences are tried exclusively by the provincial courts.

The provincial governments delegate certain bylaw-making powers to cities and municipalities. Cities may have bylaws that govern snow removal from sidewalks, licensing of pets, "poop-and-scoop" bylaws, and bylaws requiring licences for various forms of business, and so on. These bylaws are enforced in Provincial Court, usually by municipal prosecutors.

Employment Relationships Between Governments and Prosecutors

There are at least three different employment relationships between the federal and provincial governments and their prosecutors. First, the government may hire full-time employees, who have all the benefits of employment: fixed salary, holidays, pensions, some security of employment, etc. One of the advantages of this arrangement, from the government's point of view, is that some of these prosecutors will become career prosecutors and build up expertise in the area. These prosecutors are also more readily controlled by the Attorney General through management, training courses, and supervisory tactics.

Second, governments may hire full-time or part-time contract prosecutors. These prosecutors will do the same work as employees but have none of the employment benefits listed above. Their contracts may be as short as a few months, or may last for two or more years, with no guarantee that they will be renewed. This lack of employment certainty leads to a high turn-over rate, and to the continual short-term hiring of inexperienced prosecutors. In cities without federal legal offices (for example, Calgary), prosecutions are usually contracted out to a specific law firm, appointed as standing agents, often based on the political party in power and the political persuasion of the law firm. This system of agents has been attacked for years as an example of questionable political patronage.

Third, governments might hire *ad hoc* prosecutors (agents), to determine whether charges should be laid in certain circumstances, or to prosecute specific cases. These lawyers are often from the private bar, and typically have many years of experience in criminal defence work. The advantage of using *ad hoc* agents is that the government can call on a senior criminal lawyer for advice without keeping the individual on a regular salary. This approach is sometimes used to provide the government with an "independent" opinion on high-profile cases, or to avoid suggestions of conflict of interest or political influence in certain prosecutions. If the provincial Attorney General of British Columbia appoints a special prosecutor under section 7 of the *Crown Counsel Act*, R.S.B.C. 1996, c.87, the decision of the special prosecutor is final and the Attorney General is not allowed to shop around until he or she finds someone who will recommend a prosecution (*Blackmore*).

Introduction

Under the definition section of the *Criminal Code*, all indictable offences (discussed further in Chapter 1) must be prosecuted by lawyers qualified to practice law in the province (see definitions of "prosecutor" and "counsel" in section 2). Summary conviction offences, provincial and municipal offences can be prosecuted by non-lawyers, (see section 785). At one time many of the provincial offences were prosecuted by police officers, but this is no longer the case. Note that these rules do not preclude private prosecutions (discussed in Chapter 3).

THE FEDERAL COURT OF CANADA

The Federal Court of Canada, which had both trial and appeal divisions, was split into two separate courts in 2003–the Federal Court and the Federal Court of Appeal. The Federal Court consists of a Chief Justice and up to 32 other judges, who travel to the major cities across Canada and hear cases within their jurisdiction. A Federal Court judge might hear a case in Ontario one month, and another in British Columbia the next. Judges of both courts are required to live within 40 kilometres of National Capital Region.

The Federal Court system is used to sue the federal government and appeal decisions of some federally created administrative agencies, although it now has concurrent jurisdiction in many areas with the superior courts in of the provinces. Although at one time never used for criminal prosecutions, some offences under the *Competition Act* can now be prosecuted in Federal Court with the consent of the accused. Many have advocated that the Federal Court be abolished, because of unnecessary duplication; however, it appears as though its role has been expanding.

THE *CANADIAN CHARTER OF RIGHTS AND FREEDOMS*

The *Charter of Rights and Freedoms* is Part I of the *Constitution Act*, and came into force on April 17, 1982. Sections 15 and 28, the equality provisions, did not come into force until April 17, 1985. Section 52(1) of the *Constitution Act* states that:

> The Constitution of Canada is the supreme law of Canada, and any law that is inconsistent with the provisions of the Constitution is, to the extent of the inconsistency, of no force or effect.

Since the *Charter* is part of the *Constitution Act,* it is part of the supreme law of Canada, as specified in section 52. There is, however, an "opting-out" provision (section 33) which allows federal or provincial governments to declare that a certain piece of legislation "shall operate notwithstanding a provision included in section 2 or sections 7 to 15 of the *Charter*." The existence and use of this section have created much controversy.

The *Charter* applies to the federal, provincial and territorial governments of Canada, in respect of all matters within their jurisdiction (see section 32). Does the *Charter* apply outside of Canada? As a general rule, laws in one country do not apply in another country. In *Terry*, the Supreme Court of Canada found that the *Charter* did not apply to the U.S. police, in the United States, when they questioned a suspect who was charged with an offence in Canada. The U.S. police had complied with the laws in the United States. In *Cook*, the Court found that Canadian police officers are bound by the *Charter* when they take a statement from a suspect in the United States as part of their investigation of an offence committed in Canada. However, in *Hape*, the Court found that the *Charter* did not apply to Canadian police officers who were seizing documents in a foreign jurisdiction in cooperation with foreign officials to be used in a prosecution in Canada. For further discussion on this topic see Berger (2008), Roach (2007), and the Supreme Court of Canada's decision in *Khadr* (2008).

Introduction

Section 1

Another major restriction on the rights and freedoms guaranteed under the *Charter of Rights* is contained in section 1 of the *Charter*, which provides that the rights and freedoms contained in the *Charter* are subject to such "reasonable limits prescribed by law as can be demonstrably justified in a free and democratic society." Thus, the rights set out in our *Charter* are not absolute rights; some rights can be overridden by legislators using the "notwithstanding clause" (section 33), and all may be restricted, provided the restriction can be "demonstrably justified in a free and democratic society" under section 1. Decisions about the applicability of section 1 are left to the judiciary, and as a consequence, the *Charter of Rights* has changed our legal system from one in which Parliament was supreme to one in which the courts have become involved in the political process (see, for example, the discussion by Mandel 1994). If the law cannot be demonstrably justified, the court will find that it is of no force and effect under section 52 of the *Constitution Act*.

Before section 1 can become an issue, there must be a violation of rights "prescribed by law" (that is, set out in law). Such a limitation or violation can be prescribed through legislation, regulations, or orders-in-council, and in the common law or judge-made law (*Therens*). If a law limits rights and freedoms under the *Charter*, the judge asks the second question: can the limit be demonstrably justified in a free and democratic society? The Supreme Court of Canada first considered the meaning of this clause in the *Oakes* case. Chief Justice Dickson concluded that section 8 of the then *Narcotic Control Act* (which set out a two-stage process in narcotics possession trials, whereby the accused who had been found to be in possession of drugs would have to establish that she or he was not in possession for the purpose of trafficking) violated section 11(d) of the *Charter*, the right to be presumed innocent. Given legislation which violated the accused's rights under the *Charter*, the next question was whether the violation was demonstrably justified in a free and democratic society under section 1 of the *Charter*. Dickson, C.J.C., set out the test under section 1 as follows:

> To establish that a limit is reasonable and demonstrably justified in a free and democratic society, two central criteria must be satisfied. First, the objective, which the measures responsible for a limit on the *Charter* right or freedom are designed to serve, must be "of sufficient importance to warrant overriding a constitutionally protected right or freedom" . . . . It is necessary, at a minimum, that an objective relate to concerns which are pressing and substantial in a free and democratic society before it can be characterized as sufficiently important.
>
> Secondly, once a sufficiently significant objective is recognized, then the party invoking s. 1 must show that the means chosen are reasonable and demonstrably justified. This involves "a form of proportionality test." Although the nature of the proportionality test will vary depending on the circumstances, in each case courts will be required to balance the interests of society with those of individuals and groups. There are, in my view, three important components of a proportionality test. First, the measures adopted must be carefully designed to achieve the objective in question. They must not be arbitrary, unfair or based on irrational considerations. In short, they must be rationally connected to the objective. Secondly, the means, even if rationally connected to the objective in this first sense, should impair "as little as possible" the right or freedom in question. Thirdly, there must be a proportionality between the *effects* of the measures which are responsible for limiting the *Charter* right or freedom, and the objective which has been identified as of "sufficient importance" (348).

The Supreme Court of Canada found that the section of the *Narcotic Control Act* was not demonstrably justified, and the section was declared to be unconstitutional, and of no force and effect.

The analysis under section 1, the *Oakes* test, and its application, are dealt with in a number of cases in both Parts I and II of this text. It is important to consider the type of evidence the court will hear, the use (or lack of use) of social science research, and whether the courts' decisions could be improved if they properly used social science research. Kiedrowski and Webb (1993) examined the use of social science research in section 1 analyses, and predicted that it will be used more and more in the future.

Section 24

There are many circumstances under which a person's rights under the *Charter* may be violated, but where the violation or limitation is not condoned or prescribed by law, so that section 1 is not invoked. Under such circumstances, the aggrieved person can look to section 24 of the *Charter* for a remedy. Section 24(1) states that

> Anyone whose rights or freedoms, as guaranteed by this charter, have been infringed or denied may apply to a court of competent jurisdiction to obtain such remedy as the court considers appropriate and just in the circumstances.

What remedy is appropriate and just in the circumstances is up to the judge; the judge could stay the charges, order damages be paid to the person, order that costs be paid to the person whose rights were infringed, or grant other remedies.

Section 24(2) deals with the admission or exclusion of evidence obtained in a manner which violated the accused's rights. It reads:

> Where in proceedings under subsection (1), a court concludes that evidence was obtained in a manner that infringed or denied any rights or freedoms guaranteed by this Charter, the evidence shall be excluded if it is established that, having regard to all the circumstances, the admission of it in the proceedings would being the administration of justice into disrepute.

Section 24(2) is a compromise between the traditional English and Canadian common law approach, that illegally obtained evidence is admissible if it is relevant, and the general position in the United States, that illegally obtained evidence must be excluded. However, both the English and the American approaches have moved toward the middle ground. The interpretation of section 24(2), and the exclusion of evidence under it, are dealt with in greater detail in Part II of this text.

QUESTIONS TO CONSIDER

(1) Distinguish between social fact, social framework, and social authority.

(2) What are the advantages and disadvantages of codification? What are the advantages and disadvantages of strict rules, as opposed to standards?

(3) The federal government has the constitutional power to prosecute offences under the *Criminal Code*. Explain why employees of the provincial government actually conduct these prosecutions.

(4) There are three levels of government which create offences and enforce infractions. What are these levels, and give examples of the types of laws which are created by each level of government.

(5) What are the three different types of arrangements that Attorneys-General enter into with prosecutors? What are the advantages and disadvantages of each? (You should be able to think beyond the few discussed in this Chapter).

(6) What is intervener status, and why is it granted or refused?

(7) What is the distinction between factual guilt and legal guilt? What purpose does it serve in our criminal justice system?

(8) Under what circumstances might the court appoint an *amicus curiae*?

(9) Give an example of how assumptions of fact underlie and at times determine the law.

Introduction

(10)　Why were the victims not allowed to intervene in the *O'Connor* case (the bishop accused of sexual assault)?

(11)　What argument can the provincial government make if someone argues that a provincial law violates their rights under the Charter?

(12)　When is section 24(2) of the *Charter* used?

(13)　What three legal venues could deal with a lawyer who sexually assaults his client?

BIBLIOGRAPHY

Akhtar, Suhail. "Improprieties in Cross-Examination." (2004) 15 *Criminal Reports* (6th) 236.

Bakht, Natasha. "Problem Solving Courts as Agents of Change." (2005) 50 *Criminal Law Quarterly* 224.

Barrett, Joan M. *Balancing Charter Interests: Victims' Rights and Third Party Remedies.* Scarborough: Carswell, 2001.

BC Criminal Justice Reform. "How the Court Works" (www.criminaljusticereform.gov.bc.ca; accessed August 27, 2009).

Berger, Benjamin L. 2008. "The Reach of Rights in the Security State: Reflections on *Khadr v. Canada (Minister of Justice)*" (2008) 56 *Criminal Reports* (6th) 268.

Bonnycastle, Kevin and George Rigakos (eds.). *Unsettling Truths: Battered Women, Policy, Politics, and Contemporary Research in Canada.* Vancouver: Collective Press, 1998.

Bouchard, Josée, Susan B. Boyd, and Elizabeth A. Sheehy. "Canadian Feminist Literature on Law: An Annotated Bibliography." (1999) 11(1&2) *Canadian Journal of Women and the Law* 1.

Boyd, Neil (ed.). *The Social Dimensions of Law.* Scarborough, Ont.: Prentice Hall Canada, Inc., 1986.

Braiden, Patricia and Joan Brockman. "Remedying Wrongful Convictions Through Applications to the Minister of Justice Under Section 690 of the *Criminal Code.*" (1999) 17 *Windsor Yearbook of Access to Justice* 3.

Brockman, Joan. "Social Authority, Legal Discourse and Women's Voices." (1992) 21(2) *Manitoba Law Journal* 212.

Brockman, Joan. "Controlling Crimes Among the Professional Elite." in Margaret A. Jackson and Curt Griffiths (eds.) *Canadian Criminology: Perspectives on Crime and Society*, 2nd ed. Toronto: Harcourt Brace Jovanovich Canada Inc., 1995.

Brockman, Joan. "An Offer You Can't Refuse:" Pleading Guilty When Innocent (2010a) 56 *Criminal Law Quarterly* (in press).

Brockman, Joan. "Fraud Against the Public Purse by Health Care Professionals: The Privilege of Location" Chapter 2 in Janet Mosher and Joan Brockman (eds.) *Constructing Crime: Contemporary Processes of Criminalization* (Vancouver: UBC Press, 2010b).

Bryden, Phillip L. "Public Interest Intervention in the Courts." (1987) 66 *Canadian Bar Review* 490.

Burtch, Brian. *The Sociology of Law: Critical Approaches to Social Control.* Toronto: Thomson Nelson, 2003.

Carter, Ian. 2008. "A complicated friendship: the evolving role of Amicus Curiae." (2008) 54 *Criminal Reports* (6th) 89.

Chesterton, Gilbert K. "Twelve Wise Men" in *Tremendous Trifles.* New York: Dodd Mead, 1915.

Introduction

Chiodo, Anida L. "Sentencing Drug-Addicted Offenders and the Toronto Drug Court" (2001) 45 *Criminal Law Quarterly* 53.

Chopra, Sonia R. and James R. P. Ogloff. "Evaluating Jury Secrecy: Implications for Academic Research and Juror Stress." (2001) 44 *Criminal Law Quarterly* 190.

Comack, Elizabeth and Gillian Balfour. *The Power to Criminalize: Violence, Inequality and the Law.* Halifax: Fernwood Publishing, 2004.

Comack, Elizabeth and Stephen Brickey (eds.). *The Social Basis of Law: Critical Readings in the Sociology of Law,* 2nd ed. Halifax: Garamond, 1991.

Cotterrell, Roger. *The Sociology of Law: An Introduction,* 2nd ed. London: Butterworths, 1992.

Davis, Kenneth Culp. "An Approach to Problems of Evidence in the Administration Process." (1942) 55 *Harvard Law Review* 364.

Di Luca, Joseph. "Expedient McJustice or Principled Alternative Dispute Resolution? A Review of Plea Bargaining in Canada." (2005) 50 *Criminal Law Quarterly* 14.

Ericson, Richard V. and Patricia M. Baranek. *The Ordering of Justice: A Study of Accused Persons as Dependants in the Criminal Process.* Toronto: University of Toronto Press, 1982.

Faubert, Jacqueline and Ronald Hinch. "The Dialectics of Mandatory Arrest Policies." in Thomas O'Reilly-Fleming (ed.), *Post-Critical Criminology.* Toronto: Prentice-Hall Canada, 1996, 230.

Federal-Provincial-Territorial Heads of Prosecutions Committee Working Group. *Report on the Prevention of Miscarriages of Justice* (September 2004); www.justice.gc.ca/en/dept/pub/hop/toc.html; accessed September 19, 2005.

Fitzgerald, Oonagh E. *The Guilty Plea and Summary Justice.* Toronto: Carswell, 1990.

Frank, Jerome. *Courts on Trial: Myth and Reality in American Justice.* Princeton: Princeton University Press, 1949.

Freedman, Samuel. "Admissions and Confessions." in Roger E. Salhany and Robert J. Carter (eds.), *Studies in Canadian Criminal Evidence.* Toronto: Butterworths, 1972.

Gall, Gerald L. *The Canadian Legal System,* 5th ed. Scarborough, Ontario: Carswell, 2004.

Heerema, Mark. "An Introduction to the Mental Health Court Movement and Its Status in Canada." (2005) 50 *Criminal Law Quarterly* 255.

Hinch, Ronald (ed.). *Readings in Critical Criminology.* Scarborough, Ontario: Prentice Hall, 1994.

Holmes, Oliver Wendell. *The Common Law.* Cambridge: Harvard University Press, 1881, reprinted 1963.

Kiedrowski, John and Kernaghan Webb. "Second Guessing the Law-Makers: Social Science Research in *Charter* Litigation." (1993) 19(4) *Canadian Public Policy* 379.

Klein, John F. *Let's Make a Deal: Negotiating Justice.* Lexington, MA: Lexington Books, 1976.

Law Reform Commission of Canada. *Classification of Offences.* Working Paper #54. Ottawa: Law Reform Commission of Canada, 1986.

Law Reform Commission of Canada. *Our Criminal Procedure.* Report #32. Ottawa: Law Reform Commission of Canada, 1988.

Introduction

Law Reform Commission of Canada. *Plea Discussion and Agreement*. Working Paper #60. Ottawa: Law Reform Commission of Canada, 1989.

Manfredi, Christopher P. *Feminist Activism in the Supreme Court: Legal Mobilization and the Women's Legal Education and Action Fund*. Vancouver: UBC Press, 2004.

Mandel, Michael. *The Charter of Rights and the Legalization of Politics in Canada*, 2nd ed. Toronto: Thompson Educational Publishing Inc., 1994.

McBarnet, Doreen. "The Fisher Report on the Confait Case: Four Issues." (1978) 41 *Modern Law Review* 455.

Monahan, John and Laurens Walker. "Social Science Research in Law." (1988) 43(6) *American Psychologist* 465.

Mosher, Janet and Joe Hermer. "Welfare Fraud: The Constitution of Social Assistance as Crime" Chapter 1 in Janet Mosher and Joan Brockman (eds.) *Constructing Crime: Contemporary Processes of Criminalization* (Vancouver: UBC Press, 2010).

Moran, Bridget. *Judgement at Stoney Creek*. Vancouver: Tillacum Library, 1990.

Murdoch, Caroline and Joan Brockman. "Who's On First? Disciplinary Proceedings by Self-Regulating Professions and other Agencies for 'Criminal' Behaviour." (2001) 64(1) *Saskatchewan Law Review* 29.

Paciocco, David M. "Evidence About Guilt: Balancing the Rights of the Individual and Society in Matters of Truth and Proof." (2001) 80 *Canadian Bar Review* 433.

Paciocco, David M. "Why the Constitutionalization of Victim Rights Should not Occur." (2005) 49 *Criminal Law Quarterly* 393.

Peck, Richard C. C. "The Adversarial System: A Qualified Search for the Truth." (2001) 80 *Canadian Bar Review* 456.

Pilkington, Marilyn. "Equipping the Courts to Handle Constitutional Issues: The Adequacy of the Adversarial System and Its Techniques of Proof." in *Special Lectures of the Law Society of Upper Canada, 1991: Applying the Law of Evidence*. Toronto: Carswell, 1992, 51.

Ratner, R.S. and John L. McMullan. *State Control: Criminal Justice Politics in Canada*. Vancouver, University of British Columbia Press, 1987.

Roach, Kent. "Victims Rights and the *Charter*." (2005) 49 *Criminal Law Quarterly* 474.

Roach, Kent. "Editorial: *R. v. Hape* Creates Charter-Free Zones for Canadian Officials Abroad." (2007) 53 *Criminal Law Quarterly* 1.

Ross, Rupert. "Victims and Criminal Justice: Exploring the Disconnect." (2002) 46 *Criminal Law Quarterly* 483.

Stuart, Don, Ron Delisle, and Tim Quigley. *Learning Canadian Criminal Procedure*, 9th ed. Toronto: Carswell, 2008. Chapter 6 covers the adversarial system and the role of defence and Crown counsel.

Uniform Law Conference of Canada. *Report of the Federal/Provincial Task Force on Uniform Rules of Evidence*. Toronto: Carswell, 1982.

Van de Veen, Sherry L. "Some Canadian Problem Solving Court Processes." (2004) 83 *Canadian Bar Review* 91.

Victoria Times Colonist. "Friend of the court' to assist at Ruffolo trial." (22 August 2009) (www.timescolonist.com; accessed August 27, 2009).

Introduction

Young, Alan N. "Adversarial Justice and the Charter of Rights: Stunting the Growth of the 'Living Tree'–Part I." (1997a) 39 *Criminal Law Quarterly* 334.

Young, Alan N. "Adversarial Justice and the Charter of Rights: Stunting the Growth of the 'Living Tree'–Part II." (1997b) 39 *Criminal Law Quarterly* 362.

Young, Alan N. "Crime Victims and Constitutional Rights." (2005) 49 *Criminal Law Quarterly* 432.

CASES

R. v. Benji, [2006] B.C.J. No. 2837 (B.C.S.C.).

R. v. Bjelland, [2009] 2009 SCC 38.

Blackmore v. British Columbia (Attorney General), 2009 BCSC 1299.

R. v. Cook, [1998] 2 S.C.R. 597.

R. v. Hape, [2007] 2 S.C.R. 292.

R. v. Hauser (1979), 46 C.C.C. (3d) 481 (S.C.C.).

Canada (Justice) v. Khadr, [2008] 2 S.C.R. 143.

Kripps Pharmacy Ltd. and Kripps; sub nom. Wetmore (1981), 64 C.C.C. (2d) 25 (B.C.C.A.); *rev'd.* (1983), 7 C.C.C. (3d) 507 (S.C.C.).

R. v. McNeil, 2009 SCC 3.

R. v. Noël, [2002] 3 S.C.R. 433.

R. v. Oakes (1986), 24 C.C.C. (3d) 321 (S.C.C.) .

Re Regina and O'Connor et al. (1993), 82 C.C.C. (3d) 495 (B.C.C.A.).

M.N.R. v. Symes, [1993] 4 S.C.R. 695.

R. v. Terry, [1996] 2 S.C.R. 207.

R. v. Therens (1985), 18 C.C.C. (3d) 481 (S.C.C.).

R. v. X (1983), 8 C.C.C. (3d) 87 (Ont. H.C.).

Part I

The Procedures for
Bringing an Accused to Court,
Trials, Dispositions,
and Appeals

CHAPTER 1: *Classification of Offences, Elections, and Jurisdiction of the Court*

CHAPTER OBJECTIVES

In studying this chapter, you should develop an understanding of the following topics and concepts:

- the classification of offences
- a court's jurisdiction, including jurisdiction over the offence, jurisdiction over the person, territorial jurisdiction, and the effect of the passage of time on jurisdiction
- which courts have jurisdiction over which offences, and whether an accused has an election as to mode of trial for any given offence
- how the Crown can override an accused's election as to mode of trial
- what limitation periods apply to any given offence, and how limitation periods may sometimes be avoided in the case of hybrid offences
- the extent of a court's territorial jurisdiction, and exceptions to it in terms of waiver of charges and change of venue
- how delay in the criminal litigation process can affect the proceedings pursuant to the *Charter of Rights*, and the distinction between pre- and post-charge delay

CLASSIFICATION OF OFFENCES

Offences in Canada are divided into three categories: (1) **indictable**, (2) **summary conviction**, and (3) **hybrid (or dual) offences** (described in more detail below). Persons charged with criminal offences will be tried according to one of two procedures: the procedure set out for indictable offences (Parts XIX and XX of the *Criminal Code*) or the procedure set out for summary conviction offences (Part XXVII of the *Criminal Code*). These provisions govern how the trial will be conducted, what appeal procedures will be available, and to some extent, what sentencing options will exist (Chapter 6). Whether an offence is summary conviction or indictable will also determine the powers of arrest (Chapter 2), whether an accused is required to appear for the purposes of the *Identifications of Criminals Act*, R.S.C. 1985, Chap. I-1 (Chapter 2), and the waiting period for a pardon under the *Criminal Records Act*, R.S.C. 1985, Chap. C-47. The court in which an accused is tried is determined by the classification of the offence and in some cases by the election of the accused.

A person charged with an offence in Canada will be tried in one of two levels of court, a **provincial court** (where judges are appointed by the provincial government), or a "**superior court of criminal jurisdiction**" (where judges are appointed by the federal government). Superior courts have different names across Canada, such as the Newfoundland Supreme Court, the Alberta Court of Queen's Bench, and the Ontario Superior Court of Justice (see the definition of "superior court of criminal jurisdiction" in section 2 of the *Criminal Code*). The territory of Nunavut has only one court (a unified court system), the Nunavut Court of Justice, and the *Code* contains separate provisions for it. There have been discussions in the rest of Canada on the possibility of a unified criminal court system that would unite provincial and superior courts (see, for example, Friedland 2004; Healy 2004; Webster and Doob 2004; Baar 2004; Seniuk and Borrows 2004).

Indictable Offences

Indictable offences are generally considered to be the more serious offences, and only the federal government can create this type of offence. The maximum penalties (usually 2, 5, 10, or 14 years, or life imprisonment) are typically set out in the charging section. Section 124 of the *Criminal Code* is an example of an indictable offence: "every one who…[sells an office]…is guilty of an indictable offence and liable to imprisonment for a term not exceeding five years." This charging section sets out the offence, the category of offence, and the maximum penalty (other examples are found in sections 209, 236, 247, and 345 of the *Code*). Some charging sections also set out minimum penalties (see section 235). If no punishment is provided for an indictable offence, an accused is liable to a term of imprisonment of five years or less (see section 743).

Summary Conviction Offences

Summary conviction offences are the less serious offences and can be created by both the federal and provincial governments. The charging section for federal offences will indicate whether an offence is a summary conviction offence. For example, section 442 of the *Code* states "every one who wilfully pulls down, defaces, alters, or removes anything planted or set up as a boundary line or part of a boundary line of land is guilty of an offence punishable on summary conviction" (see also sections 210(2) and 439(1)). Section 787(1) states that "except where otherwise provided by law, every one who is convicted of an offence punishable on summary conviction is liable to a fine of not more than five thousand dollars [up from two thousand effective 29 May 2008] or to imprisonment for six months or to both." Some charging sections for summary conviction offences set out more serious penalties; for example, someone convicted of double ticketing under section 54(2) of the federal *Competition Act*, is liable "to a fine not exceeding ten thousand dollars or to imprisonment for a term not exceeding one year, or to both."

Hybrid or Dual Offences

Hybrid or dual offence probably emerged in England in the late 1800s (Law Reform Commission of Canada 1986, 17–8). These are offences for which the Crown (i.e., the prosecutor) has the option of proceeding either by summary conviction or by indictment. Only the federal government can create hybrid offences. For example, section 215(3) of the *Code* states "every one who commits an offence under subsection (2) [failing to provide the necessaries of life] is guilty of (a) an indictable offence and is liable to imprisonment for a term not exceeding five years; or (b) is guilty of an offence punishable on summary conviction and liable to imprisonment for a term not exceeding eighteen months." See also sections and 249(2) and 264.1(3).

The Crown usually makes its position known (whether it will elect to proceed by way of indictment or summary conviction) either before the accused appears in court or at the **arraignment** of the accused. At the arraignment, the charges are read to the accused in court (see Chapter 3). The present classification of offences, and particularly the use of hybrid offences, has been criticized by the Law Reform Commission of Canada and by others.

REFORMING THE CLASSIFICATION OF OFFENCES

The Law Reform Commission of Canada, in its Working Paper *Classification of Offences,* described the present classification of offences as "unnecessarily complex," and as "structured on accidents of history rather than any rational plan" (1986, 1). Sir James Fitzjames Stephens was a major figure in law reform in England and Canada during the late 1800s. His *English Draft Code* of 1879, originally intended as a model criminal code for India, was introduced into Canada in 1892. There was some confusion as to why Canada adopted a classification system when Stephens was of the view that the "most convenient course in practice is to have no classification at all" (quoted in the Law Reform Commission of Canada 1986, 13). Contrary to Stephens' earlier view, the Law Reform Commission of Canada thought that a classification of offences was both possible and necessary.

In 1986, the Commission proposed that all federal offences be classified as either *crimes*, for which a prison term is possible upon conviction, or *infractions*, for which a prison term is not an option on conviction. Infractions under the Commission's proposal would not be dealt with under the new code of criminal procedure, but rather under an entirely new infractions procedure act. The Commission drew the following distinctions:

> In principle the criminal law's concern is with seriously wrongful acts violating common standards of decency and humanity. In practice only a minority of criminal offences fall under this heading. The majority, which total more than 20,000, are not necessarily wrong in themselves but prohibited for expediency. Such acts have to do with commerce, trade, industry and other matters which must be regulated in the general interest of society (Law Reform Commission of Canada 1986, 23).

What types of offences or "violations" would the Law Reform Commission of Canada remove from the category of "crime"? What types of debates might revolve around "commerce, trade and industry" activities that we now treat as "crimes"? Sutherland (1940) critically suggested that white-collar criminals were being processed by administrative agencies, rather than the criminal justice system. For contemporary debates over "What is a Crime?" see Henry and Lanier 2001; Hillyard *et. al.* 2004; Law Commission of Canada 2004; and Mosher and Brockman, 2010.

In 1996, the federal government introduced a system of ticketing, through the *Contraventions Act*, which allows individuals to pay fines for less serious federal contraventions (where imprisonment is not an option) and to avoid criminal convictions. The purpose of the legislation is to distinguish between criminal offences and regulatory offences, and "to alter or abolish the consequences in law of being convicted of a contravention, in light of that distinction" (section 4). A person convicted of a contravention does not have a criminal record (section 63), and it is an offence to require disclosure of such convictions for any employment over which the federal government has jurisdiction (section 64). In any ticketing system, the offender has the right to dispute the allegations by not paying the fine and appearing in court.

In 2003, the federal government introduced Bill C-38, which would have allowed the possession of fifteen grams or less of marijuana and one gram or less of hashish to be processed under the *Contraventions Act*. Proposed fines ranged from $150-$400, and the provinces would have been allowed to set lower fines for such contraventions (see Erickson, et al. 2004 for a discussion). The Bill died on November 12, 2003, and similar provisions have not been re-introduced. It is interesting to note that this so called "decriminalization" would actually have increased the number of individuals who would have been penalized, as the government expected the legislation to be used instead of police warnings (Department of Justice 2004).

The Law Reform Commission (1986) further recommended that all hybrid offences be abolished. It criticized the present system in which the decision to proceed by way of indictment or summary conviction is within the sole discretion of the prosecutor (this point of law was decided by the Supreme Court of Canada in 1971, in *Smythe*). In England, the decision as to which way a hybrid offence will proceed is made by the judge. Other criticisms of the present practice relating to hybrid offences come from Rosenthal (1990–91), who examined a number of the *Charter* issues that arise out of this type of unregulated Crown discretion. He recommended that the sole basis for exercising this discretion ought to be the seriousness of the offence. The Law Reform Commission of Canada's recommendations have not been adopted; rather the federal government has expanded the number of hybrid offences. In the last couple of years, some summary conviction offences (e.g., keeping a cockpit under section 447) and some indictable offences (e.g., causing injury to cattle under sections 444) have been changed to hybrid offences. New offences (luring a child via the internet under section 172.1) have been added to the *Code* as hybrid offences.

ELECTIONS

Which court a trial is held in depends in some cases on the classification of the offence, and in others on the election of the accused as to how she or he wants to be tried. Historically, a list of offences developed for which an accused could be tried only by magistrate, even without the accused's consent; these offences were within the absolute jurisdiction of the magistrate. A court has **exclusive** or **absolute jurisdiction** over an offence when no other court is allowed to try the offence, and as a result, the accused does not have an election.

To determine in which court an accused may be tried, one has to know how offences are classified—see the above discussion of indictable, summary, and hybrid offences (also see the "Offence Grid" provided in Martin's *Criminal Code*, which provides an easy reference to the classification of offences, sentencing options, and so on). The easiest way of determining whether an accused has an election is to eliminate the cases in which the accused does not have an election. The following offences are within the absolute jurisdiction of a provincial court judge, therefore the accused does not have an election:

(1)   all provincially enacted offences (which are necessarily summary conviction offences),

(2)   all summary conviction offences in the *Criminal Code* and other federal statutes,

(3)   all dual or hybrid offences (that is, where the Crown can proceed either by way of summary conviction or indictment) *if* the Crown proceeds by summary conviction (in which case the offence is considered to be a summary conviction offence),

(4)   indictable or hybrid offences listed in section 553, even if the Crown proceeds by way of indictment.

The provincial court has absolute jurisdiction over these offences, and they will all be dealt with under Part XXVII of the *Criminal Code* (Summary Convictions), except for those proceeded with by way of indictment, which will be dealt with under Part XIX (Indictable Offences—Trial Without Jury). The accused does not have an election for any of these offences. There will not be a preliminary inquiry prior to trial for these offences (preliminary inquiries are discussed in Chapters 3 and 4). None of these offences will be tried by a jury.

There are some offences considered by Parliament to be so serious that the accused has no election but *must* be tried by a superior court of criminal jurisdiction (see the definition in section 2 of the *Code* for the names of these courts in each province and territory). These serious offences, such as treason and murder, are listed in section 469 and require a trial by judge and jury (section 471). The only exception is that these offences may be tried by a superior court judge without a jury if both the accused and the Attorney-General consent (section 473). If consent is given, it can be withdrawn only if both the accused and the Attorney-General agree (section 473(2)). Until 1985, this option of consenting to a non-jury trial was available only in Alberta. The section was challenged

CHAPTER 1: *Classification of Offences, Elections, and Jurisdiction of the Court*

Figure 1.1
Elections by Accused and Forum of Trial

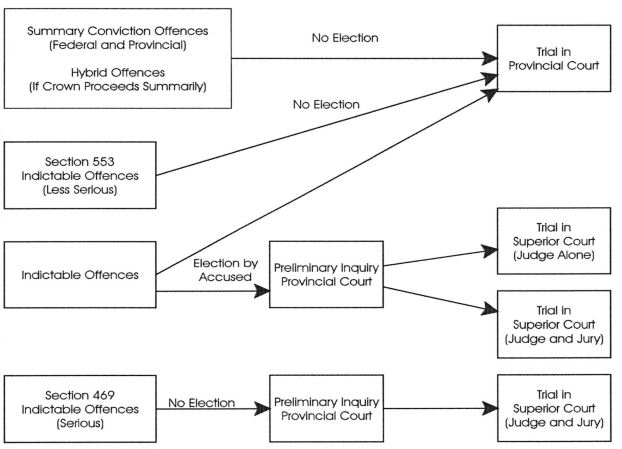

unsuccessfully under the equality provisions of the *Charter*, but Parliament subsequently amended the section to apply to all of Canada.

It is important to note that an accused does not have a right to be tried without a jury for section 469 offences; the consent of the Attorney-General is required if an accused wishes to be tried without a jury (*Turpin*). However, there may be circumstances in which the court will dispense with the Crown's consent. When the Crown in *Bird* was asked if he would consent to the accused being tried by judge alone for a murder charge, the Crown asked, "Who is the trial judge?" Having heard the reply, he refused to consent. The Court found that the Crown's decision to withhold consent "was motivated in whole or in part by a consideration of the identity of the assigned trial judge.... Such a consideration constitutes, in all the circumstances, an improperly motivated exercise of the Crown's discretion to withhold consent and an interference with the integrity of the process of the court" (para. 17). According to the Court, the Crown "is not concerned with 'winning' or 'losing'" (para. 13), and "must be motivated by a desire to achieve fundamental fairness and not by any inclination whatsoever to obtain a tactical or strategic advantage at the trial setting" (para. 15). The court dispensed with the Crown's consent, and the accused was allowed to proceed to trial by judge alone. In *Ng*, the Alberta Court of Appeal (application for leave to appeal to the Supreme Court of Canada dismissed), found that the Crown is not required to give reasons for refusing to consent to an accused's attempt to re-elect trial by judge alone, following an earlier election to be tried by judge and jury. A refusal to give reasons does not amount to evidence of abuse of process. Should the Crown be allowed to force an accused to have a trial by judge and jury, or should a jury trial be for the benefit and choice of the accused?

To summarize the election process (see Figure 1.1), the accused does not have an election when either the provincial court or the superior court of criminal jurisdiction have exclusive jurisdiction over the offence. For all other offences (there are only a limited number of indictable offences remaining), the accused has an election. Where an accused has an election, there are three choices:

(1)   trial by provincial court judge,
(2)   trial by superior court judge without a jury, or
(3)   trial by superior court judge and jury.

The accused's election takes place in provincial court, and section 536(2) contains the election that is read to the accused. (See section 536.1(2) for the election options in Nunavut). As of June 1, 2004, accused who elect trial by a superior court judge or by judge and jury must request a preliminary inquiry if they want one. The preliminary inquiry takes place in provincial court (see Chapter 4). There is no preliminary inquiry if the accused elects to be tried by a provincial court judge.

With the above information, you should be able to determine whether an accused has an election for any federal or provincial offence, and to identify which court or courts have jurisdiction over the offence. If the accused has an election, both levels of court will have jurisdiction over the offence until the election is made (see the Election Exercise in Box 1.1).

Overriding Elections and Re-Elections

The Crown or the court has authority to override an accused's election in certain circumstances. Under section 568 (569 in Nunavut), the Crown can require that the accused be tried by judge and jury for offences punishable by more than five years' imprisonment. In a case with several accused who make different elections (such as some accused electing trial in provincial court, some opting for trial in superior court with or without a jury), a provincial court judge can refuse to accept certain elections and the accused will be deemed to have elected to be tried by judge and jury (see sections 567 and 565; 565(1.1) in Nunavut). All of these options, which are outside of the control of the accused, tend to force the accused toward a trial by judge and jury.

Rules regarding re-election by the accused are set out in sections 561 to 563.1 of the *Criminal Code*. See Chapter 1 in Salhany (online) and Chapter 13 in Quigley (2006).

JURISDICTION

Jurisdiction in criminal matters refers to the legal authority of a court to hear a criminal matter, and to apply a penalty to a person whom it convicts. The definition of "person" is found in section 2 of the *Criminal Code* (under the definition of "every one"), and it includes public bodies and organizations (which are further defined under "organizations"). If a court does not have initial jurisdiction over an offence, any judgment rendered by such a court can be declared a **nullity**. A

Box 1.1 Election Exercise

For each of the listed offences, determine whether the accused has an election and which court(s) has (have) jurisdiction.

	Election?		Court With Jurisdiction	
	Yes	No	Provincial	Superior
1. Impaired driving by indictment	❑	❑	❑	❑
2. Murder	❑	❑	❑	❑
3. Sexual assault by indictment	❑	❑	❑	❑
4. Procuring	❑	❑	❑	❑
5. Theft not exceeding $5000 by indictment	❑	❑	❑	❑
6. Theft over $5000	❑	❑	❑	❑
7. Attempted murder	❑	❑	❑	❑
8. Keeping a gaming house	❑	❑	❑	❑
9. Conspiracy to commit murder	❑	❑	❑	❑
10. Insider trading	❑	❑	❑	❑

court that has jurisdiction initially can also lose its jurisdiction (see Salhany online, Chapter 2, for more information).

A court must have **jurisdiction over the offence** to try a person charged with the offence. A court in New Brunswick cannot try an offence committed in Prince Edward Island. The court also must have **jurisdiction over the person**. For example, the criminal courts could not try a child who is 11 years old.

Time Limitations

Certain time limitations restrict the jurisdiction or power of the court to try a case. At English common law, there were no time limits on a prosecution. As a prosecution was seen as an act of the king, and as the king could do no wrong, a lapse of time before a prosecution was not seen as a defect. The common law is still the law in Canada today, except as modified by legislation (see section 8(2) of the *Criminal Code*).

The common law has not been changed for most indictable offences. There are no time limitations on the court's jurisdiction to try these offences. However, there are exceptions in the *Criminal Code* and other federal statutes. For example, section 48(1)—treason—states that "no proceedings for an offence of treason as defined by paragraph 46(2)(a) shall be commenced more than three years after the time when the offence is alleged to have been committed." Given the number of historical prosecutions (e.g., sexual offences), some people have raised the question of whether there should be time limitations on all prosecutions (see, for example, Anand 2000). Although they do not recommend the imposition of time limitations, Connolly and Read (2003) discuss some of the problems with adjudicating historical child sexual abuse cases.

Section 786(2) governs summary conviction offences, and states that "no proceedings [under Part 27—Summary Convictions] shall be instituted more than six months after the time when the subject-matter of the proceedings arose, unless the prosecutor and the defendant so agree." The last clause of the section was added in 1997. Section 786(1) reads "except where otherwise provided by law, this part applies to proceedings as defined in this part." Thus, there is a six-month limitation period on summary conviction offences unless a statute provides for a different limitation period. For example, section 244(4) of the *Income Tax Act* states that "an information or complaint under the provisions of the *Criminal Code* relating to summary convictions, in respect of an offence under this Act, may be laid or made at any time within but not later than 8 years after the day on which the matter of the information or complaint arose."

According to section 786(2) of the *Criminal Code*, the prosecutor usually has six months from the time of the offence to "institute" proceedings for summary conviction offences. Charges are "instituted" when an information has been "laid" (discussed in more detail in Chapter 3). Section 504 speaks of laying "an information in writing and under oath before a justice." The swearing of a written information (see Form 2 at the back of the *Code)* can be considered the act that interrupts the time limitation; that is, proceedings are instituted or commenced by the laying or swearing of an information before a justice. So long as the information is sworn before the time limitation expires, the court will have jurisdiction (at least as far as time is concerned).

When an offence is a dual or hybrid offence, the Crown can proceed either by way of summary conviction or by way of indictment. The time limitation will depend on which way the Crown proceeds. In a sense, this choice allows the Crown to bypass the summary conviction limitation period for hybrid offences. A question arises whether the Crown can proceed by way of indictment simply because it is out of time to proceed summarily. What if it starts summarily and then switches to proceeding by indictment because it realizes the time limitation has expired?

After a number of conflicting lower court decisions on the question of whether a prosecutor who commences an action by way of summary conviction can stay the charge and recommence by way of indictment to bypass the limitation period, the Supreme Court of Canada recently resolved this issue. According to the majority in *Dudley*, if the Crown proceeds by way of summary conviction after the expiration of the limitation period, "the initial election and all subsequent proceedings are a nullity. They can therefore have no effect on the Crown's ability to proceed by indictment" (para. 43). There would be an exception to this conclusion if the evidence disclosed "an abuse of process arising from improper Crown motive, or resulting prejudice to the accused sufficient to violate the community's sense of fair play and decency"(para. 44). If the Crown proceeds by way of summary conviction on a hybrid offence, with proceedings commenced after the limitation period has expired, and if the accused is then convicted of a summary conviction offence, the accused may appeal. Such an action would result in the proceedings being declared a nullity and the Crown could then proceed by way of indictment unless the court decided the proceedings would be an abuse of process (para. 5). However, the Crown could not appeal an acquittal on the basis of the expired limitation period, because it is the Crown's responsibility "to ensure that the proceedings were properly instituted" (para. 6).

Territorial Limitations

Generally speaking, an accused cannot be convicted in Canada for an offence committed outside of Canada. Section 6(2) of the *Criminal Code* contains that general principle, with the proviso, "subject to this act or any other act." Thus, unless a statute states otherwise, an accused cannot be tried for an offence committed outside of Canada.

There were always exceptions to this general rule at common law (e.g., piracy), and a number of other exceptions have added in the *Criminal Code*. See, for example, sections 7(1) (offences committed on an aircraft), 7(3.2) (possession of nuclear material), 7(4.1) (sexual interference with a person under the age of 14–see Perrin 2009), 57 (passport forgeries), 58 (fraudulent use of citizenship papers), 83.01 "terrorist activity," 465(4) and (5) (specified conspiracies), and 290(1)(b) (bigamy).

A general rule regarding **territorial jurisdiction** is found in section 478(1) of the *Code*: "subject to this Act, a court in a province shall not try an offence committed entirely in another province." Again, there are provisions in the *Code* that deal with offences committed on territorial boundaries, and with offences that occur in more than one province or territory (see section 476).

Waiver of Charges

It is possible, under some circumstances, for an accused who has committed an offence in one province to plead guilty to the offence in another province. Section 478(3) allows for the **waiver** of charges for offences, other than those listed in section 469, from one province to another. An accused may wish to plead guilty in Vancouver to an offence committed in Winnipeg for many reasons: the accused now resides in Vancouver, the expected penalty might be less in Vancouver, family and friends may be less likely to hear about it, and so on. Section 479 allows for a waiver of charges within a province, under certain circumstances. The consent of the applicable prosecutor is required for these types of waivers.

Change of Venue

At early common law, an accused was tried by jurors who lived in the same community as the accused, and who had knowledge of the facts surrounding the offence. A rule developed that an accused could only be tried in the neighbourhood or county where the offence took place. The term for the location from which the jury was drawn was known as the **venue.** Today, the venue of a trial refers to the location where the accused is to be tried; generally speaking, this location will be the county or municipality where the offence occurred.

A **change of venue,** that is, of the location where a trial takes place, is somewhat rare, but does occur. Section 599 allows the Crown or the accused to apply to have the trial moved to another territorial division of the same province. The application might be made under circumstances, for example, where adverse comments in the local newspaper might affect the impartiality of jurors.

In the case of *Collins,* the Ontario Court of Appeal held that the decision to change the venue of a trial within a province is within the trial judge's discretion. The test in deciding whether the accused should be allowed a change of venue was whether the accused could receive a "fair trial, with an impartial jury." In the *Collins* case, the accused was charged with the first-degree murder of a police officer at a shopping centre (the accused approached a police officer, who was on his coffee break, and told the officer that his time was up). The trial judge took into account that the newspaper stories, which were sympathetic toward the police and called for a return of capital punishment, were year old, and that only four of the 24 articles published referred specifically to the accused. The application for a change of venue was denied.

Box 1.2 Changes of Venue

Dr. Joseph Charalambous and a co-accused were charged with the killing of Sian Simmonds. Simmonds was a woman who was going to testify against Charalambous at a hearing before the British Columbia College of Physicians and Surgeons. The trial was moved from Vancouver to Vancouver Island in January 1994 because the media in the Lower Mainland had made the case "a *cause célèbre*." Likewise, Paul Bernardo's trial for the murder of Kristen French and Leslie Mahaffy was moved from St. Catharine's to Toronto.

In the Newfoundland case of *English,* the accused was charged with several offences arising out of allegations of abuse at the Mount Cashel Orphanage. On April 11, 1991, the trial judge heard an application for a change of venue. One problem was the extent of pre-trial publicity emanating from the Hughes inquiry, which had begun on September 11, 1989, and had ended on June 29, 1990. Also, two authors had written about the allegations of abuse based on the Hughes' Inquiry: Michael Harris, *Unholy Orders,* and Dereck O'Brien, *Suffer the Little Children* (see Appendix C). The Hughes report was withheld from circulation until after the trials of the Christian Brothers, as was a television documentary based on the inquiry. The Newfoundland Court of Appeal was of the view that the place of trial ought not to be changed unless the court is "satisfied that a full and impartial trial cannot be held" where the charges are properly laid (526). The trial judge refused to change the venue, and the Newfoundland Court of Appeal dismissed the accused's appeal.

Early attempts to use social science research in applications to change the venue of trials were unsuccessful. A professor at Queen's University in Kingston, Ontario, who conducted a survey for a 1976 trial, concluded that there was "a firm and reasonable probability of partiality or prejudice." The judge dismissed the application for a change of venue, and the social scientist's research was not well received (see Arnold and Gold 1978–79; also see Vidmar and Judson 1981). Although this case appears to have dampened social scientists' interest in conducting such research, more recently, academics are again returning to the legal arena to offer their expertise on change of venue and other issues (Ogloff and Vidmar 1994; Freedman and Burke 1996; also see Chapter 5).

Prior to the *Charter*, the defence sought such a change of venue in the Saskatchewan case of *Threinen,* involving the gruesome murders of four children (ages four to six) near Saskatoon in the early to mid-1970s. There was a great deal of publicity about the case at the time. Defence counsel applied to have the trial moved to Winnipeg, arguing for a creative interpretation of the *Criminal Code,* as well as stressing the accused's right to a fair and impartial hearing under the *Bill of Rights*. The Saskatchewan Court of Queen's Bench decided that the *Criminal Code* did not allow for the venue of a trial to be moved from one province to another and that the *Bill of Rights* argument would require the judge to act outside the judge's legislative authority. The application accordingly did not succeed. What might the courts now do with a case similar to *Threinen*? Would the Supreme Court of Canada be more prepared to rewrite legislation today under the *Charter* than it was in 1976 under the *Bill of Rights*? Larry Fisher, accused in the 1969 rape and killing of Gail Miller in Saskatoon after David Milgaard was exonerated in 1997, requested that his trial be moved out of the province of Saskatchewan. His request was refused on February 19, 1999, and he was later convicted of murder.

THE EFFECT OF DELAY ON *CHARTER* RIGHTS

Since 1982, the *Charter* also has to be considered in terms of time limitations. Section 11(b) of the *Charter* states that "any person charged with an offence has a right to be tried within a reasonable time." While this requirement does not affect the court's jurisdiction, a violation of this right can result in the court entering a judicial stay of proceedings under section 24(1) of the *Charter*. In this instance, time plays a major role in determining whether a *Charter* right has been violated, and whether an accused can seek a remedy under section 24(1).

In examining delay under section 11(b) of the *Charter*, it is important to look at the time sequence of an offence as it makes its way through the criminal justice system:

- step 1, the offence
- step 2, the complaint to the police
- step 3, the investigation (may include search warrants or wiretap authorizations)
- step 4, charges are laid
- step 5, the preliminary inquiry (optional)
- step 6, possible adjournments
- step 7, the trial
- step 8, the decision of the trial judge or jury
- step 9, the sentencing
- step 10, the appeal
- step 11, the decision on appeal
- step 12, the retrial (if ordered)

Pre-Charge Delay

Does pre-charge delay count under section 11(b) of the *Charter*? Pre-charge delay is the time between the commission of the alleged crime and when the accused is charged. Remember that a charge is not instituted until it is laid, but the *Charter* does not say anything about the right to have a charge *instituted* in a reasonable time. It reads "a right to be tried within a reasonable time."

According to the Supreme Court of Canada in the *Carter* case, pre-charge delay is generally not relevant to a person's rights under section 11(b) to be tried within a reasonable time. In the *Carter* case, the police had problems locating the complainant (after they received her complaint, it took about two and one-half years to locate her for further information), and the laying of the information took place almost three years after the investigation started. The accused was first informed of the investigation a couple of weeks before the charges were laid.

The provincial court judge at the preliminary inquiry entered a stay of proceedings because the time delay was "totally unreasonable." The Crown then made an application to the British Columbia Supreme Court for *certiorari* and *mandamus*. An application for *certiorari* is an application to a higher court to quash (i.e., overturn) the decision of a lower court, in this case to quash the stay of proceedings entered by the Provincial Court judge. An application for *mandamus* is an application to have a higher court order a lower court to do something, in this case to conduct the preliminary hearing into the offences charged in the information.

The British Columbia Supreme Court granted the Crown's application, quashed the Provincial Court judge's decision, and ordered the judge to proceed with the preliminary hearing. The accused then appealed to the British Columbia Court of Appeal, which dismissed his appeal and upheld the Supreme Court's decision that the preliminary inquiry was to proceed. The accused further appealed to the Supreme Court of Canada, which dismissed his appeal.

Mr. Justice Lamer, who wrote for the majority, stated that "the time-frame to be considered in computing trial within a reasonable time generally runs only from the moment a person is charged." He added, "I say generally because there might be exceptional circumstances under which the time might run prior to the actual charge on which the accused will be tried" (1605). The example the court gave was that if the Crown withdrew a charge and substituted a different charge based on the same transaction, the computation of time might start from the laying of the first charge.

Carter was not even aware that there was any complaint or investigation until shortly before he was charged. What if an accused is arrested and then not charged until some time later? When should the time begin to run in calculating whether the accused is tried within a reasonable time? The Supreme Court of Canada considered this issue in *Kalanj*. The accused was arrested without warrant for theft and released the same day, but charges were not laid until eight months later because the Crown needed time to review the wiretap evidence. Should these eight months be considered in deciding whether the accused was tried within a reasonable time? Mr. Justice McIntryre, in his decision for the majority (three out of five justices), said that the delay prior to the swearing of the information (that is the laying of a charge) was not relevant under section 11(b). There was a strong dissent, but McIntryre, J. was of the view that pre-charge delay could be dealt with under the *Criminal Code*, or section 7 of the *Charter* (1610-11).

Pre-charge delay can be lengthy, especially in cases of sexual or physical abuse of children. In *L(W.K.)*, an accused was charged in 1986 with sexually abusing his stepdaughter and two daughters between 1957 and 1985. According to the Supreme Court of Canada, the passage of time alone will not save an accused from trial, especially in sexual abuse cases, where part of the reason for the delay in reporting is the effect of the abuse on its victims. If the pre-charge delay affected the fairness of the trial, the accused could argue under section 7 of the *Charter* that the delay in prosecution violated his right to a fair hearing, or constituted an abuse of process.

Post-Charge Delay

The most common form of delay considered by the courts is the delay that occurs between the laying of the charge and the trial. Carl Baar (1993), a political scientist from Brock University who swore an affidavit in the *Askov* case, based on his research on court delay, suggested that there are at least three ways to deal with the concept of "trial within a reasonable time." First, Parliament could legislate a fixed rule (a legislative approach) such that all trials would have to be held within a certain time, subject to specified exceptions (such as waiver). The Law Reform Commission of Canada suggested this approach in 1978, but Baar considers this approach to be too rigid. Second, the court could consider "whether the circumstances surrounding delay in an individual case justify dismissal" (1993, 309). According to Baar, this approach, which is now used in the United States, is far too flexible and amounts to no standard at all. The third approach, the "comparative approach," provides a compromise between the first two approaches. With this approach, judges would examine delay in comparable jurisdictions to arrive at a decision about whether an accused was tried within a reasonable time.

The Supreme Court of Canada used a comparative approach in the *Askov* case, which was highly criticized and had a severe impact on the processing of cases in Ontario (see Box 1.3). Seventeen months later in *Morin*, the Court case used a more open-ended, flexible approach. This more flexible approach was recently confirmed by the Court in *Godin* (2009, para. 18) in which Cromwell, J. (for the Court) adopted the remarks of Sopinka J. in *Morin*:

> [t]he general approach . not by the application of a mathematical or administrative formula but rather by a judicial determination balancing the interests which [s. 11 (b)] is designed to protect against factors which either inevitably lead to delay or are otherwise the cause of delay.

Box 1.3 Delay—*Askov*

Several things happened after the *Askov* case. During the next year in Ontario, it is estimated that the Crown withdrew or stayed 47,000 criminal charges, purportedly because they would not pass The *Askov* tests. The public was outraged at the number of charges being stayed, given the seriousness of some of them. Mr. Justice Cory (who wrote the majority decision in *Askov*) made a public statement to the effect that the court had no idea that the *Askov* decision would have the impact that it did. Some judges called *Askov* a public relations disaster for judges. Subsequently, the Ontario Government spent $39.2 million to hire 27 more provincial court judges, 61 new prosecutors, and 168 court staffs to reduce delays.

Mr. Justice Sopinka in *Morin* described the accused's interest under section 11(b):

The right to security of the person is protected in s. 11(b) by seeking to minimize the anxiety, concern and stigma of exposure to criminal proceedings. The right to liberty is protected by seeking to minimize exposure to the restrictions on liberty which result from pre-trial incarceration and restrictive bail conditions. The right to a fair trial is protected by attempting to ensure that proceedings take place while evidence is available and fresh (786).

He suggested an 8- to 10-month guideline for tolerable delay between charges being laid and a trial in provincial court. Baar was of the view that the courts added two months for "inherent time requirements," and concluded that given the rapid growth of cases in Oshawa and the lack of prejudice to the accused, the delay of 14 and a half months was not unreasonable (1993, 322).

Although Cromwell, J., in *Godin*, noted that "there is a strong societal interest in having serious charges tried on their merits" (para. 41), he also quoted McLachlin J. (as she then was) in *Morin*, describing problems with delayed trials in these terms:

Witnesses forget, witnesses disappear. The quality of evidence may deteriorate. Accused persons may find their liberty and security limited much longer than necessary or justifiable. Such delays are of consequence not only to the accused, but may affect the public interest in the prompt and fair administration of justice" (Para. 40).

The Supreme Court of Canada has suggested four factors that a judge should consider in determining whether a delay is unreasonable, and therefore in violation of the accused's right under section 11(b) of the *Charter* to be tried within a reasonable time:

(1) the length of the delay;
(2) the reason(s) for the delay, including:
     (a) inherent time requirements of the case,
     (b) actions of the accused [e.g., consenting to or requesting an adjournment],
     (c) actions of the Crown,
     (d) limits on institutional resources [systemic delay], and
     (e) other reasons for the delay;
(3) waiver of time periods; and
(4) prejudice to the accused (*MacDougall* para. 40, summarizing the factors as discussed in *Askov* and in *Morin*).

The length of the delay has to be sufficient to raise an issue of prejudice to the accused (*MacDougall* para. 44) before a judge moves on to consider the other factors. According to the Supreme Court of Canada, a waiver "must be clear and unequivocal, with full knowledge of the rights the procedure was enacted to protect and of the effect that waiver will have on those rights" (*Askov* para. 65). Also as a result of the *Morin* case, there is an increased requirement on the accused to establish some specific prejudice beyond that occurring from the delay itself.

Delay Between Verdict and Written Reasons

In *Teskey*, the trial judge took four months to deliver a guilty verdict, indicating that reasons would follow. Eleven months later, long after an appeal was launched, the trial judge issued written reasons. The Alberta Court of Appeal considered the written reasons in its judgment. The Supreme Court of Canada found that in the circumstances of the case, "a reasonable person would apprehend that the trial judge's written reasons, delivered more than 11 months after the verdict was rendered, did not reflect the real basis for the convictions" (2007, para. 2). The Court allowed the accused's appeal. and ordered a new trial.

Box 1.4 How Slow is Slow?

The trial of Canadian Albert Walker in England, accused of killing his business partner and taking on his identity, took 11 days (see Cairns 1998 and Schiller 1998 in Appendix C). Experts say that in Canada it would have taken six weeks. Alan Gold, a Toronto criminal lawyer, suggested that the trial of O.J. Simpson in the United States (O.J. was accused of killing his wife) would have taken half as long in Canada, but only one quarter the time in England (Ibbitson and Bindman, 1998, A6). The issue of mega-trials has recently become an issue in Canada Code, 2008).

Delay Between Conviction and Sentencing

The Supreme Court of Canada considered the impact of delay between conviction and sentencing in *MacDougall,* where the trial judge fell ill and later resigned. Madame Justice McLachlin, for the court, found that section 11(b) of the *Charter* also covers the sentencing component of a trial, and applied the factors enumerated in *Askov* and *Morin*. She found that the prosecutor had no reason to believe that the trial judge would not return to duties prior to his resignation, and therefore was not at fault. The delay was systemic, and not unreasonable in the circumstances. In addition, MacDougall suffered no prejudice, and never pressed the Crown to proceed with the sentencing hearing. She concluded that his rights under section 11(b) were not violated.

Delay Caused by Appeals

In *Potvin,* the Supreme Court of Canada decided that section 11(b) of the *Charter* did not apply to delay caused through the appellate process. The court did, however, suggest that section 7 of the *Charter* may apply in some circumstances, where a trial "would violate those fundamental principles of justice which underlie the community's sense of fair play and decency." In such circumstances, the court could use section 7 to "prevent the abuse of a court's process through oppressive or vexatious proceedings."

SUMMARY

There are three types of criminal offences in Canada. Indictable offences are generally the most serious, contain the harshest potential penalties, and can only be created by Parliament. Summary conviction offences are generally the least serious and can be created by either Parliament or the provincial legislatures. Hybrid offences are created by the federal government and can be proceeded with as either indictable or summary conviction offences, at the option of the prosecutor.

Many offences are within the absolute jurisdiction of the provincial court, including all provincial offences, all summary conviction offences, and all hybrid offences that are proceeded with by way of summary conviction. Additionally, all indictable or hybrid offences listed in section 553 of the *Criminal Code* fall within the absolute jurisdiction of the provincial court. The most serious of the remaining offences, listed in section 469, are within the exclusive jurisdiction of the superior courts. What is left are those indictable offences in respect of which an accused may elect the mode of their trial: in provincial court, in superior court by a judge alone, or in superior court by a judge and jury. Even in this situation, however, it is possible for the Crown or the court to override the accused's election in some circumstances.

Jurisdiction is a term referring to the power of a court to try a case, and it encompasses several concepts. Jurisdiction over the offence refers to the authority of a given court or level of court to hear trials of certain types of offences. Jurisdiction over the person refers to whether the court has power over a given person or class of persons to try their cases. Jurisdiction may also be restricted by time limitations. Territorial jurisdiction refers to whether a court can try offences that occurred in various locations. Except for offences in section 469, an accused may waive a charge to a court in another province that would otherwise lack territorial jurisdiction, in order to plead guilty. While a charge should ordinarily be heard in the location the offence is alleged to have been committed, it is possible for the court to order a change in the venue of the trial within the province to ensure a fair trial.

Section 11(b) of the *Charter* guarantees the right to be tried within a reasonable time. The courts have concluded that in most situations, pre-charge delay is not relevant under section 11(b), and that only delay between the laying of charges and the time of trial (through to sentencing) is important. Factors to be considered by the courts in deciding whether delay is unreasonable include the length of delay, the reasons for the delay, whether the accused waived the right in respect of any part of the delay, and any prejudice to the accused. Generally, the fact that a party avails itself of a right to appeal will not give rise to an unreasonable delay in the context of a subsequent re-trial.

QUESTIONS TO CONSIDER

(1)   Should the Crown have the authority to decide whether hybrid offences are proceeded with by way of summary conviction or indictment? What are the advantages and disadvantages to this approach? What are the alternatives?

(2)   What alternates exist for the Crown when it starts a proceeding by way of summary conviction and then discovers that it is out of time to proceed summarily?

(3)   Can an accused who is charged with murder in Montreal waive the charge to Vancouver in order to pled guilty and be sentenced in Vancouver? Why or why not?

(4)   For what reasons might an accused or the Crown be allowed to move the venue of a trial from Vancouver to Victoria?

(5)   When is a charge "instituted"?

(6)   From what time does the limitation period for a summary conviction offence generally begin?

(7)   Does the right to a trial within a reasonable time extend to sentencing? To the appeal process?

(8)   What factors will a court consider when it evaluates whether a person's right to be tried within a reasonable time was violated?

(9)   Under what circumstances would a delay of 25 years between the commission of a murder and the laying of the charge be considered a violation of an accused's right under the *Charter*? Which section would be applicable?

(10)   What are the three approaches that Carl Baar suggests the courts can take, regarding the concept "trial within a reasonable time"? Which approach does he favour, and why?

(11)   What is *certiorari*? What is *mandamus*?

BIBLIOGRAPHY

Anand, Sanjeev S. "Should Parliament Enact Statutory Limitation Periods for Criminal Offences?" (2000) 44 *Criminal Law Quarterly* 8.

Arnold, Stephen J., and Alan D. Gold. "The Use of Public Opinion Poll on a Change of Venue Application."(1978–79) 21 *Criminal Law Quarterly* 445.

Baar, Carl. "Criminal Court Delay and the Charter: The Use and Misuse of Social Facts in Judicial Policy Making." (1993) 72 *Canadian Bar Review* 305.

Baar, Carl. "Trial Court Reorganization in Canada: Alternative Futures for Criminal Courts." (2004) 48 *Criminal Law Quarterly* 110.

Barton, P.G. "Why Limitations in the Criminal Code."(1998) 40 *Criminal Law Quarterly* 188.

Béliveau, Pierre, Jacques Bellemare, and Jean-Pierre Lussier. *Our Criminal Procedure*. Cowansville: Les Èditions Yvon Blais Inc., 1982.
        Title I, Subtitle 1, Chapter II, Jurisdiction of Courts, discusses jurisdiction over the offence (indictable, summary, and provincial offences), jurisdiction over the person (including immunities, Canadian and Foreign Soldiers, and minors) and territorial jurisdiction (offences committed in Canada, in a province, and in a territorial division). Each topic is followed by a discussion of the effects of lack of jurisdiction. Tme limitations and their effects are discussed.

Code, Michael A. *Trial Within a Reasonable Time: A Short History of Recent Controversies Surrounding Speedy Trial Rights in Canada and the United States.* Toronto: Carswell, 1992.

CHAPTER 1: *Classification of Offences, Elections, and Jurisdiction of the Court*

Code, Michael. "Law Reform Initiatives Relating to the Mega Trial Phenomenon." (2008) 53 *Criminal Law Quarterly* 421.

Connolly, Deborah A. and J. Don Read. "Remembering Historical Child Sexual Abuse" (2003) 47 *Criminal Law Quarterly* 438.

Coughlan, Stephen G. "Trial Within a Reasonable Time: Does the Right Still Exist?" (1992) 12 *Criminal Reports* (4th) 34.

Department of Justice. "Backgrounder: Cannabis Reform Bill" (November 2004); www.justice.gc.ca/eng/news-nouv/nr-cp/2004/doc_31276.html (accessed November 30, 2009).

Erickson, Patricia G., Andrew D. Hathaway and Cristine D. Urquhart. "Backing into Cannabis Reform: The CDSA and Toronto's Diversion Experiment." (2004) 17 *Windsor Review of Legal and Social Issues* 9.

Freedman, J.L., and T.M. Burke. "The Effect of Pre-trial Publicity: The Bernardo Case." (1996) 38(3) *Canadian Journal of Criminology* 253.

Freidland, Martin L. "The Provincial Court and the Criminal Law." (2004) 48 *Criminal Law Quarterly* 15.

Healy, Patrick. "Constitutional Limitations Upon the Allocation of Trial Jurisdiction to the Superior or the Provincial Court in Criminal Matters." (2004) 48 *Criminal Law Quarterly* 31.

Henry, Stuart and Mark M. Lanier (eds.). *What is Crime? Controversies over the Nature of Crime and What to Do about It.* Lanham, Maryland: Rowan & Littlefield Publishers Inc., 2001.

Hillyard, Paddy, Christina Pantazis, Steve Tombs, and Dave Gordon (eds.). *Beyond Criminology: Taking Harm Seriously.* Blackpoint, Nova Scotia, Pluto Press, 2004.

Ibbitson, John, and Stephen Bindman. "Justice Probably Swifter in Britain than in Canada." (7 July 1998) *Vancouver Sun* A6.

Law Commission of Canada (ed.). *What is a Crime?* Vancouver: UBC Press, 2004.

Law Reform Commission of Canada. *Classification of Offences.* Ottawa, 1986.
      In Chapter 3, the Commission traces the origin of our present classification system and points out that it is primarily the result of constitutional law in Canada. The comparison between England (where the courts decide which way hybrid offences are to be prosecuted) and Canada (where the prosecutor makes this decision) provides good material for discussion.

      In Chapter 4, the Commission distinguishes between real crime (characterized by stigma, solemn trial, and imprisonment) and infractions and proposes separate codes for each category (the *Code of Criminal Procedure* and an *Infractions Procedure Act*). Crimes would be divided into those punishable by two years of imprisonment or less, and those punishable by more than two years imprisonment (including the possibility that first offences could fall within the first category and subsequent offences could fall in the latter category). Infractions would not be punishable by imprisonment and would be enforced under federal regulatory legislation. Hybrid offences would be abolished. Criticisms of the present system include the discretion exercised by prosecutors to proceed by indictment when the period has lapsed for proceeding by way of summary conviction.

Levesque, J.F.R. "Trial Within a Reasonable Time." (1988–9) 31 *Criminal Law Quarterly* 55.

Mosher, Janet and Joan Brockman (eds.) *Constructing Crime: Contemporary Processes of Criminalization* (Vancouver: UBC Press, 2010).

Ogloff, J.R.P., and N. Vidmar. "The Impact of Pretrial Publicity on Jurors: A Study to Compare the Relative Effects of Television and Print Media in a Child Sex Abuse Case."(1994) 18(5) *Law and Human Behavior* 507 (see Chapter 5).

Perrin, Benjamin. "Taking a Vacation from the Law? Extraterritorial Criminal Jurisdiction and Section 7(4.1) of the Criminal Code." (2009) 13 *Canadian Criminal Law Review* 175.

CHAPTER 1: *Classification of Offences, Elections, and Jurisdiction of the Court*

Quigley, Tim. *Procedure in Canadian Criminal Law,* 2nd ed. Toronto: Carswell, 2006.
> Chapter 3 examines types of offences, Chapter 9 discusses compelling the accused, Chapter 13, elections and re-elections, and Chapter 19, trial within a reasonable time.

Roach, Kent, Gary Trotter and Patrick Healy. *Criminal Law and Procedure: Cases and Materials*, 9th ed. Toronto: Emond Montgomery Publications Limited, 2004.
> Chapter 4, "The Criminal Trial Process," contains a discussion on pre-trial publicity and change of venue.

Rosenthal, Peter. "Crown Election Offences and the Charter." (1990–91) 33 *Criminal Law Quarterly* 84.

Salhany, Roger E. *Canadian Criminal Procedure,* 6th ed. Toronto: Canada Law Book Inc., (available online on Criminal Spectrum).
> Chapter 1, "Classification of Offences," sets out the history of the classification of offences in Canada. It then considers the implications of the present classification of indictable and summary conviction offences in terms of (a) mode of trial, (b) procedure on appeal, and (c) sentencing. Procedures on re-election are also discussed. Chapter 2 discusses limitations on the jurisdiction of the court: (a) limitation by time and (b) limitation by territory (extra-Canadian, interprovincial, intra-provincial, and change of venue).

Seniuk, Gerald T.G. and John Burrows. "The House of Justice: A Single Trial Court." (2004) 48 *Criminal Law Quarterly* 126.

Skelton, Chad, with Nahlad Ayed. "Fast tracking for trials wins support," *Vancouver Sun* (16 August 16 1999) A1, A2.

Stuart, Don, Ron Delisle, and Tim Quigley. *Learning Canadian Criminal Procedure*, 9th Ed. Toronto: Carswell, 2008.
> Chapter 1 on Jurisdiction considers the classification of offences and time and territorial limitations.

Sutherland, Edwin H. "White-Collar Criminality" (1940) 5(1) *American Sociological Review* 1.

Vidmar, Neil and John Judson. "The Use of Social Science in a Change of Venue Application." (1981) 59 *Canadian Bar Review* 76.

Webster, Cheryl Marie and Anthony N. Doob. "The Superior/Provincial Criminal Court Distinction: Historical Anachronism or Empirical Reality." (2004) 48 *Criminal Law Quarterly* 77.

CASES

R. v. Askov, [1990] 2 S.C. R. 1199. Discussion of factors relating to the s.11(b) right to be tried within a reasonable time, and of the method for analyzing alleged breaches of the right.

R. v. Bird (1996), 107 C.C.C. (3d) 186 (Alta. Q.B.). Court dispensed with need for Crown's consent for a trial by judge alone.

R. v. Carter (1987), 26 C.C.C. (3d) 572 (S.C.C.). The relevance of pre-charge delay in determining whether the trial has taken place within a reasonable time.

R. v. Collins (1989), 48 C.C.C. (3d) 343 (Ont. C.A.). Discusses change of venue due to pre-trial publicity.

R. v. Dudley, 2009 SCC 58.

R. v. English (1993), 84 C.C.C. (3d) 511 (Nfld. C.A.); leave to appeal to S.C.C. refused [1994] 1 S.C.R. vii. Effect of lengthy pre-charge delay, and discussion of change of venue due to pre-trial publicity. Discussion of procedure in challenging prospective jurors for cause.

R. v. Frisbee (1989), 48 C.C.C. (3d) 386 (B.C.C.A.), leave to appeal refused 50 C.C.C. (3d) vi (S.C.C.). Territorial jurisdiction of Canadian courts over offences committed on vessels in Canadian territorial waters, and the validity of the Attorney-General's consent. Scope of the special plea of *autrefois acquit.*

CHAPTER 1: *Classification of Offences, Elections, and Jurisdiction of the Court*

R. v. Godin, 2009 SCC 26. The Court confirmed the *Morin* approach to delay, and confirmed the trial judge's decision that the accused's right to be tried within a reasonable time was violated after a 30-month delay in a straightforward case where the Crown was responsible for most of the delay.

R. v. Kalanj, [1989] 1 S.C.R. 1594. Relevance of pre-charge delay in determining whether trial has occurred within a reasonable time.

R. v. L(W.K.), [1991] 1 S.C.R. 1091. Effect of lengthy pre-charge delay and the procedure to use in determining whether the trial has occurred within a reasonable time.

R. v. MacDougall, [1998] 3 S.C.R. 45. The right to be tried within a reasonable time (section 11(b) of the *Charter*) applies to the sentencing part of a trial.

R. v. Morin, [1992] 1 S.C.R. 771. Discussion of factors affecting whether the trial has occurred within a reasonable time, especially the relevance of limitations on institutional resources.

R. v. Ng, [2003] A.J. No. 489 (Alta. C.A.); application for leave to appeal to the Supreme Court of Canada dismissed; [2004] S.C.C.A. No. 33.

R. v. Potvin, [1993] 2 S.C.R. 880. Discussion of unreasonable delay, and the relevance of delay due to the appellate process.

Smythe v. R. (1971), 3 C.C.C. (2d) 366 (S.C.C.). The courts cannot interfere with the exercise of the Attorney-General's discretion under the *Criminal Code*.

R. v. Teskey, [2007] 2 S.C.R. 267.

R. v. Threinen (1976), 30 C.C.C. (2d) 42 (Sask. Q.B.). The courts cannot change the venue of a prosecution to another province.

R. v. Turpin, [1989] 1 S.C.R. 1296. Validity of the *Code* provision allowing for trial by judge alone on a murder charge only in Alberta.

CHAPTER 2: *Compelling the Appearance of the Accused and Judicial Interim Release*

CHAPTER OBJECTIVES

In studying this chapter, you should develop an understanding of the following topics and concepts:

- the various means by which a person can be compelled to attend court
- the legal criteria for arrest, with or without warrant
- the meaning and use of investigative detention
- the applicability of the *Charter* rights to counsel, and to not be arbitrarily detained or imprisoned
- the ways in which a person who has been arrested can be released before appearing in court
- the forms of bail (judicial interim release) available to the court, the procedure governing the show cause hearing, the grounds to be considered in deciding to grant bail, and the manner and order in which the forms of bail must be considered
- reverse onus situations
- ways of altering, appealing or reviewing bail
- the consequences of breaches of conditions of release

COMPELLING THE APPEARANCE OF THE ACCUSED

Prior to the bail reform amendments to the *Criminal Code* in the early 1970s, there were three ways that the police could compel an accused to attend court: by arrest under the authority of a warrant, arrest without a warrant, or by summons. A summons was (and still is) issued only after an information had been laid before a justice, and the justice has determined that there is sufficient reason to summon the accused to appear in court. When the alternative of release on appearance notice was added as a means of compelling a suspect to attend court, it allowed the police to take steps to compel court attendance of an accused even before they had sworn an information. An appearance notice provides greater efficiency in two ways: first, a person given an appearance notice does not have to be served with a summons following the swearing of an information; second, the police can more readily release accused immediately following arrest, without having to keep them in custody as the only alternative to a summons. The bail reform also allowed police officers and officers-in-charge to release suspects upon their signing of different documents, depending on the circumstances and the offences.

After examining the multiple procedures used to compel a person to attend court in light of seven principles (fairness, efficiency, restraint, protection of society, clarity, accountability and participation), the Law Reform Commission of Canada concluded that the procedures are a "'tangled web' that reduces the fairness and efficiency of the process" (1988, 27-32). This section of the chapter examines these various means of compelling an individual to attend court—by **appearance notice, summons, arrest without warrant** by private citizens ad peace officers, and **arrest with warrant.**

Appearance Notice

Section 496 allows a peace officer to issue an appearance notice, under certain circumstances, for three types of offences: offences listed in section 553, hybrid offences, and summary conviction offences. Section 501 specifies that an appearance notice must contain the name of the accused, the nature of the alleged offence, and the requirement that the accused attend court at a particular time and place. If the accused is alleged to have committed an indictable offence, the appearance notice may require the accused to attend (usually at a police station) for the purpose of the *Identification of Criminals Act*, R.S.C. 1985, Chap. I-1 (i.e., for fingerprinting and photographs–"pictures and prints"). At this stage of the proceedings, hybrid offences are treated as indictable (see section 34(1)(a) of the *Interpretation Act*, R.S.C. 1985, Chap. I-21 and *Connors* para. 69). The Supreme Court of Canada has recently recognized that there are contrary views, but has not yet resolved the issue (*Dudley* paras. 23, 73, 74).

Section 501 also requires that the appearance notice recite the text of sections 145(5) and (6), and section 502, which create the offences of failing to appear in court or for fingerprinting. Section 501(4) further requires a peace officer to request that the accused sign the appearance notice; however, if the accused refuses to sign it, or it is not signed by the accused, it is still valid.

Form 9, at the back of the *Criminal Code*, specifies the form and contents of an appearance notice "issued by a police officer to a person not yet charged with an offence." The practice is for a peace officer to:

(1)   issue an appearance notice to the accused (which means writing out what looks like a "ticket," and giving it to the accused. Note: An appearance notice requires the accused to attend court to respond to the charges, unlike the ticketing system under the *Contraventions Act*).

(2)   write up a police report.

(3)   convey the report to the prosecutor's office (directly or through a superior). Note: in British Columbia, New Brunswick, and Quebec, the prosecutor has to approve the charge(s) before they can be laid. In the other provinces, prosecutors would generally not see the report until after the charges are laid (discussed further in Chapter 3).

(4)   take the appearance notice to a Justice of the Peace, who will confirm the appearance notice (i.e., sign it).

(5)   (at the same time) present to the Justice an Information (Form 2), which sets out the charge.

(6)   swear to the truth of the contents of the Information before the justice (see Chapter 3).

In some locations, the police officer who issues the appearance notice passes the paper work onto another police officer who appears before the justice.

Section 145(5) states that an appearance notice must be confirmed by a justice; otherwise, a person who does not attend court as required by the appearance notice cannot be charged with failure to appear. So while police officers are allowed to compel the appearance of a suspect in court, it is not an offence to fail to attend unless the appearance notice has been confirmed by a justice. In practice, of course, a suspect will not know whether the appearance notice has been confirmed prior to appearing (or failing to appear) in court.

Summons

Another means of compelling the attendance of an accused is for a justice to issue a summons (see Form 6 in the *Criminal Code*). Section 509 describes the required contents of a summons. Similar to an appearance notice, the summons is directed to the accused, states the alleged offence, requires the accused to attend court at a specified time, recites the text of sections 145(4) and 510 of the *Code* regarding failure to appear, and may require the person to appear under the *Identification of Criminals Act* (for fingerprinting) if the accused is alleged to have committed an indictable offence.

In this case, the peace officer who encounters a suspect, or who investigates an allegation, will:

(1)   take or ascertain the person's name and address.
(2)   write up the report.
(3)   have the charges approved by a prosecutor (where required by provincial practice—see Chapter 3).
(4)   appear before a justice and swear an information. Where the justice considers that a case is made out for compelling the attendance of the accused, the justice will issue a summons.
(5)   serve the summons on the accused.

Section 509 requires that a summons be served by a peace officer, who shall deliver it to the accused named in it, or "if that person cannot be conveniently found, shall leave it for him at his last or usual place of abode with some inmate thereof who appears to be at least sixteen years of age." Some judges want proof that the summons was personally served on the accused before they will issue a warrant for the arrest of an accused who fails to appear, unless there is evidence that the accused is evading service. This proof or evidence is presented in an *affidavit of (attempted) service*–a document in which the person who tried to serve or who actually served the summons lists the particulars of service, or the efforts made to find the accused, and any evidence that might suggest the accused was evading service.

Arrest Without Warrant by a Private Citizen Or a Peace Officer

Section 494(1) states that:

(1) Anyone may arrest, without warrant
    (A) a person whom he finds committing an indictable offence; or
    (b) a person who, on reasonable grounds, he believes
        (i) has committed a criminal offence, and
        (ii) is escaping from and freshly pursued by persons who have lawful authority to arrest that person.

A person making such an arrest is entitled to use reasonable force (*Asante-Mensahm*). For the purposes of arrest, "indictable offence" includes hybrid offences (section 34(1)(a) of the *Interpretation Act*). A "criminal offence" includes all offences, whether enacted by the federal or a provincial government.

Section 494(2) of the *Criminal Code* allows the owner of property, or a person authorized by the owner of property, to arrest a person "whom he finds committing a criminal offence on or in relation to that property." Following an arrest, section 494(3) requires that anyone other than a peace officer "shall forthwith deliver the person to a peace officer."

The Law Reform Commission of Canada (1986, 25-26) pointed out that it is unrealistic to expect a private citizen to sort out all of these distinctions before making an arrest, and has suggested that in the absence of a police officer, any person should be allowed to arrest without warrant, "a person whom he believes on reasonable grounds is committing or has just committed a criminal offence." The Commission also recognized the fact that most arrests by "private citizens" are really made by private security guards, who are increasingly being employed by owners of shopping malls, industrial companies, apartment blocks and office towers (27). Given the uneven training received by private security guards, the Commission recommends that their powers of arrest be limited to those of private citizens, but that their duty to inform a suspect be equivalent to those of police officers (27, 51).

In *Lerke*, the Alberta Court of Appeal ruled that a tavern owner, who arrested a person under section 494(1) for re-entering a tavern after being forbidden to do so without proof of age, was exercising a government function to which the *Charter* applied. However in *Buhay*, the Supreme Court of Canada, without mentioning *Lerke*, stated that private security guards would only be subject to the *Charter* (section 8, in this case) if they "can be categorized either as 'part of government' or as performing a specific government function, or if they can be considered state agents" (para. 25, citations omitted). In *Buhay*, private security guards who searched a locker at a bus station in Winnipeg were not acting as state agents, and so section 8 of the *Charter* did not apply. In *Dell*, the Alberta Court of Appeal, although recognizing that three appellate courts in Canada had decided that the *Charter* did not apply to citizens' arrests, held that *Buhay* had not overrule *Lerke*. However, on the facts in *Dell* (investigative detention), the Court ruled that the *Charter* did not apply. In 2006, the Law Commission of Canada recommended that legislation be passed to clarify the duties of private security guards in light of the increasingly blurred line between public and private policing in Canada (2006, 74-75).

Arrest Without Warrant by A Peace Officer Only

A peace officer (see definition under section 2) usually has more options than a private citizen when it comes to arresting a suspect. Under section 495(1), a peace officer may arrest without a warrant:

> (a) a person who has committed an indictable offence or who, on reasonable grounds, he believes has committed or is about to commit an indictable offence;
> (b) a person whom he finds committing a criminal offence; or
> (c) a person in respect of whom he has reasonable grounds to believe that a warrant of arrest or committal . . . is in force within the territorial jurisdiction in which the person is found.

Section 495(1)(a) allows a protective or preventive arrest—an arrest to prevent the commission of an indictable offence. The Law Reform Commission of Canada suggested that such an arrest power should be restricted to circumstances "where there is a risk of personal injury or property damage" (1986, 23).

The term "finds committing" in section 495(1)(b) was clarified by the Supreme Court of Canada in the *Biron* case. Biron was charged with causing a disturbance in a public place by shouting, a summary conviction offence under what is now section 175. Therefore, in order for the peace officer to arrest him without warrant, he had to find him "committing" the offence. Biron was later acquitted of creating a disturbance by shouting because there was no evidence that he had been shouting. The issue before the Supreme Court of Canada was whether Biron could be convicted of resisting a peace officer in the execution of his duty (now section 129). Biron's counsel argued that Biron's acquittal on the charge of disturbing the peace meant that the peace officer did not find him committing an offence, and consequently had no authority to arrest him. If the peace officer had no authority to arrest

Biron, then Biron could hardly be convicted of resisting a peace officer in the execution of his duty, because the peace officer would have been acting beyond the scope of his duty. The Supreme Court of Canada interpreted "finds committing" to turn on "what was apparent" to the peace officer at the time of the arrest. Thus, "finds committing" means "apparently committing" (*Biron* 526). In the later case of *Roberge*, the Supreme Court of Canada held that "apparently committing" means "it must be apparent to a reasonable person" acting in the circumstances (314).

In 2001, section 83.3(4) was added to the *Criminal Code*, allowing a peace officer to arrest a person if the peace officer "suspects on reasonable grounds that the detention of the person in custody is necessary in order to prevent a terrorist activity." This power could be exercised only if there were grounds for laying an information but exigent circumstances made it impracticable to do so, or if an information has already been laid (section 83.3(4)). In 2007, the House of Commons voted not to extend the life of this section and it is no longer in effect.

The Duty Not to Arrest

Section 495(2) imposes a duty on a peace officer **not** to arrest a person without a warrant for three types of offences (those found in section 553, hybrid offences and summary conviction offences), in any case where:

> (d) he believes on reasonable grounds that the public interest, having regard to all the circumstances including the need to
>> (i) establish the identity of the person,
>> (ii) secure or preserve evidence of or relating to the offence, or
>> (iii) prevent the continuation or repetition of the offence or the commission of another offence,
>>> may be satisfied without so arresting the person, and
> (e) he has no reasonable grounds to believe that, if he does not so arrest the person, the person will fail to attend in court in order to be dealt with according to law.

Section 496 allows the police officer to issue an appearance notice, if the accused is not arrested under section 495(2).

Reasonable Grounds to Make an Arrest

What are **"reasonable grounds"** for making an arrest? In *Storrey*, the accused was charged with aggravated assault under section 268 of the *Criminal Code,* an indictable offence punishable by a term not exceeding fourteen years. At the trial, the judge found that the arrest was arbitrary, thus constituting a violation of section 9 of the *Charter* (the right not to be arbitrarily detained or imprisoned), and accordingly entered a stay of proceedings under section 24(1) of the *Charter*. The Ontario Court of Appeal found that there were reasonable grounds for the arrest, and directed a new trial.

The Supreme Court of Canada dismissed Storrey's appeal. Mr. Justice Cory, who delivered the judgment of the Court, said:

> the arresting officer must subjectively have reasonable and probable grounds on which to base the arrest. Those grounds must, in addition, be justifiable from an objective point of view. That is to say a reasonable person placed in the position of the officer must be able to conclude that there were indeed reasonable and probable grounds for the arrest. . . [The police] are not required to establish a *prima facie* case for conviction (324).

Prima facie means "on the face of it," or "presumably." In *Feeney*, the Supreme Court of Canada confirmed that the police officer must have subjective grounds; the fact that objective grounds exist

---

Box 2.1 Feeney's Fate

At Feeney's second trial, his blood-stained clothes and the $350.00 in cash were not presented as evidence. However, a discarded cigarette butt, which contained Feeney's DNA, was admitted as the police would have found it without a *Charter* breach, and the accused had no reasonable expectation of privacy regarding it (*Feeney* 1999a para. 64). On February 2, 1999, after seven days of deliberation, the jury convicted Feeney of second degree murder. In sentencing Feeney to life without parole for 12 years, Mr. Justice Oppal noted that Mr. Boyle was an "85 year old defenceless man," and commented, "This was a senseless act of violence and there is no evidence of remorse" (*Feeney* 1999b para. 10). A further appeal by Feeney to the British Columbia Court of Appeal on seven grounds was dismissed (*Feeney* 2001).

---

are not sufficient. Madame Justice L'Heureux-Dubé, in dissent, suggested that judges should not assume an officer did not have reasonable grounds simply because "a skillful cross-examination elicits the desired responses" (para. 122).

Entering a Private Dwelling to Make an Arrest

Under what circumstances may a peace officer enter a private dwelling in order to arrest someone? In the 1975 civil case of *Eccles and Bourque*, the Supreme Court of Canada decided that police officers could enter a private dwelling to make an arrest pursuant to an arrest warrant. In that case, the Court said that normally, prior to using force to enter, police officers should give:

> i) notice of presence by knocking or ringing the doorbell, ii) notice of authority, by identifying themselves as law enforcement officers, and iii) notice of purpose, by stating a lawful reason for entry. Minimally they should request admission and have admission denied although it will be recognized that there will be occasions on which, for example, to save someone within the premises from death or injury or to prevent destruction of evidence or if in hot pursuit notice may not be required (summarized in *Landry* 157).

The law in this area changed with the Supreme Court of Canada's decision in *Feeney*. Michael Feeney was charged with the 1991 bludgeoning death of his 86-year-old neighbour, Frank Boyle. Without sufficient grounds to make an arrest, the police entered Feeney's trailer to investigate their suspicion that they might find evidence against Feeney. They entered woke him, and found that his clothes were stained with blood and that he had $350 in cash under his mattress, allegedly stolen from Boyle. This evidence was found to be admissible at his trial, and Feeney was convicted. The British Columbia Court of Appeal dismissed Feeney's appeal; however, the Supreme Court of Canada allowed his appeal and ordered a new trial. Mr. Justice Sopinka noted that the emphasis on privacy in the home gained considerable importance with the introduction of the *Charter* (para. 42). The purpose of section 8 of the *Charter* is to prevent unjustified searches, and therefore police officers must have prior authorization to enter dwelling homes to make arrests. As a general rule, "warrantless arrests in private dwellings are prohibited" (para. 47). Sopinka, J. did acknowledge the "hot pursuit" exception at common law, which had been recognized by the Supreme Court of Canada in *Macooh*; however, he was not prepared to say that any other exigent circumstances would justify a warrantless entry to execute an arrest (para 47).

In *Godoy*, the Supreme Court of Canada considered the propriety of police officers entering a private dwelling following a 911 "unknown trouble call," in which the call was disconnected before the caller spoke. Chief Justice Lamer distinguished this case from that of *Feeney* and found that the police had a common law duty to respond to 911 calls. In the circumstances, they were justified in entering the dwelling house even though the accused, who answered the door, said there was "no problem." Pringle (1999) suggests that this case "represents a reversion to the pre-*Charter* Dark Ages" (1999, 227; also see Stuart 1999).

In response to the *Feeney* decision, Parliament introduced what is now section 529.3(2). Peace officers may enter a private dwelling, without a warrant, for the purpose of arresting a person in exigent circumstances. Exigent circumstances include preventing "imminent bodily harm or death to any person," and loss or destruction of evidence related to indictable offences. Section 529.4(3) requires that the peace officer not enter the dwelling without prior announcement unless he or she has reasonable grounds to believe that such announcement would expose someone to bodily harm or result in the loss of evidence.

Arrest with Warrant

Form 7 lists the various reasons and sections under which a warrant for arrest may be issued. Warrants are directed to all police officers within the territorial jurisdiction of the judicial officer issuing the warrant, under section 513 of the *Code*. The contents of the warrant are stipulated in section 511, and must include the name or description of the accused, a brief statement of the offence alleged, and an order that the accused be arrested and brought before the justice or judge issuing the warrant, or another within the same territorial division. Where applicable, the warrant will authorize entry into a specified dwelling-house. The justice is supposed to operate in a neutral manner in deciding whether to issue the arrest warrant.

Following the *Feeney* decision by the Supreme Court of Canada, and its widespread and critical media coverage, the federal government also introduced legislation to allow for what are sometimes called "Feeney warrants" or "entry warrants," that allow police officers to enter a private dwelling to make an arrest. Section 529 provides for a conventional arrest warrant to authorize entry (see Form 7); however, immediately before entering the dwelling, the police officer must still have "reasonable grounds to believe that the person to be arrested or apprehended is present in the dwelling house" (section 529(2)).

Section 529.1 allows for a "free-standing" arrest warrant, where authority to arrest pre-exists and the judge or justice is satisfied that "there are reasonable grounds to believe that the person is or will be present in the dwelling house" (see Form 7.1). Although the information for both types of warrants must be under oath, there is no requirement that the Information to Obtain a section 529.1 warrant be in writing. Both types of warrants can be obtained as telewarrants (section 529.5).

Section 529.2 allows the judge or justice to impose any terms and conditions on section 529 or section 529.1 warrants that are "advisable to ensure that the entry into the dwelling-house is reasonable in the circumstances." Either warrant may authorize a peace officer to enter a dwelling-house

> without prior announcement if the judge or justice is satisfied by information under oath that there are reasonable grounds to believe that prior announcement of the entry would
> (a) expose the peace officer or any other person to imminent bodily harm or death, or
> (b) result in the imminent loss or imminent destruction of evidence relating to the commission of an indictable offence (section 529.4(1)).

In addition, the peace officer must hold these beliefs prior to entering a dwelling-house without announcement (section 529.4(2)).

Execution of the warrant is governed by section 514, and permits arrest by the peace officers to whom the warrant is directed, either within the territorial jurisdiction of the issuing official or court, or anywhere in Canada in the event of fresh pursuit.

Arrest for Breach of Judicial Interim Release, With or Without Warrant

Section 524(1) allows a justice to issue a warrant for the arrest of an accused if the justice:

> is satisfied that there are **reasonable and probable grounds** to believe that an accused
> (a) has contravened or is about to contravene any summons, appearance notice, undertaking or recognizance that was issued or given to him or entered into by him, or
> (b) has committed an indictable offence after any summons, appearance notice, promise to appear, undertaking or recognizance was issued or given to him or entered into by him.

Section 524(2) further allows a peace officer to arrest an accused without a warrant if she or he believes on reasonable grounds that those same conditions exist. The rest of that section covers the procedure for dealing with the accused's application for judicial interim release in such a case. Section 524(4) states that if the accused was released under subsection 522(3)(a) by a judge of a superior court, the accused will be detained unless he or she shows cause why detention is not justified under 515(10).

CHARTER CONSIDERATIONS

Two sections of the *Charter* are particularly relevant when a person is arrested or detained: section 9, "the right not be arbitrarily detained or imprisoned," and section 10, the right "on arrest or detention (a) to be informed promptly of the reasons therefor; and (b) to retain and instruct counsel without delay and to be informed of that right."

The Meaning of Detention

Prior to the introduction of the *Charter*, the Supreme Court of Canada said that "detain" implies some form of "compulsory restraint," which would not include a demand for a sample of breath (*Chromiak*). After the introduction of the *Charter*, this definition of **detention** was replaced by the Supreme Court of Canada in the *Therens* case, where Mr. Justice LeDain wrote:

> In my opinion, it is not realistic, as a general rule, to regard compliance with a demand or direction by a police officer as truly voluntary, in the sense that the citizen feels that he or she has the choice to obey or not, even where there is in fact a lack of statutory or common law authority for the demand or direction and therefore an absence of criminal liability for failure to comply with it. Most citizens are not aware of the precise legal limits of police authority. Rather than risk the application of physical force or prosecution for wilful obstruction, the reasonable person is likely to err on the side of caution, assume lawful authority and comply with the demand. The element of psychological compulsion, in the form of a reasonable perception suspension of freedom of choice, is enough to make the restraint of liberty involuntary. Detention may be effected without the application or threat of application of physical restraint if the person concerned submits or acquiesces in the deprivation of liberty and reasonably believes that the choice to do otherwise does not exist (para. 57; approved of by the Supreme Court of Canada in *Grant* paras. 25, 28 and 30).

In 2004, the Supreme Court of Canada in *Mann* clarified the concept of detention:

> . . .the police cannot be said to "detain", within the meaning of ss. 9 and 10 of the *Charter*, every suspect they stop for purposes of identification, or even interview. The person who is stopped will in all cases be "detained" in the sense of "delayed", or "kept waiting". But the constitutional rights recognized by ss. 9 and 10 of the *Charter* are not engaged by delays that involve no significant physical or psychological restraint (para. 19; approved of by the Supreme Court of Canada in *Grant* para. 26).

According the Supreme Court of Canada in *Grant*, it follows that a person is detained when they are legally required to comply with a direction or demand from a police officer (2009 para. 30). A person is also detained where "a reasonable person would conclude by reason of the state conduct that he or she had no choice but to comply" (para. 44). Where there is no physical restraint or legal obligation imposed on an individual's encounter with the police, the court will examine all of the circumstances to determine whether the person is detained. The court may consider the following factors:

> a) The circumstances giving rise to the encounter as they would reasonably be perceived by the individual: whether the police were providing general assistance; maintaining general order; making general inquiries regarding a particular occurrence; or, singling out the individual for focussed investigation.
> b) The nature of the police conduct, including the language used; the use of physical contact; the place where the interaction occurred; the presence of others; and the duration of the encounter.
> c) The particular characteristics or circumstances of the individual where relevant, including age; physical stature; minority status; level of sophistication (para. 44).

For commentaries on the Court's decision in *Grant* and *Suberu* (discussed below), see Coughlan (2009) and Quigley (2009).

Investigative Detention

Historically, police officers did not have any right to detain someone for the purposes of investigating a crime (investigative detention); they could, however, arrest the person if they had grounds for the arrest. In 1993, the Ontario Court of Appeal suggested that police officers had a common law authority to detain for investigative purposes "if the detaining officer has some 'articulable cause' for the detention" (*Simpson* 500). After much confusion over, and criticism of, these expanded police powers (see for example, Gorham 2004; McCoy 2002; Patel 2001; and Striboupolos, 2003), the Supreme Court of Canada considered the issue of investigative detention in the *Mann* case.

Iacobucci, J., for the majority in *Mann*, recognized that investigative detention was a "complex area," in which changes in the law were "better accomplished through legislative deliberation than by judicial decree." Nevertheless, he said, the courts cannot "shy away from the task where common law rules are required to be incrementally adapted to reflect societal change" (para. 17). He found that while there is no "general power of detention for investigative purposes," police officers may detain for investigative purposes provided they have "reasonable grounds to detain," which he described as follows:

> The detention must be viewed as reasonably necessary on an objective view of the totality of the circumstances, informing the officer's suspicion that there is a clear nexus between the individual to be detained and a recent or on-going criminal offence. Reasonable grounds figures at the front-end of such an assessment, underlying the officer's reasonable suspicion that the particular individual is implicated in the criminal activity under investigation. The overall reasonableness of the decision to detain, however, must further be assessed against all of the circumstances, most notably the extent to which the interference with individual liberty is necessary to perform the officer's duty, the liberty interfered with, and the nature and extent of that interference (para. 34).

The police are not allowed to detain "on the basis of a hunch, nor can it become a *de facto* arrest" (para. 35). It also appears from the above passage that there has to be some "recent or on-going" criminal activity that is "under investigation." Despite this clarification of investigative detention, the decision has been the subject of criticism, leaving many unanswered questions (see, for example, Berger 2004; Fiszauf, 2007, 2008; Latimer 2007; McCoy 2004; Quigley 2004; Skiblinsky 2006; Stribopoulos 2007; Tanovich 2004).

Investigative Detention and Racial Profiling

Confusion over the role of racial profiling and investigative detention is illustrated by a suggestion from Toronto City Councillor Michael Thompson (a Black man) that the police should randomly target and question black youth because of the many shootings and killings taking place in the black community (D'Andrea 2005a). Having thought through the implications of his statement in the light of criticisms of racial profiling, Councillor Thompson withdrew his comments the next day (D'Andrea 2005b).

In a number of cases, the Ontario Court of Appeal has quoted the following definition of racial profiling, initially offered by the African Canadian Legal Clinic:

> Racial profiling is criminal profiling based on race. Racial or colour profiling refers to that phenomenon whereby certain criminal activity is attributed to an identified group in society on the basis of race or colour resulting in the targeting of individual members of that group. In this context, race is illegitimately used as a proxy for the criminality or general criminal propensity of an entire racial group (*Brown* para. 7).

The existence of racial profiling (Gorham 2004; Tanovich 2002, 2006, 2008) introduces an element of discrimination into law enforcement. In 2005, the federal government announced an initiative to further enhance "bias-free policing" (Thompson 2005), although efforts to introduce anti-racial profiling legislation have been unsuccessful (Tanovich, 2008, 660).

Arbitrary Detention—Section 9 of the *Charter*

According to the Supreme Court of Canada, "a detention not authorized by law is arbitrary," and violates section 9 of the *Charter* (*Grant* para. 54). In such a case, the court will determine whether the evidence is admissible under section 24(2) of the *Charter* (discussed further in Chapter 7).

There are some legislative provisions that provide for detention that might be considered arbitrary. If such a law violates a *Charter* right, the government will have to show that the law is "demonstrably justified in a free and democratic society," under section 1. In *R. v. Hufsky*, Mr. Justice LeDain, for the Supreme Court of Canada, found that spot checks of drivers for sobriety were arbitrary, in that there were no criteria (expressed or implied) for "the selection of drivers to be stopped and subject to the spot check procedure" (407). The Ontario *Highway Traffic Act* authorized arbitrary detention. Having decided that the legislation violated the accused's rights under section 9 of the *Charter*, LeDain, J. considered whether this violation was demonstrably justified under section 1 of the *Charter*. In arriving at the decision that the legislation was justified, he relied upon the ten volumes of material that had been filed in the Ontario Court of Appeal in an earlier decision (*Seo*), previously used by the Supreme Court of Canada in *Thomsen* (discussed below).

Right to Counsel—Section 10 of the *Charter*

Section 10 of the *Charter* states that:

> Everyone has the right on arrest or detention
>> (a) to be informed promptly of the reasons therefor;
>> (b) to retain and instruct counsel without delay and to be informed of that right.

The right to counsel is applicable to both arrest and investigative detention (*Suberu* para. 2). According to the Supreme Court of Canada, "the immediacy of this obligation is only subject to concerns for officer or public safety, or to reasonable limitations that are prescribed by law and justified under s. 1 of the Charter" (*Suberu* para. 2).

The right to counsel under section 10(b) of the *Charter* has received considerable attention in drinking and driving cases. Section 254(2) of the *Criminal Code*, which now also applies to driving while under the influence of drugs, states:

> If a peace officer has reasonable grounds to suspect that a person has alcohol or a drug in their body and that the person has, within the preceding three hours, operated a motor vehicle or vessel, operated or assisted in the operation of an aircraft or railway equipment or had the care or control of a motor vehicle, a vessel, an aircraft or railway equipment, whether it was in motion or not, the peace officer may, by demand, require the person to comply with paragraph (a), in the case of a drug, or with either or both of paragraphs (a) and (b), in the case of alcohol:

>> (a) to perform forthwith physical coordination tests prescribed by regulation to enable the peace officer to determine whether a demand may be made under subsection (3) or (3.1) and, if necessary, to accompany the peace officer for that purpose; and
>> (b) to provide forthwith a sample of breath that, in the peace officer's opinion, will enable a proper analysis to be made by means of an approved screening device and, if necessary, to accompany the peace officer for that purpose.

Note that this section covers a demand for evidence used to screen people who may later be asked to provide a sample of breath under section 254(3), to acquire further evidence of impairment under section 254(3.1) or 254(3.4). In cases prior to these 2008 changes, the Supreme Court of Canada court found that a screening device which required breath samples "forthwith" implicitly did not allow for the suspect to consult counsel. This limitation violated a person's right under section 10(b) of the *Charter* but, according to the Court, the limitation was demonstrably justified under section 1 of the *Charter* (*Thomsen*). One of the factors the Court considered was that evidence from a sample under section 254(2) (the screening sample) was not admissible in court (para. 22).

The legislative objective in section 254(2) was to reduce drinking and driving through increased detection and increased public perception of police efficiency at detection (*Thomsen* para. 21). Regarding the problem of drinking and driving in Canada, Mr. Justice LeDain (for the Court) restated ten conclusions drawn by the Ontario Court of Appeal in *Seo*, including the direct relationship between 1) drinking drivers and automobile accidents, 2) the severity of accidents and the quantity of alcohol, and 3) the number of accidents and the blood level of the drivers. He also noted that detection of impaired drivers was difficult simply through observation and that the "most effective deterrent is the strong possibility of detection" (para. 21).

Continuing with the requirements to invoke section 1, LeDain further quoted from *Edwards Books,* saying that the government must also show that:

[2.] . . . the means chosen to attain those objectives must be proportional or appropriate to the ends. The proportionality requirement . . . has three aspects: [a.] the limiting measures [that is the restrictions on the right to counsel in the requirement that a person provide a sample of breath forthwith] must be carefully designed, or rationally connected, to the objective; [b.] [the limiting measures] must impair the right as little as possible; and [c.] [the effects of the limiting measures] must not so severely trench on individual or group rights that the legislative objective, albeit important, is, nevertheless, outweighed by the abridgement of the rights (para. 20).

If legislation violates the *Charter*, and if it is not demonstrably justified under section 1, then the legislation will be found to be of no force and effect under section 52 of the *Constitution Act*.

Section 254(3) allows a peace officer to demand a sample "as soon as practicable," thus allowing time to consult with counsel prior to providing the sample, and avoiding a violation of section 10(b) of the *Charter*. Sections 254(3.1), (3.2), and (3.4) allow for demands that a suspect submit to evaluations of whether their ability to drive is impaired by alcohol or drugs "as soon as practicable." Again, this implies that the suspect first has a right to consult with counsel, and if the proper procedures are followed and there are no *Charter* violations, these tests are admissible against the accused.

PROTECTION AND LIABILITY OF PERSONS ENFORCING THE LAW

A number of sections of the *Criminal Code* protect those enforcing the law from criminal and civil liability. Generally, persons required or authorized by law to enforce the law are justified in using "as much force as is necessary," so long as they act on reasonable grounds (see Section 25 of the *Criminal Code*, and *Asante-Mensahm*, for a discussion of the common law). Any one who uses excessive force is criminally liable (section 26). Section 28 provides some protection for a person who arrests the wrong person, and section 29(2) states that anyone who arrests a person is required, where it is feasible, to tell that person the reasons for the arrest. Sections 32 and 33 deal more specifically with riots.

There are other means of deterring unlawful arrest. Those who unlawfully arrest someone may be subject to a civil suit, and could be required to pay damages to the person arrested. Police officers are also subject to discipline, through their own internal disciplinary mechanisms or by local police boards.

RELEASING A SUSPECT

Section 497(1) requires a peace officer who arrests a person without a warrant for offences listed in section 496 (offences in section 553, hybrid offences, and summary conviction offences) to release the person "as soon as practicable." However, the peace officer shall not release the person if the peace officer believes on reasonable grounds that it is necessary, in the public interest, to detain the person, in order to:

(i) establish the identity of the person,
(ii) secure or preserve evidence of or relating to the offence,
(iii) prevent the continuation or repetition of the offence or the commission of another offence, or
(iv) ensure the safety and security of any victim of or witness to the offence (section 497(1.1)).

In addition, the peace officer should not release the person if the peace officer believes that the person will fail to attend court if released (section 497(1.1)(b)).

If the peace officer does not detain the person, the peace officer can issue the suspect an appearance notice, or may record the necessary information with the intention of compelling the person's appearance in court by way of summons. Section 497(2) provides for certain exceptions to this requirement to release suspects in cases where the suspect is arrested without a warrant in a province other than the province within which the offence took place.

Where a peace officer does not release a suspect under section 497, but instead takes the suspect into custody, section 498(1) requires the **officer in charge** (see section 493 for a definition) or another peace officer (this option was added in 1997) to release the suspect "as soon as practicable" (for offences under section 553, hybrid offences, summary conviction offences, and indictable offences punishable by imprisonment for five years or less). However, under section 498(1.1), the person is not to be released if the officer in charge or other officer has reasonable grounds to believe that it is necessary to detain the person for the same reasons set out in section 497(1.1). Again, there is an exception (section 498(2)) for individuals arrested without a warrant, if they are arrested outside the province where the offence took place.

Under section 498(1), the officer in charge or another peace officer has a number of options for releasing a suspect. Subject to the exceptions set out in section 498(1.1), the officer shall

a) release the person with the intention of compelling their appearance by way of summons;
b) release the person on their giving a promise to appear;
c) release the person on the person's entering into a recognizance before the officer in charge or another peace officer without sureties in an amount not exceeding $500. . . but without deposit of money or other valuable security; or
d) if the person is not ordinarily resident in the province . . . or does not ordinarily reside within 200 kilometres of the place in which the person is in custody, release the person on the person's entering into a recognizance before the officer in charge or another peace officer without sureties in an amount not exceeding $500 that the officer directs and, if the officer in charge so directs, on depositing with the officer a sum of money or other valuable security not exceeding in amount or value $500, that the officer directs.

Section 493 defines the various terms. A **promise to appear** (see form 10) is a document signed by the accused, and section 501 specifies its contents. A **recognizance** is an acknowledgement that the accused owes Her Majesty a sum of money, which will be forfeited to the Crown if the person fails to attend court. The obligation is discharged if the accused meets the conditions in the recognizance. One condition of the recognizance, of course, is to show up for trial and all other scheduled appearance dates (See Form 11). Section 501 also provides the contents of a recognizance. **No deposit** means the person does not have to leave any money or personal property with the officer in charge. A **surety** is someone who undertakes to ensure that the accused will appear in court at the required time, and who pledges an amount of money or property to support their promise. If the accused does not appear, the surety may be required to pay that amount. Sureties should be informed of any changes to the bail conditions of the accused. If they are not, they may not be liable for the amount posted (*U.S.A. v. MacFarlane*).

Section 499 provides a procedure similar to that in section 498, whereby the officer in charge can release a suspect who has been arrested under a warrant (if the warrant is endorsed by a justice under section 507(6) in Form 29). In such a case the officer in charge has the option to ("may") release the accused, whereas under section 498, the officer in charge or another peace officer must ("shall") release the accused, unless the criteria of the exceptions in the section are met. In 1997 (with amendments in 1999), section 499(2) was added to allow an officer in charge to require a person to enter into an undertaking in Form 11.1, with conditions as allowed by paragraphs (a) through (h). The conditions include depositing the person's passport with the police, abstaining from alcohol or drugs, abstaining from communicating with victims, witnesses or other persons, etc. In 1999, the section was amended to allow the officer to require the person to "comply with any condition specified in the undertaking that the officer in charge considers necessary to ensure the safety and security of any victim or witness to the offence." The 1997 and 1999 amendments give more discretion and power over the release of suspects to the police. Prosecutors and persons who enter into these undertakings required by officers in charge may apply to a justice to have them replaced (sections 499(3) and (4) and 503(2.2) and (2.3)).

FAILURE TO APPEAR

The forms of appearance notice, summons, promise to appear and recognizance all refer to sections 145(5) and (6), and section 502 of the *Criminal Code*. These sections provide the consequences for **failing to appear** (sometimes abbreviated to **FTA**).

    Section 145(5) is the charging section for a person who fails to appear "without lawful excuse the proof of which lies on the person," at the time and place set out in the appearance notice, promise to appear or recognizance. This offence includes both failing to appear in court and failing to attend at the police station for photos and fingerprinting under the *Identification of Criminals Act*. Section 145(6) states that a defect in how the offence is worded is not a lawful excuse for not appearing. Failure to appear is a hybrid offence, and the penalty is a maximum of two years imprisonment if the Crown proceeds by indictment, or six months if the Crown proceeds by way of summary conviction. While this may not appear to be a serious offence, failings to appear are often treated harshly by the court because they are essentially breaches of court orders. They are also considered very seriously when an accused is seeking bail (now referred to as **Judicial Interim Release**, discussed below) in respect of a new charge. Section 502 allows a justice to issue a warrant for the arrest of the accused if the accused does not appear for the purposes of the *Identification of Criminals Act*.

    Section 512 allows a justice, under certain circumstances, to issue a warrant for the arrest of an accused who has been released on an appearance notice, promise to appear, or recognizance, or who has been served with a summons. Section 597(1) allows for a warrant to be issued if the accused does not remain in attendance throughout his or her trial. Note that the appearance notice, promise to appear, or recognizance entered into before an officer in charge must have been confirmed (i.e., signed) by a justice before it is an offence to fail to appear in response to it.

JUDICIAL INTERIM RELEASE

Before the bail reforms of the early 1970s, police officers did not have the power to release persons whom they arrested. In 1968, the Canadian Committee on Corrections (the Ouimet Report), concerned about the unnecessary pre-trial detention of accused in Canada, recommended reforms, some of which were enacted in 1970 (discussed in Kiselbach, 1988-89, and Salhany, online). The purpose of the bail reform legislation of 1970 was to reduce the number of unnecessary pre-trial arrests and detentions, through release of suspects by police officers. Today, a person arrested for certain offences may be released by a peace officer or by an officer in charge, under sections 497-499 of the *Criminal Code* (discussed above).

    Because of public dissatisfaction with the new, "easy" release of accused, additional amendments, passed in 1975, retreated somewhat from the legislation of 1970. A **reverse onus** clause was introduced, requiring accused charged with certain offences, or under certain circumstances, to show why they should be released. More offences were added to this list over the years, most recently in 2008 (discussed later under "Reverse Onus Clause").

Section 469 Offences

A peace officer or an officer in charge cannot release someone who is charged with an offence under section 469; only a superior court judge (defined in section 2) has jurisdiction to deal with bail. If an accused charged with a section 469 offence appears before a justice or a provincial court judge, the accused must be detained in custody until dealt with by a superior court judge (see sections 515(11) and 522).

Section 522 imposes a reverse onus on an accused detained and charged with a section 469 offence. The accused shall be detained by the superior court judge unless the accused, on a balance of probabilities, shows cause why detention is not justified within the meaning of 515(10) (discussed below under "Grounds for Judicial Interim Release").

The Law Reform Commission of Canada has recommended that all **show cause hearings** be held in provincial court, because in many respects the judges in provincial court have more experience in dealing with judicial interim release than the judges who sit in superior court (1988, 35). To date, that recommendation has not been implemented.

Offences Other Than Section 469 Offences

For all offences other than those in section 469, if the accused is not released by the officer who made the arrest, the officer in charge, or another peace officer, section 503(1) requires that the person be brought before a **justice** "without unreasonable delay," and within 24 hours of her or his arrest if a justice is available, or "as soon as possible" where a justice is not available within 24 hours. "Justice" is defined as a justice of the peace or a provincial court judge. Excessive delay at this stage could amount to a violation of the accused's rights under section 9 of the *Charter*. However, sections 503(2) and (2.1) also allow the peace officer or officer in charge to release the person under certain conditions, rather than bring the person before a justice.

Forms of Judicial Interim Release

Section 515 provides the basic rule that the justice shall release the accused on an **undertaking to appear** (UTA) without conditions (see Form 12 in the back of the *Code* for the contents of a UTA), unless the prosecutor shows cause why the accused should be detained, or why conditions should attach to the release of the accused. This requirement to "show cause" is why a bail hearing is often referred to as a "show cause hearing." An accused must be present for a show cause hearing; however in 1997, section 515(2.2) was added, so that an accused may now appear by telecommunication under certain circumstances satisfactory to the justice.

A show cause hearing is not a trial, and so the prosecutor does not have to show cause beyond a reasonable doubt, but only on the balance of probabilities. Strict rules of evidence do not apply, as section 518(1)(e) states the justice may hear and base the decision "on evidence considered credible or trustworthy. . .in the circumstances of each case." Usually, there are no witnesses, although the Crown and the accused are both entitled to call witnesses. Section 518 allows for the leading of evidence under oath at a show cause hearing, but the accused cannot be asked any questions about the offence, unless questioning is initiated by counsel for the accused.

A common practice (although the practice will vary from jurisdiction to jurisdiction) is for the Crown to read the circumstances of the alleged offence from the police report, and the contents of any criminal record alleged, and then for the accused or defence counsel to respond. Section 518(1)(c) provides that the prosecutor can lead evidence on the circumstances of the offence, the criminal record alleged, and so on, and section 518(1)(d) allows for the judge to take into account "any matters agreed on by the prosecutor and the accused."

The prosecutor is given a reasonable opportunity to show cause, and under section 516 can ask the justice to adjourn the bail hearing. The adjournment may not be for more than three clear days, except with the consent of the accused.

Section 515(2)) sets out the available forms of release, in the order in which they must be considered. Section 515(3) states that "the justice shall not make an order under any of paragraphs 2(b) to (e) unless the prosecution shows cause why an order under the immediately preceding paragraph shall not be made." The presumption is in favour of release, and on the least stringent conditions.

The list of forms of release, in the order they must be considered, from lightest to most stringent, are as follows:

(1)   an undertaking to appear without conditions, under section 515(1);
(2)   an undertaking to appear with conditions, under section 515(2)(a);
(3)   a recognizance without sureties or deposit, and with or without conditions, under 515(2)(b);
(4)   a recognizance with sureties, with or without conditions, but without deposit, under 515(2)(c);
(5)   a recognizance without surety, but with deposit of money or valuable security, and with or without conditions, under 515(2)(d) (note, this is "cash bail" and can normally be imposed only with the consent of the prosecutor);
(6)   a recognizance with or without sureties, but with deposit, if the accused is not ordinarily resident in the province or does not reside within 200 kilometres of where he is in custody, under 515(2)(e); or
(7)   a detention order.

Section 515(4) lists the types of conditions that a justice can attach to any of the above orders. In addition to various forms of reporting (such as to a Bail Supervisor), and the relinquishing of a passport, the justice can order any other "reasonable conditions . . . as the justice considers desirable." This last clause has been used to restrict the accused's movements. For example, people accused of trafficking in narcotics in Vancouver have often been given a "No-Go Granville" condition, which requires them not to be found on the Granville Street Mall. Similar conditions have been imposed to restrict the movements of those charged with prostitution or shoplifting. Section 515(4.1) requires a justice to consider extra conditions (surrender of firearms) where a person is charged with an offence of violence against persons, a terrorism offence, criminal harassment, or with certain offences under the *Controlled Drugs and Substances Act* and the *Security of Information Act*. Section 515(4.2) requires the justice to consider whether the accused should be prohibited from speaking to victims and witnesses, if charged with the offences listed in section 515(4.3).

There is case law to the effect that it is wrong to fix a cash deposit or a surety in an amount that it is so high that the accused will not be able to raise it (*Cichanski,* and *Garrington*). In such cases, the deposit or surety is an illusory form of bail, in that it really amounts to a detention order, but without the safeguards that accompany detention orders (discussed below under "Section 525 Review").

Grounds for Judicial Interim Release

In 1992, the Supreme Court of Canada found that an accused could not be detained in the "public interest," as stated in section 515(10) of the *Code,* because it violated section 11(e) of the *Charter,* the "right not be denied reasonable bail without just cause" (*Morales*). The concept of "public interest" was too vague, in that it did not give citizens fair notice of what the law is, and did not impose any limits on law enforcement discretion (*Morales* 100). In this same decision, however, the Court found that detention for "public safety" was justified.

Section 515(10), which was amended following decision in *Morales*, allows for the detention of an accused under one or more of the specified grounds: a) to ensure the accused's attendance in court; b) for the protection or safety of the public or witnesses (including factors such as the likelihood of re-offending if released); and c) if detention is "necessary to maintain confidence in the administration of justice." In deciding the last ground under section 515(10)(c) the following circumstances should be considered:

> (i) the apparent strength of the prosecution's case,
> (ii) the gravity of the offence,
> (iii) the circumstances surrounding the commission of the offence, including whether a firearm was used, and
> (iv) the fact that the accused is liable, on conviction, for a potentially lengthy term of imprisonment or, in the case of an offence that involves, or whose subject-matter is, a firearm, a minimum punishment of imprisonment for a term of three years or more [see Stuart 2008 for a discussion of this section].

In the 2002 case of *Hall*, the Supreme Court of Canada decided that refusing bail where it is "necessary in order to maintain confidence in the administration of justice" does not violate the *Charter*. Hall was charged with murder, having inflicted 37 wounds on the deceased and having attempted to cut off her head. The judge hearing the judicial interim release had found there was no reason to detain Hall under 515(10) (a) or (b), but detained him in order to "maintain confidence in the administration of justice." According to the majority in the Supreme Court of Canada, there was no error in the bail judge's reasoning. However, the Court decided that denying bail under an earlier version of section 515(10)(c), which allowed the denial of bail "on any other just cause being shown" and "without limiting the generality of the foregoing," violated the presumption of innocence and section 11(e) of the *Charter* and could not be demonstrably justified under section 1 of the *Charter*.

Reverse Onus Provisions at the Show Cause Hearing

There are several reverse onus clauses respecting judicial interim release. Section 522(2) imposes a reverse onus when the accused is charged with an offence listed in section 469; the accused is required in such cases to show cause why detention is *not* justified. In addition, section 515(6) provides that the onus at the bail hearing lies on the accused if the accused is charged with certain offences, including the commission of an indictable offence "while at large after being released in respect of another indictable offence," specified offences relating to criminal organizations, terrorism, and the *Security of Information Act*, and "an offence punishable by imprisonment for life under subsections 5(3), 6(3) or 7(2) of the *Controlled Drugs and Substances Act* or the offence of conspiring to commit such an offence." In 2008, the government added a number of firearms and weapons offences to the list of reverse onus offences.

Do the reverse onus provisions violate section 11(e) of the *Charter*, which states that "any person charged with an offence has a right not to be denied reasonable bail without just cause"? A number of lower courts have said no, and the Supreme Court of Canada has dealt with two of these provisions, in *Morales* (discussed above), and in *Pearson*.

In *Pearson*, the Supreme Court of Canada upheld the reverse onus in section 515(6)(d), in relation to certain offences under the *Narcotic Control Act* (now the *Controlled Drugs and Substances Act*). Mr. Justice Lamer found that there was just cause for these special bail rules for traffickers, who had special characteristics.

According to Lamer, drug importers and traffickers "have access both to a large amounts of funds and to sophisticated organizations which can assist in flight from justice." Traffickers "pose a significant risk that they will abscond rather than face trial" (669). Trafficking is "often a business and a way of life . . . it is highly lucrative, creating huge incentives for an offender to continue criminal behaviour even after arrest and release on bail . . . . [Therefore] special bail rules are required in order to establish a bail system which maintains the accused's right to pre-trial release while discouraging continuing criminal activity" (695). What do you think of this argument? Would Lamer have made a different decision if this had been a low-level trafficker? Lamer, J. was of the view that the section applies to the "small fry" as well as the major traffickers, and even to the "generous smoker" who shares a single joint of marijuana at a party (698).

Madame Justice McLachlin wrote a strong dissent. She objected to the section applying to all traffickers, and to Lamer's rationale, writing that much trafficking has nothing to do with making money, and that "the lowly street vendor, the person most likely to be arrested, cannot count on the distant drug lord to run the risk of stealing him out of the country" (706-07). Lamer was of the view that the small fry could easily establish they were not part of a major drug ring, but McLachlin wrote, "Criminal organizations, unlike unions and service organizations, do not distribute lists of their members. How does one prove that one is not a member?" (708).

The Supreme Court of Canada also stressed, in both *Morales* and *Pearson,* that section 11(d) of the Charter (the right to be presumed innocent) is not relevant at bail hearings—guilt or innocence is not being determined, and there is no punishment being imposed.

Release Pending Appeal

Section 679(3) allows a judge of the Court of Appeal to release a convicted person pending the determination of an appeal. The onus is on the appellant to establish that 1) the appeal is not frivolous, 2) the appellant will be available to return to custody as required, and 3) "detention is not necessary in the public interest." Both the Ontario Court of Appeal and the British Columbia Court of Appeal have distinguished this situation from that in *Morales,* deciding that the reference to "public interest" in respect of bail pending appeal does not violate the appellant's rights under the *Charter* (*Farinacci* and *Branco*). Both courts decided that section 11(e) of the *Charter* did not apply to bail pending appeal, since the appellant in such cases is not "accused" of a crime, but has been convicted. The presumption of innocence no longer applies. The Supreme Court of Canada dismissed (without reasons) leave to appeal in *Branco*.

BAIL VARIATION

Section 523 stipulates that bail conditions continue until the trial ends. If the accused is convicted, bail conditions continue until the accused is sentenced. If there is reason to believe the accused will not show up for sentencing, bail can be revoked immediately upon conviction.

The conditions of bail can be *varied* at any time, upon cause being shown. The trial court judge or the judge at the preliminary hearing can vary bail upon either cause being shown, or by the consent of both the prosecutor and the accused (section 523(2)).

BAIL REVIEW

A **bail review** is different from a variation of bail. A bail review, which takes place under section 520, is essentially an appeal, heard by a judge (defined in section 493 to mean a judge of a superior court

Box 2.2: The Public Interest

Following his conviction after his second trial, Guy Paul Morin applied for bail pending appeal. The only issue was whether his detention was necessary in the public interest. In support of their application for release, the defence filed the affidavit of an official with a public opinion research company, detailing the results of a telephone survey of 561 people asked whether they thought Morin had received a fair trial, and whether they believed he should be released pending appeal. While the Ontario Court of Appeal rejected that evidence on the issue of "public interest," it did accept the affidavits of four residents of the area, "deposing to doubts in that community about the appellant's guilt and to the lack of any concern in the community if he were to return there pending the determination of his appeal" (at 402). Although this decision is limited to its own unique circumstances, it is interesting that the personal opinions of four residents, presumably chosen for their particular views, as to what they believed the community felt, were more persuasive than a representative sample survey conducted by a professional polling agency (*Morin*).

of criminal jurisdiction). Sections 520 and 521 allow the accused or the Crown respectively to apply to a judge of a superior court of criminal jurisdiction for a review of bail set by a justice or a provincial court judge. Where bail was set for a section 469 offence by a judge of a superior court of criminal jurisdiction, the review is carried out by a judge of the Court of Appeal in the province, under section 680.

SECTION 525 REVIEW

Another review of bail is automatic under section 525, where a trial is delayed. It is referred to as the **"90 day bail review"** if the offence is an indictable one, and a **"30 day bail review"** if the matter is a summary conviction matter. The purpose is to prevent the accused from languishing in jail awaiting trial, and being forgotten. It is only available if there has been a detention order, and not, for example, if an accused was given a $200 cash bail but remains in custody because she or he cannot come up with the $200. Upon the review, the judge takes into account who is responsible for the delays (section 525(3)).

The general rule under section 525(4) is that the judge shall release the accused on conditions or other forms of bail unless the judge is satisfied that the continued detention of the accused is justified within section 515(10), which refers to the grounds for detention.

ESTREATMENT AND FORFEITURE

If an accused has put up cash bail and breaches the bail terms, the Crown can apply for **forfeiture** of the money to the Crown. If an accused has entered into a recognizance and breaches the terms of bail, or if a surety has signed for a specified amount of money and the accused breaches the bail terms, the Crown can apply for the bail to be **estreated**. This means that the Crown will have a judgment against the accused or the surety, and can enforce it like any other civil judgment (i.e., have the sheriff seize assets, or have the judgment registered against real property, such as a house).

SUMMARY

An accused may be compelled to attend court to answer charges in response to an appearance notice or a summons. Anyone, whether a peace officer or not, may arrest without warrant a person found committing an indictable offence, or who is believed on reasonable grounds to have committed a criminal offence and to be fleeing lawful pursuit. Property owners or their agents may also arrest anyone found committing an offence in relation to their property. A police officer has additional powers to arrest without warrant anyone reasonably believed to have committed or to be about to commit an indictable offence, anyone found committing a criminal offence, and anyone in respect of whom it is reasonably believed an arrest warrant exists. The police have a concomitant duty **not** to arrest without a warrant persons suspected of certain categories of less serious offences unless it is necessary to do so to identify them, to preserve evidence, to prevent the continuation or repetition of the offence, or where it is believed necessary to arrest them to ensure their attendance in court. Police officers may also arrest anyone who is the subject of an arrest warrant issued by a justice.

The courts have defined the standard of "reasonable grounds" to require the person conducting the arrest to subjectively have reasonable grounds for believing an offence has been committed, when those grounds are also objectively justifiable. The police must now obtain warrants to enter a home to make an arrest, although there are exceptions.

Section 9 of the *Charter* guarantees freedom from arbitrary detention or imprisonment, and section 10 gives everyone who is arrested or detained the right to be informed of the reason, and the right to counsel.

A police officer who arrests an accused in respect of certain less serious offences may immediately release that person by issuing an appearance notice—which requires the accused to attend court—or may release the person outright, intending to have a summons issued later. If the arresting officer decides not to release an arrested suspect, the officer in charge or another peace officer must then release the suspect as soon as practicable, unless it is believed necessary to continue detention in order to identify the accused, to preserve evidence, to prevent the continuation or repetition of the offence, or to ensure the attendance of the accused in court. The officer may release the person outright, intending to have a summons issued later, or may release the suspect on a promise to appear or a recognizance.

If a suspect is not released by the arresting officer, the officer in charge, or another peace officer, the suspect will normally appear in court for the purpose of a show cause hearing. Usually, the onus will be on the Crown to show why the accused's detention is necessary, or why a relatively stringent form of bail is necessary. In some cases, a reverse-onus applies, and the accused must show why his or her release is justified. There are several steps, ranging from outright release on an Undertaking to Appear without conditions, to a Detention Order, and the court must consider them in sequence. In the normal case, a judge must be satisfied by the Crown that no lower form of bail is adequate before making a higher-level order. Bail continues, unless varied, until the completion of the trial proceedings.

It is an offence to fail to appear as required by, or to breach any other term or condition of, the imposed form of release. Additionally, any cash deposit may be ordered forfeited, and any sureties posted may be ordered estreated.

Provisions exist for varying bail by consent, or on review. If an accused is ordered detained, provisions exist to ensure that their custodial status is automatically reviewed after 90 days in the case of indictable offences, or 30 days in the case of summary conviction offences.

QUESTIONS TO CONSIDER

(1)   An Appearance Notice must be_____ by a justice before a person can be charged with failing to appear as set out in the Appearance Notice.

(2)   An Information is sworn (in time) before_____, but after_____.

(3)   What sections of the *Charter* affect the law regarding detention and arrest in Canada?

(4)   Can you (as a private citizen) arrest someone who is causing damage to your neighbour's fence if you see it happening? Why or why not?

(5)   Can you arrest the person in Question 4 if you didn't actually see him do the damage, but you believe that he did it? What if you see the person causing damage to the fence, but at trial under section 430, he is acquitted—was your arrest lawful?

(6)   Under what circumstances could you, as a private citizen, arrest someone for procuring (section 212)? For murder (section 235)?

(7)   When might a private security guard be governed by the *Charter*?

(8)   A peace officer may arrest a person without warrant whom he finds committing a criminal offence. What does "finds committing" mean?

(9)   On what grounds can a police officer stop someone for an investigative detention?

(10)   Does a person stopped for investigative detention have a right to counsel?

(11)   How does the Supreme Court of Canada define "detention" in *Grant*? Provide an example of circumstances that might be consider detention and circumstances that would not be considered detention.

(12)   On what grounds can a police officer arrest someone?

(13)   What does it mean to say that a police officer needs reasonable grounds to arrest a person?

(14)   What is a recognizance? What is a surety?

(15)   Which court will set bail for an accused charged with murder? Who will have the onus of proof?

(16)   What is the difference between bail variation and a bail review?

BIBLIOGRAPHY

Allman, Anthony. "Further Perspectives on Section 10(b) of the *Charter*: A Reply to Gold." (1994) 25 *Criminal Reports* (4th) 280.

Berger, Benjamin L. "Race and Erasure in *R. v. Mann*." (2004) 21 *Criminal Reports* (6th) 58.

Coughlan, Stephen G. "Great Strides in Section 9 Jurisprudence." (2009) 66 *Criminal Reports* (6th) 75.

D'Andrea, Armando. "'Question Black Youths': Councillor." (17 August 2005a) *National Post* A1.

D'Andrea, Armando. "Race Row: Councillor Backtracks: Response to gun deaths" (18 August 2005b) *National Post* A8.

Fiszauf, Alec. "Articulating Cause–Investigative Detention and Its Implications." (2007) 52 *Criminal Law Quarterly* 327.

Fiszauf, Alec. *Investigative Detention*. Markam, Ontario: Lexis Nexis Canada, 2008.

Gorham, Nathan J.S. "Police Discretion, Racial Profiling and Articulable Cause." (2004) 49 *Criminal Law Quarterly* 50.

Gorman, Wayne. "Arbitrary Detentions and Random Stops." (1999) 41 *Criminal Law Quarterly* 41.

Hill, S. Casey, David M. Tanovich and Louis P. Strezos. *McWilliams' Canadian Criminal Evidence*. Aurora, ON: Canada Law Book Limited (available online on Criminal Spectrum).
        Chapters 32 talks about bail hearings.

Kiselbach, Daniel. "Pre-Trial Procedure: Preventive Detention and the Presumption of Innocence." (1988-89) 31 *Criminal Law Quarterly* 168.

Latimer, Scott. 2007. "The expanded scope of search incident to investigative detention." (2007) *Criminal Reports* (6th) 48.

Law Commission of Canada. *In Search of Security: The Future of Policing in Canada*. Ottawa: Law Commission of Canada, 2006.

Law Reform Commission of Canada. *Compelling Appearance, Interim Release and Pre-Trial Detention (Working Paper #57)* Ottawa: Law Reform Commission of Canada, 1988.

Law Reform Commission of Canada. *Arrest* (Working Paper #41) Ottawa: Law Reform Commission of Canada, 1985.

Law Reform Commission of Canada. *Arrest* (Report #29) Ottawa: Law Reform Commission of Canada, 1986.

McCoy, Lesley A. "Liberty's Last Stand? Tracing the Limits of Investigative Detention." (2002) 46 *Criminal Law Quarterly* 319.

McCoy, Lesley A. "Some Answers from the Supreme Court of Canada on Investigative Detention. . . and Some More Questions." (2004) 49 *Criminal Law Quarterly* 268.

Patel, Aman S. "Detention and Articulable Cause: Arbitrariness and Growing Judicial Deference to Police Judgment." (2001) 45 *Criminal Law Quarterly* 198.

Pomerance, Renee M. "Parliament's Response to *R. v. Feeney*: A New Regime for Entry and Arrest in Dwelling Houses." (1994) 13 *Criminal Reports* (5th) 84.

Pringle, Heather. "The Smoke and Mirrors of *Godoy*: Creating Common Law Authority While Making Feeney Disappear." (1999) 21 *Criminal Reports* (5th) 227.

Pringle, Heather. "Kicking the Castle Doors: The Evolution of Exigent Circumstances." (1999) 43 *Criminal Law Quarterly* 86.

Quigley, Tim. *Procedure in Canadian Criminal Law,* 2nd ed. Toronto: Carswell, 1997 (Updates available in looseleaf).
        Quigley covers compelling the attendance of the accused (Chapter 9), arrest and detention (Chapters 5 and 6), bail (Chapter 11).

Quigley, Tim. "*Mann*, It's a Disappointing Decision." (2004) 21 *Criminal Reports* (6th) 41.

Quigley, Tim. 2009. "Was it Worth the Wait? The Supreme Court's New Approaches to Detention and Exclusion of Evidence." (2009) 66 *Criminal Reports* (6th) 88.

Rigakos, George S. and David R. Greener. "Bubbles of Governance: Private Policing and the Law in Canada." (2000) 15(1) *Canadian Journal of Law and Society* 145.

Salhany, Roger E. *Canadian Criminal Procedure,* 6th ed. Toronto: Canada Law Book Inc., (available online on Criminal Spectrum).
> Chapter 3, Arrest and Seizure of Property, covers the appearance of an accused, the definition of detention, *Charter* rights upon detention, arrest powers with and without a warrant, and duties upon arrest. It also examines the law regarding appearance notices and summonses.
> Chapter 4, Bail, discusses the history of bail, and the law surrounding release by an officer in charge and by a justice of peace, as well as the *Charter* issues.

Skiblinsky, Christina. 2006. "Regulating Mann in Canada." (2006) 69 *Sask. Law Review* 197.

Stribopoulos, James. "A Failed Experiment? Investigative Detention: Ten Years Later." (2003) 41 *Alberta Law Review* 335.

Stribopoulos, James. "The Limits of Judicially Created Police Powers: Investigative Detention after *Mann*." (2007) 52 *Criminal Law Quarterly* 299.

Stuart, Don. "*Feeney*: New *Charter* Standards for Arrest and Undesirable Uncertainty." (1998) 7 *Criminal Reports* (5th) 175.

Stuart, Don. "*Godoy*: The Supreme [*sic.*] Reverts to the Ancilliary Powers Doctrine to Fill a Gap in Police Power." (1999) 21 *Criminal Reports* (5th) 225.

Stuart, Don. "Bail: Decisions on Effect of New Amendments in Firearms Cases." (2008) 61 *Criminal Reports* (6th) 136.

Stuart, Don, Ron Delisle, and Tim Quigley. *Learning Canadian Criminal Procedure*, 9th ed. Toronto: Carswell, 2008.
> Chapter 3 covers charging, arrest and detention, and Chapter 7 discusses bail hearings.

Tanovich, David M. "Using the Charter to Stop Racial Profiling: The Development of Equality-Based Conceptions of Arbitrary Detention." (2002) 40 *Osgoode Hall Law Journal* 149.

Tanovich, David M. "The Colourless World of *Mann*." (2004) 21 *Criminal Reports* (6th) 47.

Tanovich, David M. *The Colour of Justice: Policing Race in Canada* (Toronto: Irwin Law Inc. 2006).

Tanovich, David M. "The Charter of Whiteness: Twenty-Five Years of Maintaining Racial Injustice in the Canadian Criminal Justice System (2008) 40 *Supreme Court Law Reports* (2d) 655.

Thompson, Elizabeth. "Ottawa Moves to Combat Racial Profiling by Police." (29 March 2005) *Vancouver Sun* A3.

Trotter, Gary T. *The Law of Bail in Canada*, 2nd ed. Ontario: Carswell, 1999.

CASES

R. v. Asante-Mensahm, [2003] 2 S.C.R. 3. A citizen is entitled to use reasonable force in making a lawful arrest.

R. v. Beare, [1988] 2 S.C.R. 387. The section of the *Identification of Criminals Act* that allows for the taking of fingerprints does not violate the *Charter*.

R. v. Bielefeld (1981), 64 C.C.C. (2d) 216 (B.C.S.C). Area restrictions for the purpose of preventing further offences are valid bail conditions.

R. v. Biron (1975), 23 C.C.C. (2d) 513 (S.C.C.). Respecting the powers of arrest, "finds committing" in the *Code* means "apparently committing," so that the validity of the arrest is determined with respect to the circumstances apparent to the police officer at the time.

R. v. Branco (1994), 87 C.C.C. (3d) 71 (B.C.C.A.); leave to appeal to the Supreme Court of Canada dismissed (without reasons); [1994] S.C.C.A. No. 134. An accused who is appealing conviction may be granted bail under section 679, but only upon establishing that the appeal is not frivolous and that detention is not necessary in the public interest. This does not violate section 11(e) of the *Charter*, because the presumption of innocence is lost upon conviction.

R. v. Brown, (2003) 173 C.C.C. (3d) 23 (Ont. C.A.).

R. v. Buhay, [2003] 1 S.C.R. 631. Private security guards who searched a locker at a bus station were not acting as state agents, and so section 8 of the *Charter* did not apply.

Chromiak v. The Queen (1979), 49 C.C.C. (2d) 257 (S.C.C.). A driver subjected to a demand for a breath sample is not detained, and therefore a denial of the right to counsel under section 2(c)(ii) of the *Bill of Rights* does not constitute a reasonable excuse for refusing to provide the sample.

R. v. Cichanski (1976), 25 C.C.C. (2d) 84 (Ont. H.C.). Where bail is granted, it should not be in such a large amount as to be equivalent to a detention order.

R. v. Connors, [1998] B.C.J. No. 41.

R. v. Dell, [2005] A.J. No. 867 (Alta. C.A.).

R. v. Dudley, 2009 SCC 58.

Eccles v. Bourque, [1975] 2 S.C.R. 739.

R. v. Farinacci (1994), 86 C.C.C. (3d) 32 (Ont. C.A.). The requirement in section 679 that an applicant for bail pending appeal establish that detention is not necessary in the public interest is not unconstitutionally vague, and is not contrary to section 11(e) of the *Charter*.

R. v. Feeney, [1997] 2 S.C.R. 13. Police officers are not allowed to enter private dwellings without authorization.

R. v. Feeney, [1999a] B.C.J. No. 688 (B.C.S.C).

R. v. Feeney, February 5, 1999b; CC980028, (B.C.S.C.).

R. v. Feeney, [2001] B.C.J. No. 311 (B.C.C.A.).

R. v. Garrington et al. (1972), 9 C.C.C. (2d) 472 (Ont. H.C.). Where bail is granted, it should not be in such a large amount as to be equivalent to a detention order.

R. v. Godoy, [1999] 1 S.C.R. 311. Police officers may enter a private dwelling following a disconnected "no voice" 911-call, where the call was disconnected before the caller spoke, under their obligation to protect life and safety.

R. v. Grant, 2009 SCC 32. The Supreme Court of Canada clarifies the definition of detention.

R. v. Hall, [2002] 3 S.C.R. 309. Permitting detention under section 515(10)(c) of the *Criminal Code* "on any other just cause being shown," and "without limiting the generality of the foregoing" violate section 11(e) of the *Charter*, and are not justified under section 1 of the *Charter*. Denying bail in order "to maintain confidence in the administration of justice," does not violate the *Charter*.

R. v. Hufsky (1988), 40 C.C.C. (3d) 398 (S.C.C.). The random stopping of cars for "spot checks" is an arbitrary detention, but is justified under section 1 of the *Charter*. The demand for production of driver's licence and insurance for inspection does not constitute a search or seizure.
R. v. Landry, [1986] 1 S.C.R. 145. A warrantless arrest under section 450(1)(a) may be made on private property, assuming the requirements of the section are met.

R. v. Lerke (1986), 24 C.C.C. (3d) 129 (Alta. C.A.). An arrest by a private citizen is a governmental function exercised on behalf of the Crown, to which the *Charter* applies. While there is a right to search incidental to arrest, that right is subject to *Charter* limits.

R. v. Macooh (1993), 82 C.C.C. (3d) 481 (S.C.C.). Police acting without a warrant, but in hot pursuit, may enter private premises to arrest an accused, even for a provincial offence.

R. v. Mann, [2004] 3 S.C.R. 59.

R. v. Morales (1992), 77 C.C.C. (3d) 91 (S.C.C.). In respect of the bail provisions in section 515, the "public interest" portion of the secondary ground is impermissibly vague, and unconstitutional under section 11(e) of the *Charter*. The "public safety" component of the secondary ground is valid. The reverse onus provisions of section 515(6)(a) and (d) are also valid.

R. v. Morin (1993), 19 C.R. (4th) 398 (Ont. C.A.).

R. v. Pearson, [1992] 3 S.C.R. 665. The reverse onus provision of section 515(6)(d) (narcotics importing or trafficking) is constitutionally valid.

R. v. Roberge (1983), 4 C.C.C. (3d) 304 (S.C.C.). Discussion of scope of section 25 of the *Code,* authorizing the use of reasonably necessary force to effect an arrest.

R. v. Seo (1986), 25 C.C.C. (3d) 385 (Ont. C.A.). Extrinsic evidence may be introduced on issues under section 1 of the *Charter,* provided the evidence is not inherently unreliable or contrary to public policy, and that it is not tendered to aid in statutory construction. Such evidence may include statistics, surveys, studies, research papers, and program evaluations.

R. v. Simpson (1993), 79 C.C.C. (3d) 482 (Ont. C.A.).

R. v. Storrey (1990), 53 C.C.C. (3d) 316 (S.C.C.). Arrest without warrant requires reasonable and probable grounds, both subjectively and objectively.

R. v. Suberu, 2009 SCC 33. An accused has a right to counsel under section 10(b) of the *Charter* when subjected to an investigative detention.

R. v. Therens, [1985] 1 S.C.R. 613. Discussion of what constitutes a detention, triggering the right to counsel, and to be informed of that right. Exclusion of evidence for a breach of a *Charter* right may only be under section 24(2), not under 24(1) as a remedy that is "appropriate and just in the circumstances."

R. v. Thomsen, [1988] 1 S.C.R. 640. Discussion of what constitutes a detention. A demand for a breath sample under section 234.1 (roadside screening) of the *Code,* without the opportunity to consult with counsel, constitutes a detention and a limitation on section 10 of the *Charter,* but is justified under section 1 of the *Charter*.

U.S.A. v. MacFarlane, [1989] B.C.J. No. 2047 (B.C.C.A). The residence and reporting terms of release were so "markedly and unilaterally changed" by the United States government that MacFarlane was no longer obliged to pay on a promissory note (signed to secure the release of an accused).

CHAPTER 3: *Informations and Indictments, Arraignment and Plea*

CHAPTER OBJECTIVES

In studying this chapter, you should develop an understanding of the following topics and concepts:

- the nature and purpose of, and the distinction between, informations and indictments, and how they are used to commence and support criminal proceedings
- private prosecutions
- the ways in which the Crown may discontinue a prosecution
- direct indictments
- means of controlling prosecutorial discretion
- available pleas

INFORMATIONS AND INDICTMENTS

The Historical Development of Prosecutions

At early common law in England, prosecutions were private. Their purpose was to compensate the victim for a loss, rather than to punish the wrongdoer. These prosecutions were more like present day torts, whereby an aggrieved person might sue someone for a personal injury or loss. Even after the concept of public prosecutions developed, it was still possible for wrongdoers to compensate their victims, and thereby avoid being prosecuted. However, while wrongdoers could "buy off" the victim's family in the case of a murder, they were not allowed to purchase their freedom from a victim of theft (Law Reform Commission of Canada 1986, 33-35).

This practice of "prosecution or payment" does not exist in our modern criminal justice system. In fact, it is an offence in Canada under section 139 of the *Code* (obstructing justice) to offer money to a witness or a victim of crime in order to influence them. It is a summary conviction offence under section 143 to publicly advertise a reward for the return of something lost or stolen with "no questions asked." Section 141 makes it an indictable offence to agree to conceal an indictable offence in exchange for valuable consideration. The section, however, provides an exception where such an agreement is entered into with the consent of the Attorney General, or as part of a diversion programme approved of by the Attorney General. In a sense, one may officially be spared prosecution in some instances by agreeing to compensate one's victim, thus reverting to a compensatory scheme. In some respects, restorative justice re-introduces the victim of crime into the process.

Informations

In most cases, criminal proceedings in Canada are formally commenced by the laying of an **information**, usually (but not necessarily) by a police officer, before a justice (see definition in section 2 of the *Criminal Code* and Form 2 at the back of the *Code*). An information is a written complaint, sworn under oath (see Box 3.1). All trials in Provincial Court are tried on an information.

If the accused has an election, and elects to be tried in superior court by judge alone or judge and jury and requests a **preliminary inquiry**, the inquiry will be held in provincial court. The information is the charging document on which proceedings are based through to the end of the preliminary inquiry (see Chapter 4). If the accused is ordered to stand trial following the preliminary inquiry, the Crown will then prefer an indictment (discussed later).

In 1997, section 508.1 was added to allow a peace officer to lay an information by means of a telecommunications device that produces a written document. (Rather than swearing the information, section 508.1(2) provides that the peace officer is to state that the information is true, and that such a statement is deemed to have been made under oath.) Charges of indictable offences "may" be in Form 2 (see section 506), while charges of summary offences "shall" be in Form 2 in the back of the *Criminal Code* (see section 788(1)). Section 849 requires that the preprinted parts of the form of the Information be in both English and French. There are lower court decisions to the effect that an information is still valid, and can be amended, if it is not in both official languages.

Box 3.1 Sample Information.

Canada,
Province of British Columbia,
City of Vancouver.

This is the information of Constable Jane Smith, of Vancouver, British Columbia, police officer, hereinafter called the informant.

The informant says that she has reasonable grounds to believe, and does believe, that

John Doe, on or about the 28th day of May, A.D. 2010, at or near the City of Vancouver, in the Province of British Columbia, did unlawfully traffic in a narcotic, to wit: diacetylmorphine (heroin), contrary to Section 5 of the *Controlled Drugs and Substances Act.*

Sworn before me this 29th day of
May, A.D. 2010, at the City of
Vancouver, British Columbia

....
(Signature of Informant)

....
A Justice of the Peace in and for
the Province of British Columbia.]

Section 504 governs informations regarding indictable offences, and section 788 informations regarding summary conviction offences. The person swearing the information under oath must have "personal knowledge," or believe on "reasonable grounds" that the accused has committed an offence. Anyone may swear an information (see the discussion below on "Private Prosecutions Today"), but most prosecutions in Canada are commenced by an information sworn by a police officer. A police officer may swear an information based on the allegations of a complainant, or on the information

Box 3.2 *R. v. Radbourne* (1787), 1 Leach 456, 168 E.R. 330.

An example of the extreme technicality of 18th century English indictments:

Henrietta Radbourne, late of the parish of St Mary-le-bone, in the county of Middlesex, widow, late servant of Hannah Morgan, widow, her mistress, not having the fear of God before her eyes, but being moved and seduced by the instigation of the devil, and of her malice aforethought, contriving and intending her the said Hannah Morgan, her mistress, to deprive of her life, and feloniously and traitorously to kill and murder on the 31st May, in the 27th year, &c. with force and arms, at the parish aforesaid, in the county aforesaid, in and upon the said Hannah, the mistress of the said Henrietta, feloniously, traitorously, wilfully, and of her malice aforethought, did make an assault; and that the said Henrietta, with a certain stick having a bayonet fixed at the end thereof, of the value of two shillings, which stick she, the said Henrietta, in both her hands then and there had and held, in and upon the top of the head of her the said Hannah, did then and there feloniously, traitorously, wilfully, and of her malice aforethought, her the said Hannah Morgan strike, cut, stab, and penetrate, giving to the said Hannah, by such striking, cutting stabbing, and penetrating of the said Hannah, with the bayonet so fixed at the end of the stick aforesaid, in and upon the top of the head of her the said Hannah, one mortal wound, the length of one inch and of the depth of half an inch, of which mortal wound the said Hannah, from the said 31st May in the year aforesaid, until the 11th day of July in the year aforesaid, in and at the parish aforesaid, in the county aforesaid, did languish, and languishing did live, on which said 11th day of July in the year aforesaid, at the parish aforesaid, in the county aforesaid, of the mortal wound aforesaid, she the said Hannah died; And so the Jurors aforesaid, upon their oath aforesaid, do say, that the said Henrietta Radbourne, otherwise Henrietta Gibbons, her the said Hannah Morgan, her said mistress, in manner and by the means aforesaid, feloniously, traitorously, wilfully, and of her malice aforethought, did kill and murder, against the peace of our said Lord the King, his crown and dignity.

contained in a police report prepared by another police officer. Swearing an information in this manner should not be a mere "rubber stamp" process; the officer swearing the information "second hand" must be satisfied there are reasonable grounds to believe the person named has committed the offence alleged (section 504).

Particular provisions apply where an accused has been issued an appearance notice, or has been arrested and released by a police officer under section 497, or by an officer in charge or another peace officer under section 498. Section 505 requires that an information shall be laid before a justice "as soon as practicable thereafter and in any event before the time stated in the appearance notice, promise to appear or recognizance issued to or given or entered into by the accused for his attendance in court."

Public prosecutions are governed by sections 507 and 508, which require a justice receiving an information from a peace officer or a prosecutor to "hear and consider" the allegations, and any evidence of witnesses considered "desirable or necessary." This hearing is *ex parte*, which means that the accused named in the information is not in attendance. There have been a number of challenges to this closed hearing, but the constitutionality of the section (in terms of freedom of the press) has been upheld by the Ontario Court of Appeal, under section 1 of the *Charter* (see *Southam Inc.*). It follows from the fact that the justice is supposed to "hear and consider" the evidence that the justice cannot sign a blank summons or warrant; this is expressly forbidden by section 507(5).

Where a justice "considers that a case for so doing is made out," the justice shall compel the accused to attend a specified provincial court in the territory where the alleged offence took place to deal with the charge. This may be done by issuing a summons or a warrant, if the allegations "disclose reasonable grounds to believe that it is necessary in the public interest to issue a warrant for the arrest of the accused" (section 507(4)). Section 507(6) allows the justice to pre-authorize the officer in charge to release an arrested accused under section 499 (except for offences mentioned in section 522); this is known as an **endorsed warrant.** If an appearance notice, promise to appear, or recognizance was used by a peace officer, the justice may confirm that document, or may cancel it and instead issue a summons or a warrant for the arrest of the accused (section 508(1)). If no case is made out for compelling the accused to attend at court, the justice shall "cancel the appearance notice, promise to appear or recognizance…and cause the accused to be notified forthwith of such cancellation" (section 508(1)(c)).

The Law Reform Commission of Canada expressed the view that the test of whether "a case for so doing is made out" is too vague, and that a more precise standard for issuing process was needed (1988 at 32-33). It recommended that a justice be required to have "reasonable grounds to believe that the person named . . . has committed a crime" (1988, 98).

Indictments

Section 566 states that an accused's trial for an indictable offence, other than a trial before a provincial court judge, shall be on an indictment (see Form 4). The indictment is the charging document on which proceedings are based to the conclusion of the trial. Section 566.1 deals with indictments in Nunavut. An indictment is a written accusation of crime against a person or several persons. It is usually **preferred** (i.e., presented to a court of superior jurisdiction) by the Attorney General, or more typically by a prosecutor as agent of the Attorney General, pursuant to section 574. No one else may prefer an indictment except with the written order of a judge of the court in which it is preferred (section 574(3)).

Section 2 of the *Code* defines indictment to include an information. The two concepts are different, however; an information is on oath, sworn by a police officer or anyone else, whereas an indictment is not on oath, but is a document preferred by the Crown, and lodged with the trial court at or before the opening of the accused's trial.

Historically, indictments were extremely technical documents (see **Box 3.2, *R. v. Radbourne*** for an example of such an indictment), and had to be very precise as to the offence. These detailed and technical indictments, and their strict (narrow) interpretation at common law, were developed to mitigate against the harsh punishments (typically death) that were imposed even for relatively less serious offences. The courts would strictly interpret the indictments to relieve many accused of a trip to the gallows.

Today, section 581 sets out the rules regarding what should be contained in an indictment. By section 795, these rules also apply to informations in summary conviction proceedings. Section 581(1) states that "each count in an indictment shall in general apply to a single transaction and shall contain the substance of a statement that the accused or defendant committed an offence therein specified." An indictment may contain one or many **counts**. For example, if someone committed an offence by filing 52 false claims for employment insurance, the information might contain 52 separate counts, one for each false statement or claim. Prosecutors appear to have some discretion in this matter, as charges against physicians for defrauding medicare (OHIP in Ontario) are often "bulked," such that rather than facing thousands of fraud charges, the amounts defrauded by numerous claims are all added into one charge (Brockman 2010b).

The count may be stated in popular language, in the words of the enactment, or in words "that are sufficient to give to the accused notice of the offence with which he is charged" (section 581(1)(c)). Section 581(3) requires that there be sufficient detail to give the accused reasonable information to enable him or her to identify the transaction or event referred to in the count. Section 583 provides that the lack of details, such as the name of the person injured, the owner of property damaged or stolen, and so on, will not render the indictment insufficient if the requirements of section 581 are otherwise met. The guiding principle regarding the sufficiency of informations is "for the accused to be reasonably informed of the transaction alleged against him, thus giving him the possibility of a full defence and a fair trial" (*Coté*). The kind of information necessary will depend on the nature of the offence. The indictment should not mislead the accused, but it need not be as technical as was required historically.

In *Coté*, the information omitted a reference to "without reasonable excuse" in a charge of failing to provide a sample a breath, but did refer to the proper section of the *Criminal Code*. The Supreme Court of Canada held that the charge was not defective.

Box 3.3 Sample Indictment (Contemporary)

Canada,
Province of British Columbia,
In the Supreme Court of British Columbia

Her Majesty the Queen
against
John Doe

John Doe stands charged

That he, between May 28, 2007 to June 30, 2010, at the City of Vancouver, in the Province of British Columbia, unlawfully by deceit, falsehood or other fraudulent means, did defraud the Ministry of Health for the Province of British Columbia of $100,000, by submitting false claims to the Ministry of Health, in violation of section 380(1) of the *Criminal Code* of Canada thereby committing an offence contrary to Section 380(1)(b)(i) of the *Code*.

Dated this 11th day of November, A.D. 2011, at Vancouver, British Columbia

....

(Signature of signing officer, Agent of
Attorney General, etc., as the case may be)]

Amending An Information Or Indictment

Section 601 states that an objection to an information should be made before plea, presumably because, by entering a plea, the accused is conceding the validity of the information. After a plea is entered, an objection to the information can be made only by leave of the court.

Section 601(3) gives wide powers to the courts to amend an indictment (and also an information). During the course of a trial, the Crown may apply to remedy a defect in the indictment, or to amend a charge to conform to the evidence. The test the court will apply is one of fairness. Where such an amendment is allowed, the court may grant the accused an adjournment to remove any prejudice occurring as a result of the amendment. This section removes many of the technical arguments that used to be made regarding defective informations.

PRIVATE PROSECUTIONS TODAY

Section 504 of the *Criminal Code* states that "any one who, on reasonable grounds, believes that a person has committed an indictable offence may lay an information in writing and under oath before a justice." The definition of **prosecutor** in section 2 includes those individuals who lay a private information. In some instances, where government prosecutors do not, or refuse to, proceed with a case, a private individual can prosecute a case, or hire a private lawyer to conduct the prosecution.

In 2002, the federal government introduced a new procedure for justices to deal with private informations (previously dealt with under section 507), which gives prosecutors more supervisory powers over private prosecutions. Section 507.1 provides that informations laid under section 504 by a private citizen shall be referred to a provincial court judge (in Quebec, a judge of the Court of Quebec), and if the judge "considers that a case for doing so is made out," shall issue a summons or warrant. The judge can do this only after considering the allegations and evidence, and after being satisfied that the provincial Attorney General has received a copy of the information and was given

---

Box 3.4 The Attempt to Prosecute George W. Bush

In 2004, Gail Davidson, who worked with Lawyers Against the War, swore a private information in Vancouver, British Columbia, accusing George W. Bush, President of the United States of America, of torture and other offences under the *Criminal Code*. When Davidson appeared to fix a date for a hearing under section 507.1 of the *Code*, a Provincial prosecutor successfully applied to have the Information declared a nullity, "based on the diplomatic immunity of Mr. Bush." Davidson appealed to the British Columbia Supreme Court and asked for a Writ of Certiorari quashing the provincial court judge's decision, and various other declarations; however, she did not ask for a Writ of Mandamus requiring a summons or warrant to compel Mr. Bush to attend court. In fact, she stated, "Lawyers Against the War and myself are not asking at any time for process to issue" (*Davidson* para. 7). The judge dismissed her appeal on the basis of abuse of process, because the only reasonable inference to be drawn from Davidson's statement was "that she intends to use the criminal procedure under the *Criminal Code* as a forum to express her political views" (para. 8).

---

an opportunity to attend and cross-examine and call witnesses. These hearings are referred to as **pre-inquiries** (*Grinshpun* para. 6). The person named in the information is not entitled to appear at the pre-inquiry (*Robb*). The Attorney General of the province may still intervene in such a case and take control of the prosecution under section 579.

Should private prosecutions be allowed? Should the Crown have the power to stay charges that are commenced by way of private prosecution? What are the advantages and disadvantages of allowing private prosecutions? The Law Reform Commission of Canada favoured retaining private prosecutions, as they reinforce democratic values (1986, 22).

Bridget Moran, in her book *Judgement at Stoney Creek* (see Appendix C), describes a case in which the family of the victim initiated a private prosecution. The charge was against Richard Redekop, who was driving the car that struck and killed Coreen Thomas, a 21 year old pregnant Carrier Native from the Stoney Creek reservation. Generally, as happened in this case, the Crown takes over private prosecutions, and either stays the charge or prosecutes the case itself as a conventional prosecution.

DISCRETION TO LAY CHARGES AND CHARGE SCREENING

The Law Reform Commission of Canada described the criminal law as a "blunt and costly instrument," that should be "an instrument of last resort" (1976, 24). The Law Commission of Canada has re-iterated this concern, calling for "a more equitable and accountable process for defining crime and enforcing criminal law" (Des Rosiers and Bittle 2004, xxiii). The blunt force of the law is softened by the discretion granted to many of the participants in the criminal justice system.

Unlike the United States, where prosecutors often get involved in the investigation of a crime, in Canada there is a relatively clear division of functions between the prosecutors and the police (*Regan* para. 66). The practice in most provinces is for the police to lay informations, and then to bring the sworn informations to the Crown for prosecution. Once the Crown has the information, the prosecutor in charge can decide whether to proceed with or to stay the charge. Some provinces (British Columbia, Quebec and New Brunswick) have adopted a procedure whereby prosecutors review (or screen) the charges and pre-approve them *before* the information is laid by the police. In some instances, the prosecutor will interview witnesses prior to the charges being laid (*Regan* para. 82). In addition, some sections of the *Criminal Code* require the consent of the Attorney General before charges are laid (see section 83.3 regarding certain terrorist activities).

The Martin Report from Ontario discussed the advantages and disadvantages of both pre-charge screening (the system used in British Columbia, Quebec and New Brunswick) and post-charge screening (used in the other provinces). It recommended that Ontario continue to use post-charge screening, in order not to undermine "an important system of checks and balances represented by current independent rights of the police to lay charges, and of the Attorney General to stay charges" (1993, 123). It rejected the suggestion of the Law Reform Commission of Canada (1990) that the police be required to consult with a prosecutor before laying a charge, or be required to explain to a justice why it was impractical to do so (Martin Report 1993, 127).

The Martin Report also noted that the tests used to decide whether a charge should proceed vary across the country, and that British Columbia provincial prosecutors use the highest standard, whether there is a "substantial likelihood of conviction." The Martin Report recommended that the Crown proceed with charges when there is a "reasonable prospect of conviction." Federal prosecutors now use this standard, and if this standard is met, they then ask whether "the public interest requires a prosecution to be pursued." Public interest factors include the seriousness of the offence and resources available for prosecution, as well as factors specific to the case (Department of Justice, Chapter 15). Provincial prosecutors in British Columbia consider both the "substantial likelihood of conviction" test and whether the prosecution is in the public interest (Attorney General 2009). Also see Layton (2002).

ATTORNEY GENERAL'S CONTROL OVER PROSECUTIONS

There are three different methods for discontinuing a prosecution. Section 579(1) allows the Attorney General or an agent of the Attorney General to intervene and enter a **stay of proceedings,** whether it is a private prosecution or one commenced by the police. A stay simply stops the proceedings. Once a charge is stayed, the Crown can reactivate the charge under section 579(2) within one year. If the Crown does not recommence the proceedings within one year, they are treated as if they were never commenced.

The Crown also has the option of **withdrawing** the charges. If the charges are withdrawn, new charges must be laid in order to recommence proceedings against the accused. A third method of discontinuing a charge is for the prosecutor to **call no evidence** at trial. In such a case, the court would *dismiss* the charges (making a finding of "not guilty"). The Crown could not then recommence the proceedings, but could appeal the acquittal.

The Law Reform Commission of Canada (1990) has recommended that these methods of discontinuing proceedings be replaced with legislation that would allow the Crown to discontinue proceedings by entering either a temporary or permanent discontinuance.

DIRECT INDICTMENTS

Section 577 allows the Crown to dispense with the preliminary inquiry and proceed directly to trial, by preferring a **direct indictment**. The Crown can take this action before or during a preliminary inquiry, or after an accused has been discharged at a preliminary inquiry. Note that section 577 requires "the personal consent in writing" of either the Attorney General or the Deputy Attorney General. This requirement means the decision comes from the Director of Public Prosecution Service of Canada for federal prosecutions, and from provincial capitals for provincial prosecutions. The preliminary inquiry can also be bypassed by the "written order" of a superior court of criminal jurisdiction.

There have been several challenges to this section under sections 7, 9, and 15 of the *Charter*, but so far the Supreme Court of Canada has not dealt with the issue. Leave to appeal to the Supreme Court has been refused in at least three cases where courts of appeal have found that the section did not violate the *Charter*.

A prosecutor might want to bypass a preliminary inquiry for a variety of reasons. The Law Reform Commission of Canada (1990, 91–2) listed the following, gleaned from a number of cases:

1. The notoriety of the case is such that a quick trial of the merits is essential;
2. The case is long and complex, and involves many accused;
3. The accused intends to disrupt the preliminary hearing;
4. The fear that the security of the Crown's witnesses, or of other persons involved in the prosecution, is jeopardized and a speedy disposition of the case is required;
5. The need to try the case as soon as possible to preserve the Crown's case;
6. The need to avoid a multiplicity of proceedings; and
7. The need to avoid unconscionable delay which cannot otherwise be remedied.

The Law Reform Commission of Canada (1990, 904) recommended that the power of Attorneys General to prefer indictments without a preliminary inquiry be retained, but that they follow guidelines that preferred indictments are to be used only in "rare and extraordinary circumstances." The Federal Department of Justice's Deskbook states that direct indictments will only be used when it is in the public interest to do so. In addition to some of the reasons listed by the Law Reform Commission cited above, the Deskbook suggests that a decision to proceed by direct indictment might also be influenced by "an error of law, jurisdictional error, or palpable error on the facts of the case" at the preliminary inquiry, new evidence, and whether "the holding of a preliminary inquiry would unreasonably tax the resources of the prosecution, the investigative agency or the court" (Department of Justice, Chapter 17.2).

Box 3.5 Bypassing Bernardo's Preliminary Inquiry

Shortly before Paul Bernardo's preliminary inquiry on charges of sexual assault and murder was scheduled to commence, Ontario Attorney-General Marion Boyd signed a direct indictment. Defence counsel were critical of this move. The direct indictment protected Karla Homolka from being cross-examined at the preliminary inquiry (denying the accused the opportunity of disclosure through such examination), and it severely reduced the amount of time that defence counsel had to prepare for the trial (Tyler 1994, A1).

CONTROL OVER PROSECUTORIAL DISCRETION AND MISCONDUCT

There are at least three ways to control the discretion exercised by prosecutors. An accused in a criminal trial might ask for a remedy from the court (for example, a stay of proceedings), because the Crown's behaviour amounted to an abuse of process. For example, Nova Scotia's former Premier, who was charged in 1995 with numerous sexual offences dating back to the 1950s, asked for a stay of proceedings alleging an abuse of process by the Crown. The allegations of abuse included judge-shopping by the prosecutor, and the fact that following the police investigation, the police and Crown jointly re-interviewed 16 of the 22 complainants. The purpose of the joint interviews was "to provide information about the Court process to potential complainants so that they could make an informed decision as to their involvement in these proceedings; and secondly, to make assessments of credibility about these witnesses, including their capacity for recall and general demeanor issues, and to prepare for a preliminary inquiry" (*Regan* para. 13). Although the Supreme Court of Canada stated that "judge shopping is unacceptable both because of its unfairness to the accused, and because it tarnishes the reputation of the justice system," evidence of it in the *Regan* case was insufficient to enter a stay of proceedings. Regarding the pre-charge interviews conducted by the Crown, the Supreme Court of Canada found that this practice, that existed in a number of provinces to varying degrees, was not an abuse of process, but had many benefits (para. 84-86). Where there is an abuse of process, the Court went on to say that a stay of proceedings is appropriate only where: "(1) the prejudice caused by the abuse in question will be manifested, perpetuated or aggravated through the conduct of the trial, or by its outcome; and (2) no other remedy is reasonably capable of removing that prejudice" (para. 54).

A less invasive remedy is for the judge to terminate inappropriate questioning by Crown counsel (see Walsh 2007 and Code 2007). This, of course, also applies to inappropriate questions by defence counsel. See Morgan (1986-87) for a historical examination of the role of the prosecutor and Layton (2002), for a more contemporary discussion.

A second way to control prosecutorial discretion is for an accused to sue prosecutors for malicious prosecution. The broad immunity that historically protected prosecutorial discretion was altered (at least theoretically) by the Supreme Court of Canada in *Nelles* (a nurse who was wrongly charged with the death of babies in a hospital, and who settled her law suit out of court). One subsequent successful litigant was Proulx, who sued the Crown for malicious prosecution in the murder of his former girlfriend, and was awarded more than a million dollars in damages (*Proulx* 2001, para. 54). When the case got to the Supreme Court of Canada, the majority found that the Crown's improper motive included the Crown's decision to recruit a retired police officer to work on the file, notwithstanding the fact he was a defendant in the Proulx's well-publicized million dollar defamation law suit (para. 38). The Crown had also improperly distorted the accused's words so that they looked like a confession (para. 41). In 2009, there were 107 ongoing malicious prosecution lawsuits against Ontario's 800 prosecutors (Makin 2009). However, it is thought that a recent decision from the Supreme Court of Canada might make it much more difficult to succeed in such lawsuits (Makin 2009). In *Miazga v. Kvello Estate,* the Supreme Court of Canada overturned a finding of civil liability against a Saskatchewan prosecutor because there was no malice (improper purpose) on his part. The Court summarized the four requirements for a malicioius prosecution complaint to succeed: (1) the prosecutor must initiate the prosecution; (2) the prosecution must be terminated in favour of the plaintiff; (3) the prosecution must have been "undertaken without reasonable and probable cause;" and (4) the prosecutor must have been "motivated by malice or a primary purpose other than that of carrying the law" (para. 3). The Court then expanded on the concept of malice: "the malice element of the tort of malicious prosecution ensures that liability will not be imposed in cases where a prosecutor proceeds, absent reasonable and probable grounds by reason of incompetence, inexperience, poor judgment, lack of professionalism, laziness, recklessness, honest mistake, negligence, or even gross negligence" (para. 81). Prior to this case, Roach (2000) suggested that civil lawsuits might be useful to the wealthy or the determined, but are not that useful in holding prosecutors accountable.

A third way to control prosecutorial discretion is through the self-governing law societies. Prosecutors are members of self-regulating provincial law societies, and can be disciplined by their society for unethical conduct in the course of their work—see Layton (2002) and Code (2007) for a discussion of various ethical codes of conduct for lawyers. For example, an accused in Alberta complained to the Law Society of Alberta about a prosecutor who was tardy about disclosing the fact that the DNA at the scene of a murder implicated someone other than the accused. The issue before the Supreme Court of Canada was whether the Law Society had jurisdiction to discipline prosecutors, and the Court found that it did (*Krieger*).

ARRAIGNMENT

An accused may make several appearances in court, following a number of adjournments, prior to being **arraigned** in court. Arraignment is the process of calling the accused in court by name, reading the charge(s), and asking whether the accused pleads guilty or not guilty. A case may be adjourned (remanded) so that the accused can consult with a lawyer, or for a variety of other reasons, before the accused enters a plea.

An accused who elects to be tried by a superior court judge, or by a judge and jury, and who requests a preliminary inquiry, will not be arraigned until after the preliminary inquiry in provincial court. The subsequent arraignment will take place in a superior court, unless the accused re-elects before committal to be tried by a provincial court judge. With respect to summary conviction offences and offences under section 553, the arraignment takes place in a provincial court.

Prior to 2002, an accused had to appear in person for arraignment on indictable offences (section 650(1)), unless the conditions set out in section 650(1.1) were met, allowing for "closed-circuit television or any other means that allow the court and the accused to engage in simultaneous visual and oral communication." Today, section 650.01 also allows an accused to appear by counsel, after filing a form with the court designating counsel. Despite such designation, the court may still require the accused to attend (section 650.01). Section 650.02 allows such appearances by designated counsel to take place through "any technological means satisfactory to the court that permit the court and all counsel to communicate simultaneously." Section 800 allows an accused charged with a summary conviction offence to appear by counsel or agent. However, the Provincial Court judge may require the attendance of the accused. Section 800(2.1) allows for closed-circuit television appearances similar to those in section 650(1.1).

PLEAS

On arraignment, the accused can enter a variety of possible pleas, as provided for in section 606: guilty, not guilty, or one of the **special pleas** in section 607. The special pleas include *autrefois acquit* (a plea that the accused has already been acquitted of the alleged offence), *autrefois convict* (a plea that the accused has already been convicted of the offence), and **pardon** (whereby the accused pleads that a pardon has already been granted in respect of the offence). There is a common law rule against multiple prosecutions for the same act (see Salhany, online, or Law Reform Commission of Canada, 1991, for further details).

The fact that many accused plead guilty in our criminal justice system has been a cause of concern for some writers. Fitzgerald (1990, 82) argues that the judge-made law surrounding the acceptability of guilty pleas in criminals matters is "lacking in principle and consistency," is a threat to the integrity of our criminal justice system, and is in need of re-evaluation and reform (see Law Reform Commission of Canada 1989). In 2002, section 606(1.1) was added, requiring that a court accept a guilty plea only if satisfied that the plea is voluntary, and that the accused understands "the nature and consequences of the plea, and that the court is not bound by any agreement made between the accused and the prosecutor." However, the "failure of the court to fully inquire into whether [these] conditions. . . are met does not affect the validity of the plea" (section 606(1.2); see Brockman 2010a). A person may be allowed to withdraw a guilty plea if it is later established that his or her rights under the *Charter* were violated (*Taillefer*). Tanovich (2004) raises some of the ethical issues defence counsel face when representing accused who maintain their innocence, but who want to plead guilty for reasons of efficiency, or to avoid risking a more serious conviction.

Plea Bargaining

Plea bargaining, largely ignored for years in Canada, was the subject of an entire issue of the *Criminal Law Quarterly* in 2005. Di Luca (2005) suggests that the *Askov* case provided the impetus for openly accepting plea bargaining in Canada. The Martin Committee in Ontario also placed its stamp of approval on plea bargaining. Topics in the special *Criminal Law Quarterly* issue include: the history and controversy surrounding the concept of plea bargaining in Canada, and the pros and cons of such a process (Di Luca 2005); the history of plea bargaining in the United States and how Canada might avoid some of the pitfalls of plea bargaining found in that country (McCoy 2005); "the institutionalization of plea bargaining and its impact upon the routine of the criminal case as it makes its way through the courts and consequently, on the routine of the criminal defence lawyer" (Lafontaine and Rondinelli 2005, 108); the ethical responsibilities of Crown counsel in plea bargaining (Dickie 2005), and a comparison of plea bargaining in Canada and England (Waby 2005). Although Walby suggests that Crown and defence counsel "seek to canvass judicial approval" of an agreement before a guilty plea is entered (2005, 155), this practice, which exists in Ontario, is not followed in British Columbia.

SUMMARY

Criminal prosecutions are normally commenced by the swearing of an information, before a justice, alleging a criminal offence has been committed by an accused. This information is the charging document that supports the prosecution through trial or preliminary inquiry in provincial court. If, following a preliminary inquiry, the accused is ordered to stand trial, the Crown will then prefer an indictment. An indictment is a formal, written allegation that the accused committed the specified offence. It is filed with the superior court in which the trial is to take place. In this case, the indictment is the charging document that supports the proceedings through to the conclusion of the trial. A relatively rarely used procedure allows the Crown to proceed by direct indictment, which avoids a preliminary inquiry and takes the prosecution directly to trial in superior court.

Compared to the extreme technicality of the earlier common law, the current law in Canada allows informations and indictments to be worded simply and gives the courts wide latitude to amend them to correct defects in wording rather than reject them entirely.

Private prosecutions, once the predominant form of criminal proceeding, are now relatively rare in Canada. Although anyone may lay an information, and although the informant may prosecute the allegation if the Crown chooses not to, the Crown may step in at any point in the proceedings, either to take over and conduct the prosecution, or to stay it. Since 2002, the Crown must be given notice of, and an opportunity to participate in, a pre-inquiry to determine if a private prosecution should proceed.

Arraignment is the process of reading the information or indictment to the accused in court, and calling on the accused to enter a plea. An accused who elects to have a preliminary inquiry in provincial court will not be arraigned until he or she appears in superior court after being committed to stand trial. An accused upon arraignment may plead guilty, not guilty, or enter one of the special pleas (*autrefois acquit, autrefois convict*, or pardon).

The Crown may end criminal prosecutions by entering a stay of proceedings, which temporarily suspends them. The Crown may also apply to withdraw charges. If the court allows the withdrawal, it is as if the prosecution had never been commenced. A third alternative is for the Crown to go to trial but to call no evidence, resulting in an acquittal.

Wrongful behaviour on the part of the Crown can be a reason to ask the court in a criminal trial for a remedy such as a stay of proceedings, the subject of a malicious prosecution civil law suit, or the subject of a disciplinary action by a law society against the prosecutor.

QUESTIONS TO CONSIDER

(1)   Informations are sworn *ex parte*. What does this mean?

(2)   How many counts can an information contain?

(3)   What is the general principle used to determine if an information is sufficient?

(4)   The Department of Justice has asked you to prepare a brief on private prosecutions. It would like to know the advantages and disadvantages of private prosecutions, and whether you would recommend their continuation or their abolition. What changes, if any, would you make to the present system? If you don't think that any changes are necessary, defend your position.

(5)   Evaluate the present system of pre-charge screening that takes place in British Columbia, Quebec and New Brunswick.

(6)   What are the advantages and disadvantages of allowing the Crown to prefer direct indictments? Should there be any limits placed on this power?

(7)   Should judges participate in plea bargaining? Why or why not?

(8)   What are the three ways that the Crown can discontinue criminal proceedings?

(9)   What is an arraignment?

(10) Discuss three different legal forums or venues that exist to control prosecutorial misconduct.

BIBLIOGRAPHY

Attorney General (British Columbia). "The Role of Crown Counsel" (2009). (/www.ag.gov.bc.ca/prosecution-service/BC-prosecution/crown-counsel.htm; accessed September 10, 2009).

Brockman, Joan. "Review of Oonagh E. Fitzgerald, *The Guilty Plea and Summary Justice: A Guide for Practitioners.*" (1993) 3 *International Criminal Justice Review* 130.

Brockman, Joan. "An Offer You Can't Refuse:" Pleading Guilty When Innocent (2010a) 56 *Criminal Law Quarterly* (in press).

Brockman, Joan. "Fraud Against the Public Purse by Health Care Professionals: The Privilege of Location" in Janet Mosher and Joan Brockman (eds.) *Constructing Crime: Contemporary Processes of Criminalization.* Vancouver: UBC Press, 2010b.

Code, Michael. "Counsel's Duty to Civility: An Essential Component of Fair Trials and an Effective Justice System." (2007) 11 *Canadian Criminal Law Review* 97.

Department of Justice. *The Federal Prosecution Service Deskbook*; www.justice.gc.ca/eng/dept-min/pub/fps-sfp/fpd/toc.html; accessed September 10, 2009.

Des Rosiers, Nathalie and Steven Bittle. "Introduction" in Law Commission of Canada (ed.). *What is a Crime?* Vancouver: UBC Press, 2004.

Dickie, Mary Lou. "Through the Looking Glass–Ethical Responsibilities of the Crown in Resolution Discussions in Ontario." (2005) 50 *Criminal Law Quarterly* 128.

Di Luca, Joseph. "Expedient McJustice or Principled Alternative Dispute Resolution? A Review of Plea Bargaining in Canada." (2005) 50 *Criminal Law Quarterly* 14.

Fitzgerald, Oonagh E. *The Guilty Plea and Summary Justice: A Guide for Practitioners.* Toronto, ON: Carswell, 1990.
   Fitzgerald traces the historical development of guilty pleas and discusses the problems with how the courts deal with them. Problems today include the fact that accused plead guilty to offences of which they consider themselves innocent. He concludes that the guilty plea needs to be re-evaluated in light of the *Charter* and the way judicial investigations are conducted in Europe. A more active judicial investigation results in improved summary justice.

Forsyth, Frederick L. "A Plea for *Nolo Contendere* in the Canadian Criminal Justice System." (1998) 40 *Criminal Law Quarterly* 243.

LaFontaine, Gregory and Vincenzo Rondinelli. "Plea Bargaining and the Modern Criminal Defence Lawyer: Negotiating Guilt and the Economics of 21st Century Criminal Justice." (2005) 50 *Criminal Law Quarterly* 108.

Law Reform Commission of Canada. *Our Criminal Law.* Working Paper #3. Ottawa, 1976.

Law Reform Commission of Canada. *Private Prosecutions.* Working Paper #52. Ottawa, 1986.
   The Commission reviews the reasons for retaining private prosecutions and supports the use of public interest groups to initiate private prosecutions in areas where public prosecutions are less likely to be initiated, such as crimes against consumers and the environment. According to the Commission, availability of the private prosecution "reinforces democratic values and public perceptions…within a system that has public prosecutions as the linchpin of the prosecution process" (at 22). Of six possible methods for controlling private prosecutions within a public prosecution system, the Commission recommends that the Attorney-General have the right to intervene to take over the case, or to stay or withdraw the charges. The Appendix reviews the history of how private prosecutions in England were supplemented with public prosecutions, and compares our system to that in some of the civil law systems (France and Germany) and other common law systems (Scotland, New Zealand, Australia, United States, and Contemporary England).

Law Reform Commission of Canada.. *Compelling Appearance, Interim Release and Pre-Trial Detention.* Working Paper #57. Ottawa, 1988.

Law Reform Commission of Canada. *Plea Discussions and Agreements.* Working Paper #60. Ottawa, 1989.
   The Commission reviews the need for reform of plea negotiation in terms of six principles: openness, judicial supervision, protection from improper inducements, accuracy and appropriateness, equality, and enforcement. It makes 23 recommendations for reform. Appendix A reproduces some of the rules and guidelines from other reports in Canada and the United States that the Commission cites. Appendix B provides the results of a survey by Professor Anthony Doob on "Public Attitudes Towards Plea Bargaining," including the scenarios and questions used in the survey.

Law Reform Commission of Canada. *Controlling Criminal Prosecutions: The Attorney-General and the Crown Prosecutor.* Working Paper #62. Ottawa, 1990.
   This Working Paper examines the role of the Attorney-General and the need for reform in the structure of the Department of Justice and the Department of the Solicitor General, and the powers of the Attorney-General (consent to prosecutions, initiation of charges, control over the form of trial, preferred indictments, discontinuation of a prosecution). The 45 recommendations are summarized in Chapter 3. Appendix A contains a description of the institutional arrangements in England; Appendix B discusses the distribution of powers of federal and provincial ministers of Justice, Attorneys-General, and Solicitor General; Appendix C describes the organizational structure of the Department of Justice; and Appendix D contains a comparison chart of direction of public prosecutions arrangements in other countries.

Law Reform Commission of Canada. *Double Jeopardy, Pleas and Verdicts.* Working Paper #63. Ottawa, 1991.
   This Working Paper discusses the law regarding double jeopardy, pleas, and verdicts. Thirty-six recommendations are examined in light of the need for reform in terms of lack of comprehensiveness, confusing procedures, anachronistic provisions, procedures that produce delay, and shortfalls in the protection of the accused.

Layton, David. "The Prosecutorial Charging Decision." (2002) 46 *Criminal Law Quarterly* 447.

Makin, Kirk. "Ruling gives Crown Stronger Shield Against Lawsuits" (7 November 2009) *The Globe and Mail* (online).

Martin, Arthur G. *Charge Screening, Disclosure, and Resolution Discussion.* Toronto: Report of the Attorney General's Advisory, 1993.

    The Report reviews the pros and cons of charge screening by prosecutors, and the timing of such screening (before or after the charges are laid). The Report favours post-charge screening, on the grounds that it allows anyone to lay an information (as set out in the *Criminal Code,* and provided for by common law), provides checks and balances on the power of the Attorney-General, makes the Attorney-General more accountable, and allows for the operational independence of the police. (The Marshall Report also reached this conclusion.) The Report discusses the desirability of the suggested threshold test—"that there is a reasonable prospect of conviction, and that the prosecution is otherwise in the public interest" (at 130), as compared with other tests. Other aspects of post-charge screening involve controlling overcharging, which is "an irresponsible exercise of charging discretion by the police" (at 137).

McCoy, Candace. "Plea Bargaining as Coercion: The Trial Penalty and Plea Bargaining Reform." (2005) 50 *Criminal Law Quarterly* 67.

Morgan, Donna C. "Controlling Prosecutorial Powers—Judicial Review, Abuse of Process, and Section 7 of the Charter." (1986–87) 29 *Criminal Law Quarterly* 15.

Quigley, Tim. *Procedure in Canadian Criminal Law,* 2d ed. Toronto: Carswell, 2006 (looseleaf).

Roach, Kent. "The Attorney General and the Charter Revisited." (2000) 50 *University of Toronto Law Journal* 1.

Salhany, Roger E. *Canadian Criminal Procedure,* 6th ed. Toronto: Canada Law Book Inc., (available online on Criminal Spectrum).

    Chapter 6 discusses private prosecutions, prosecutorial discretion, the preferring of an indictment, and direct indictments.

Stuart, Don, Ron Delisle, and Tim Quigley. *Learning Canadian Criminal Procedure*, 9th ed. Toronto: Carswell, 2008.

    The authors discuss plea bargaining in Chapter 10, formal attacks on the indictment or the information in Chapter 11.

Tanovich, David M. "*Taillefer*: Disclosure, Guilty Pleas and Ethics." (2004) 17 *Criminal Reports* (6th) 149.

Tyler, Tracey. "Bernardo Trial Move 'Unfair': Defence Lawyer Attacks Speeding up of Murder Case." (31 March 1994) *Toronto Star* A1.

Walby, Michael. "Comparative Aspects of Plea Bargaining in England and Canada: A Practitioner's Perspective." (2005) 50 *Criminal Law Quarterly* 148.

Walsh, John J. "Cross-examination by the Prosecutor: Stopping Transgressions. It is also a Trial Judge's Responsibility." (2007) 11 *Canadian Criminal Law Review* 301.

CASES

R. v. Coté (1978), 33 C.C.C. (2d) 353 (S.C.C.). The guiding principle regarding the sufficiency of informations will be "for the accused to be reasonably informed of the transaction alleged against him, thus giving him the possibility of a full defence and a fair trial."

Davidson et al. v. Attorney General of British Columbia et al., 2005 BCSC 1765 (B.C.S.C.).

R. v. Grinshpun, [2004] B.C.J. No. 2371 (B.C.C.A.).

Krieger v. Law Society of Alberta, [2002] 3 S.C.R. 372.

Miazga v. Kvello Estate, 2009 SCC 51.

Nelles v. Ontario, [1989] 2 S.C.R. 170.

Proulx v. Quebec (A.G.), [2001] 3 S.C.R. 9.

R. v. Regan, [2002] 1 S.C.R. 297.

Robb v. York (Region), [2005] O.J. No. 198 (Ont. S.C.).

Southam Inc. v. Coulter (1990), 60 C.C.C. (3d) 267 (Ont.C.A.). The hearing conducted by a justice who receives an information, in considering whether to issue or confirm process, must be held *in camera*, and not in public. While this procedure contravenes section 2(b) of the *Charter*, it is justifiable under section 1.

R. v. Taillefer; R. v. Duguay, [2003] 3 S.C.R. 307.

CHAPTER 4: *Crown Disclosure and the Preliminary Inquiry*

CHAPTER OBJECTIVES

In studying this chapter, you should develop an understanding of the following topics and concepts:

- the extent of the duty on the Crown to provide pre-trial and ongoing disclosure
- the distinction between particulars and circumstances
- the notion of defence disclosure
- the nature and purpose of the preliminary inquiry, and the test applied
- the procedure followed at a preliminary inquiry
- the exclusion of the public from a trial
- perpetuated evidence
- arguments favouring the retention or abolition of the preliminary inquiry

CROWN DISCLOSURE

Despite calls for comprehensive Crown **disclosure** laws in Canada (see Kaiser 2009), there is little legislation governing such disclosure. Section 603 of the *Code* states that the accused is entitled "to inspect…his own statement, the evidence and the exhibits," while section 650(3) provides that the accused is entitled to "make full answer and defence." The Crown has a duty at common law to disclose all material evidence, even if it is favourable to the defence (*C.(M.H.)*). However, prior to the Supreme Court of Canada's decision in *Stinchcombe*, the practice of disclosure varied widely. Some prosecutors were of the view that the amount of information they disclosed to the defence was within their discretion, so long as the exercise of that discretion was not an abuse of process. To establish such an abuse, the defence would have to show that there had been "an affront to fair play and decency" (*Jewitt*).

The Supreme Court of Canada in *Stinchcombe* found that an accused has a right to disclosure under section 7 of the *Charter,* because the right to make full answer and defence is a "principle of fundamental justice." This requirement is consistent with the purpose of a criminal prosecution, which is "not to obtain a conviction; [but] to lay before a jury what the Crown considers to be credible evidence relevant to what is alleged to be a crime" (*Boucher*, quoted in *Stinchcombe* para. 11). The Court's decision was influenced by the fact that the Crown's failure to disclose witnesses' prior inconsistent statements to the defence was an important factor in the wrongful conviction of Donald Marshall, Jr. in 1970. The Royal Commission on the Marshall prosecution stated that "anything less than complete disclosure by the Crown falls short of decency and fair play" (quoted in *Stinchcombe* para. 17).

In *Stinchcombe*, the Court placed a heavy onus on the Crown to disclose all relevant information, whether it plans to use it or not, and whether the information supports the position of the Crown or the defence. One indication of relevance is whether the information would be useful to the accused. If the information "can reasonably be used by the accused either in meeting the case for the Crown, advancing a defence or otherwise in making a decision which may affect the conduct of the defence such as, for example, whether to call evidence" (*Egger* 467), it is relevant, and the Crown should disclose it.

According to the Supreme Court of Canada, disclosure should be made before the accused makes an election or enters a plea. Further, the obligation to disclose continues, should the Crown receive additional evidence (*Stinchcombe* para. 28), and continues through the appellate process (*McNeil* para. 17) . The Crown has some discretion with the timing of disclosure, if early disclosure would jeopardize a witness or, in some "rare" cases, a continuing investigation (*Stinchcombe* para. 28; confirmed in *McNeil* para. 18). The Crown must provide witness statements, either verbatim notes taken by police officers, or "will say" statements summarizing the evidence a witness may give. Statements by civilians who are going to be called as witnesses must also be disclosed. If there is any dispute about whether the information ought to be disclosed, the trial judge will consider the issue and rule on it in a *voir dire*.

Box 4.1 Crown Disclosure

The issue of disclosure requires a balancing of competing interests. Accused persons should not be "tried by ambush." Conversely, the public wants obviously culpable persons to be convicted.

In the *Morin* case, the defence was not advised before the first trial of the following: logs of telephone calls (the tapes of which had been accidentally destroyed) in which the victim's family speculated that their own son was concealing information; the existence of a shallow pit found near the body; the statement of a witness who spoke to a man near the site where the body was found, on the morning after the abduction; the statements of neighbours near the body site that they had heard screams on the night of the abduction; and a test conducted by the prosecution that discredited certain of its own evidence relating to the matching of clothing fibres between Morin and the victim. Morin was acquitted on his first trial in 1986, but was ordered re-tried, following appeal. When these instances of non-disclosure were raised at the second trial (1992), the Crown's response was reportedly that it is up to the defence to precisely request what it wants before the Crown is required to provide it ("What was missed at first autopsy," *The Globe and Mail* (27 January 1995, A5). Morin was finally exonerated by DNA evidence in 1995. See the Kaufman Report in Appendix A.

The Crown's obligation to disclose (also referred to as "first-party disclosure" obligation) extends to information that the Crown has in its possession or control that is relevant to the accused's case or his or her defence. According to the Supreme Court of Canada, it follows that police have a corollary duty "to disclose to the Crown the fruits of the investigation"(*McNeil* paras. 14, 24, 52), including any information that might affect the credibility of Crown witnesses, such as misconduct by investigators that bears on their credibility in the case (para. 59). The Crown also has an obligation to make "reasonable inquiries of other Crown entities and other third parties, in appropriate cases" (paras. 13 and 48-49). It should be remembered that "the Crown and the defence in a criminal proceeding are not adverse in interest for the purpose of discovering relevant information that may be of benefit to an accused" (para. 13). This does not mean that the Crown cannot redact information that attracts a reasonable expectation of privacy, although the overriding concern will be the accused's right to make full answer and defence (paras. 38 and 44).

The disclosure of third party records that are relevant, but beyond the possession or control of the prosecutor under the first-party disclosure obligations, are governed by the common law developed by the

Supreme Court of Canada in *O'Connor*. Third-party records in sexual assault cases are, however, dealt with differently under section 278.1 of the *Code* (discussed in Chapter 12). The Supreme Court elaborated on the so-called *O'Connor* application in its decisions in *McNeil*. Such applications by the accused for third party records are accompanied by an affidavit showing that the records are likely relevant to the case. If the records are not privileged, the judge decides whether the records are "likely relevant" to the accused's case–that is, is there "a reasonable possibility that the information is logically probative to an issue at trial or the competence of a witness to testify" (para. 33). If so, the judge may order the records produced for inspection by the court (para. 28). The next step is for the judge to decide whether the records can reasonably be used by the defence. The Court described this second stage test as similar to the obligation on the Crown to produce records under *Stinchcombe* (para. 47). Again, redactions can be made to protect privacy interests, so long as they do not affect the accused's right to make full answer and defence (para. 46). Paciocco (2009) suggests that the decision in *McNeil* "bulks up" the *Stinchombe* disclosure requirements.

The remedy for late disclosure is usually to adjourn the trial so that defence counsel can deal with the material that was the subject of tardy disclosure. In exceptional circumstances, the court might exclude the evidence under section 24(1) of the *Charter*. In reversing a trial judge's decision to exclude evidence that was disclosed late in the process, the Supreme Court of Canada stated:

> Thus, a trial judge should only exclude evidence for late disclosure in exceptional cases: (a) where the late disclosure renders the trial process unfair and this unfairness cannot be remedied through an adjournment and disclosure order or (b) where exclusion is necessary to maintain the integrity of the justice system. Because the exclusion of evidence impacts on trial fairness from society's perspective insofar as it impairs the truth-seeking function of trials, where a trial judge can fashion an appropriate remedy for late disclosure that does not deny procedural fairness to the accused and where admission of the evidence does not otherwise compromise the integrity of the justice system, it will not be appropriate and just to exclude evidence under s. 24(1) (*Bjelland* para. 24).

If non-disclosure is established at trial, the trial court can order production and, if necessary, an adjournment (*Dixon* para. 33). If the non-disclosure is established by the accused on appeal, "the accused bears the additional burden of demonstrating on a balance of probabilities that the right to make full answer and defence was impaired as a result of the failure to disclose" (*Dixon* para. 33). The accused must show that there is a "reasonable probability the non-disclosure affected the outcome of the trial or the overall fairness of the trial process" (*Dixon* para. 34).

A judicial stay of proceedings, considered by the Court to be an "extraordinary remedy," should only be granted where the accused has demonstrated "irreparable prejudice" to the right to make full answer and defence. A new trial can be ordered where the accused shows that non-disclosure affected the outcome or the overall fairness of the trial (para. 35). In considering the overall fairness of the trial, the court must examine defence counsel's diligence in obtaining disclosure. Defence counsel must "diligently pursue disclosure" (para. 37). This had not occurred in the *Dixon* case, and the Supreme Court of Canada considered this factor in dismissing the accused's appeal (see Davison 1998 and Mitchell 1998, for comments on this case).

The two-step test, when disclosure is raised on appeal, was summarized in *Taillefer* and *Duguay*. In assessing the impact that the undisclosed evidence might have had on the trial, the accused must first "demonstrate that there is a reasonable possibility that the verdict might have been different but for the Crown's failure to disclose all of the relevant evidence" (para. 81). The task for the appellate court is to determine whether "there was a reasonable possibility that the jury, with the benefit of all of the relevant evidence, might have had a reasonable doubt as to the accused's guilt" (para. 81). Even if the accused is unsuccessful in this first stage, the appellate court "must then inquire as to whether there is a reasonable possibility that the failure to disclose affected the overall fairness of the trial process" (para. 81). This second stage includes examining the possible uses that the accused might have made of the undisclosed evidence in mounting a defence (e.g., evidence that might have allowed the accused to impeach a Crown witness).

In 2007, the Supreme Court of Canada found it "unnecessary to consider the issue of post-conviction disclosure" because it was "entirely moot," and therefore the issue of post-conviction disclosure was "left for another day" (*Trotta* para. 17).

Box 4.2 Mammoth Trials, Mammoth Disclosure

In the case of accused Robert Pickton, charged with the first-degree murder of 26 Vancouver sex trade workers, the Crown disclosed some 750,000 pages of material to the defence (Baron 2006). The preliminary inquiry for Pickton, who had been in custody since February, 2002, started in January 2006. Other mammoth trials have included the Air India bombing prosecution, and some of the organized crime prosecutions.

Particulars

One way of discovering, or learning about, the Crown's case is through **particulars**. These may be informal particulars (sometimes called **"circumstances"**), given by phone or mailed to defence counsel, or simply discussed between Crown and defence counsel. Circumstances typically involve a narrative account of the facts alleged, any statements made, and details of the accused's alleged criminal record. Informal particulars or circumstances of the offence given to defence counsel in this manner do not form part of the indictment.

Formal particulars, however, can be ordered by the court under section 587 of the *Criminal Code*. If formal particulars are ordered, section 587(3) states that the trial shall proceed "as if the indictment had been amended to conform with the particular[s]." This means that the Crown has to prove whatever particulars are ordered, as if they were specified in the original indictment. Formal particulars are much narrower than circumstances, and may include, for example, particularizing or naming a person or place referred to in an indictment. Their purpose is to allow the accused to know with certainty what offence is alleged.

In *Thatcher*, the defence had applied for formal particulars as to whether the Crown was alleging that the accused had personally killed his wife, or that the accused had had someone else do the killing. The court refused to order the Crown to provide formal particulars on this issue. The Supreme Court of Canada later decided that it was not even necessary for the jury to have been unanimous as to whether the accused killed his wife personally or had someone else do it in order to support the murder conviction (see Chapter 5).

Pre-Trial Conference

Section 625.1(2), which was added to the *Criminal Code* in 1985, requires a **pre-trial conference** to be held prior to all jury trials, to "promote a fair and expeditious trial." In non-jury cases, or even for lengthy preliminary inquiries, a pre-trial conference may also be arranged for the same purpose, on the initiative of either the Crown or the accused. Since 1997, the presiding judge has been allowed to order such a pre-trial conference, even without the consent of the Crown or the accused.

Disclosure Court

After the Law Reform Commission of Canada made several recommendations regarding pre-trial disclosure, a voluntary disclosure project was set up in Montreal in 1975. The effect was an increase in guilty pleas, an increase in the number of charges withdrawn, and a significant saving of time and money in the administration of justice (Law Reform Commission of Canada 1984, 7–9). Other experiments took place in Vancouver, Montreal, and Ottawa. They varied in approaches and results, but the overall conclusion was that a more formal disclosure process would reduce delays and inefficiencies in the criminal justice system.

     Disclosure courts now exist across Canada in order to reduce delays and to ensure that the defence obtains full details of the Crown's case at the earliest possible stage. The judge may also canvass whether the defence is prepared to make admissions to reduce the amount of time required for a trial. The accused may also plead guilty in disclosure court.

DEFENCE DISCLOSURE

The Supreme Court of Canada in *Stinchcombe* observed that the development of comprehensive rules for Crown disclosure was likely resisted because the law reform proposals "did not provide reciprocal disclosure by the defence" (para. 10). The Court speculated that it might consider defence disclosure in the future, but also confirmed that the defence has no obligation to assist the prosecutor in the case against the accused (para. 12). It is unclear how the Court might consider or require defence disclosure. Stalker (2002) makes the argument that since the *Charter* does not apply to the accused, it could not be used to require defence disclosure. In addition, she concludes that the common law does not lend itself to require defence disclosure. The lack of defence disclosure in Canada is in sharp contrast to the laws in many American states and in England, imposing an obligation on the defence to disclose certain aspects of their case (see Box 4.3).

     There are a number of arguments for and against defence disclosure (see Costom 1996; Davison 1996; McKinnon 1995; Maude 1999; Tomljanovic 2002; Tochor and Kilback 2000). Arguments for defence disclosure include:

(1)   it prevents the defence from ambushing the Crown;
(2)   it is more efficient and will result in more guilty pleas or shorter trials through the narrowing of issues;
(3)   it assists in ascertaining the truth;
(4)   disclosure should be a two-way street;
(5)   a fair trial includes the perspective of the public, represented by the Crown;

Box 4.3 Defence Disclosure

"A criminal trial should not be a game. The Crown should bear the burden of proof but not the handicap of carrying it in the dark" (Speech of the Lord Chief Justice of England quoted in Glynn at 841–2). The result of this sentiment has been a move to defence disclosure rules in various jurisdictions. In California, reciprocal discovery was introduced by the *Crime Victims Justice Reform Act* of 1991. The provisions require the defence to inform the prosecution in advance of trial of all the witnesses and experts it intends to call, and of all the physical evidence and reports it intends to introduce. In 2003, England added the requirement of a defence statement to its earlier defence disclosure requirements of 1996. A defence statement is a written statement:

> (a) setting out the nature of the accused's defence, including any particular defences on which he intends to rely,
> (b) indicating the matters of fact on which he takes issue with the prosecution,
> (c) setting out, in the case of each such matter, why he takes issue with the prosecution, and
> (d) indicating any point of law (including any point as to the admissibility of evidence or an abuse of process) which he wishes to take, and any authority on which he intends to rely for that purpose (Criminal Justice Act 2003, Chapter 44, section 6A).

---

(6) it would simply be an extension of a number of existing requirements for defence to disclose (see Tomljanovic 2002);

(7) this slight move from the adversarial to the inquisitorial system would be a positive move.

Arguments against defence disclosure include:

(1) the Crown has more resources, and this is one way to balance things out;

(2) the Crown might tailor or alter its case after hearing defence disclosure (i.e., evidence might be fabricated to fill the holes in the Crown's case identified by defence disclosure);

(3) it violates the right to remain silent and the right to make full answer and defence (see Stalker 2002);

(4) a criminal trial is not a search for truth; it is "a systematic testing of the prosecution's case to determine whether it has proven to the legal standard required, that a particular wrongful act for which there is no legal excuse or justification has been committed by this particular accused person" (Davison, 1996 at 108);

(5) a partial move to the inquisitorial system is a negative move;

(6)   in our system, the defence takes an adversarial position; the Crown has an overriding duty to ensure that justice prevails;

(7)   defence counsel typically may not make their final decisions about defences to be raised, witnesses to be called, or even whether to call evidence at all, until they can assess the Crown's case as a whole on its completion.

Tomljanovic (2002) makes the argument that rules surrounding defence disclosure already exist, are legitimate, and should be codified. Since 2002, section 657.3(3) requires both the Crown and the accused to give each other 30 days notice where they plan to call an expert witness. The notice must include the name of the proposed witness, and the witness's area of expertise and qualifications. In addition, the Crown must provide the accused with a copy of the expert's report (if one is prepared), or a summary of the opinion evidence "within a reasonable period before the trial." The accused must provide a copy or a summary of the evidence of any expert to the Crown "not later than the close of the case for the prosecution" (section 657(3)(c)). The procedures prior to a preliminary inquiry (discussed below) require defence disclosure, as do other sections of the *Criminal Code*.

Tanovich and Crocker (1994, 346) argue that further disclosure by the defence (names and statements of all its potential witnesses, and notice of any defences) would enhance the search for truth, without jeopardizing the accused's right to a fair trial (340). Defence disclosure would not violate an accused's rights under sections 7 and 11(d) of the *Charter* (Tanovich and Crocker 1994, 341–2 discuss American cases that have held that reciprocal disclosure does not violate the accused's right under the Fifth Amendment—the right to remain silent), because it involves an "accelerated disclosure." Tomljanovic (2002 para. 35) also argues in favour of "accelerated" defence disclosure where the defence plans to call evidence. Davison suggests that Tanovich and Crocker's proposal, not accompanied by a move of our adversarial system toward an inquisitorial system, will result in "more risks for the innocent accused than potential benefits for the state or complainants" (1996, 121).

EVIDENCE IN POSSESSION OF DEFENCE COUNSEL

Although historically, defence counsel who came into possession of evidence of a crime (the "smoking gun," the "bloody shirt") saw themselves in a controversial dilemma (see, for example, Cooper 2003), today it is quite clear that defence counsel who conceal evidence of a crime are themselves committing the crime of obstructing justice. More recently, Kenneth Murray, Paul Barnardo's lawyer, was tried on charges of wilfully attempting to obstruct justice by concealing videotapes that showed two girls, who were later killed, being forced to participate with Bernardo and his wife Karla Homolka "in the grossest sexual perversions" (*Murray* para. 19). In finding that Murray committed the *actus reus* of the offence, Gravely J. stated "The tapes were the products and instrumentalities of crime and were far more potent 'hard evidence' than the often-mentioned 'smoking gun' and 'bloody shirt'" (para. 109). However, the trial judge found that Murray did not have the required *mens rea*, and acquitted him of the charge (for a discussion of this case, see Cooper 2003; Scott 2003; Renke 2003). Following the decision, the Law Society of Upper Canada proposed a Rule that would provide lawyers with guidance in relation to property relevant to a crime (see Brauit and Argitis 2003).

PROCEDURE BEFORE A PRELIMINARY INQUIRY

Changes to the *Criminal Code* in 2004 established new disclosure procedures that take place prior to a preliminary inquiry. Section 536.3 provides that the party requesting a preliminary inquiry shall (within specified time periods) "provide the court and the other party with a statement that identifies (a) the issues on which the requesting party wants evidence to be given at the inquiry; and (b) the witnesses that the requesting party wants to hear at the inquiry." The justice, before whom a preliminary inquiry is to be heard, may order a hearing to "(a) assist the parties to identify the issues on which evidence will be given at the inquiry; (b) assist the parties to identify the witnesses to be heard at the inquiry, taking into account the witnesses' needs and circumstances; and (c) encourage the parties to consider any other matters that would promote a fair and expeditious inquiry" (section 536.4(1)). Any agreement to limit the scope of the preliminary inquiry that arises out of this hearing, or any agreement reached within this hearing, is recorded and filed with the court. Such hearings have come to be known as "focus hearing" (*Gill* para. 5).

THE PRELIMINARY INQUIRY

Although the accused may obtain disclosure of the Crown's case through a preliminary inquiry, there is no constitutional right to a preliminary inquiry (*S.J.L.* 2009 para. 21). Since 2004, a preliminary inquiry is held only if the accused or prosecutor requests one. A preliminary inquiry can be requested for offences under section 469, or if the accused elects trial by a superior court judge, or judge and jury (see section 536). All preliminary inquiries are held in provincial court. The provincial court judge will not arraign the accused in such a case (that is, the accused will not be asked to enter a plea), because the superior court will be the trial court and will take the plea. If, after the preliminary inquiry, the provincial court judge commits the accused to stand trial, the arraignment will take place in a superior court of criminal jurisdiction. See section 536.1 for the procedure to be followed in Nunavut.

Historically, the purpose of a preliminary inquiry was quite different from its purpose today. In early England, it was "designed as an instrument of the prosecution for finding the culprit and preparing the evidence against him" (Devlin, quoted in *Skogman* at 171). The preliminary inquiry was used as an investigative tool by Justices of the Peace, who performed a role similar to what the police do today. The modern preliminary inquiry is primarily designed to protect the accused from being put on trial unnecessarily, and it is used as a tool for discovery. Mr. Justice Estey discussed the purpose of the preliminary inquiry in the Supreme Court of Canada's decision in *Skogman*:

> The purpose of the preliminary hearing is to protect the accused from a needless, and indeed, improper, exposure to public trial where the enforcement agency is not in possession of evidence to warrant the continuation of the process. In addition, in the course of its development in this country, the preliminary hearing has become a forum where the accused is afforded an opportunity to discover and to appreciate the case to be made against him at trial where the requisite evidence is found to be present (171).

Section 537 sets out the powers of a provincial court judge at a preliminary inquiry. Since 2004, the judge may permit the accused to be absent for part or all of the preliminary inquiry, and may order the cessation of any questioning of witnesses that is "abusive, too repetitive or otherwise inappropriate." At the end of the preliminary inquiry, the judge will either (a) order the accused to stand trial, if there is "sufficient evidence to put the accused on trial for the offence charged or any other indictable offence in respect of the same transaction;" or (b) discharge the accused if there is insufficient evidence to put the accused on trial (section 548(1)). What is "sufficient evidence to put the accused on trial?" The Supreme Court of Canada in *U.S.A. v. Sheppard* formulated the test as: "The justice is required to commit an accused person for trial in any case in which there is admissible evidence which could, if it were believed, result in a conviction" (427).

Section 540 outlines how a preliminary inquiry is to be conducted. The Crown calls witnesses to give evidence under oath. The accused or counsel has the right to cross-examine the Crown's witnesses. The judge is not to consider the credibility of the witnesses; the weighing of evidence is left for the trier of fact (either the judge, or the jury if trial by judge and jury) (*Arcuri*). The Crown need only establish a **prima facie case** (meaning "on its face"), but the Crown must present at least some evidence on each essential element of the offence. If there is no evidence on a particular element of the offence, the evidence as a whole cannot be sufficient to order the accused to stand trial (*Skogman*).

In 2004, subsections 540(7) (8) and (9) were added to the *Code* to reduce the time required for preliminary hearings and to minimize the number of times complainants, particularly those in sexual assault trials, are subjected to examination and cross-examination. Subsection 540(7) allows the preliminary inquiry judge to "receive as evidence any information that would not otherwise be admissible but that the justice considers credible or trustworthy in the circumstances of the case, including a statement that is made by a witness in writing or otherwise recorded." Subsection 540(9) allows the judge, where appropriate, to require any person to appear for examination or cross-examination with respect to evidence admitted under subsection 540(7). Early lower court decisions on these provisions have admitted such evidence under subsection 540(7), but judges seem reluctant to refuse to allow defence counsel to cross-examine witnesses on such evidence. For example, in *S.P.I.*, the preliminary inquiry judge allowed the Crown to introduce audio and video statements of the child victims of alleged sexual assault, but also decided under section 507(9) that defence counsel could cross-examine the complainants (for a critique of these provisions, see Paciocco 2004, 173-183).

Section 541 states that at the end of the Crown's evidence, the accused is to be addressed by the provincial court judge in the words set out in section 541(2). Subsection (3) says that anything the accused says is to be recorded as evidence. Subsection (4) gives the accused the option of calling witnesses. Most accused or defence counsel do not call witnesses at the preliminary inquiry, although it is done occasionally. In such a case, the defence would likely call only a potential Crown witness it wanted to hear from, but whom the Crown did not call. There would be little point for the defence to call witnesses to help the accused, since the court at the preliminary inquiry cannot weigh the evidence.

Section 549 allows for committal to stand trial by consent at any stage of preliminary inquiry, and **consent committals** happen frequently. Defence counsel often want to hear, and perhaps to cross-examine, only certain of the Crown witnesses, and then move on to a trial. A consent committal allows for this, without the court having to sit through the evidence of additional witnesses. Alternatively, an accused may intend to plead guilty in superior court, in which case the committal is just a necessary formality.

A provincial court judge hearing a preliminary inquiry is not a "court of competent jurisdiction" within the meaning of section 24(1) of the *Charter,* and therefore cannot exclude evidence under section 24(2) (*Mills;* and *Hynes*), or declare that a law is of "no force and effect" under section 52 of the *Constitution Act* (*Moore*).

Section 574 allows the Crown to prefer an indictment against any person who has been ordered to stand trial on a charge following a preliminary inquiry, or on "any charge founded on the facts disclosed by the evidence taken by on the preliminary inquiry, in addition to or in substitution for any charge on which that person was ordered to stand trial." This section appears to allow the Crown to lay whatever charge it likes, so long as it is based on the facts disclosed. Section 574 does not, however, allow the Crown to ignore a discharge by the preliminary inquiry judge. In *Tapaquon*, the Supreme Court of Canada considered a case in which the accused was charged with assault causing bodily harm. The judge at the preliminary inquiry ordered the accused to stand trial on the lesser included offence of assault. The Crown then preferred an indictment for the original, more serious charge of assault causing bodily harm. The Supreme Court of Canada said that this was not permissible. Once the accused is discharged, the only recourse the Crown has is to proceed by direct indictment.

In *Skogman,* the Court considered whether an order to stand trial could be appealed. It held that a provincial court judge's decision that there is sufficient evidence for an accused to stand trial cannot be reviewed on appeal, unless the judge was acting outside of or exceeded her or his statutory jurisdiction, such as by breaching the principles of natural justice. The appellate court will not interfere with a committal merely because it disagrees with the merits of the decision.

Exclusion of the Public and Publication Ban

The public can be excluded from the preliminary inquiry under section 537(1)(h), "where the ends of justice will be best served by doing so." Section 539 allows for a ban on publication of proceedings at the preliminary inquiry. This order must be made if requested by the accused, and remains in force until the accused is discharged at the preliminary inquiry or, if ordered to stand trial, until the end of the trial. These provisions exist to ensure a fair hearing if the accused is committed to stand trial. This type of publication ban is not the same as was made (for example) in the Karla Homolka case, where a publication ban was imposed respecting details of one accused's trial to protect the rights of another party (the co-accused) to a fair trial.

An Absconding Accused

Section 544 allows a preliminary inquiry to be conducted in the absence of the accused, if the accused does not appear at the designated time. This procedure is rarely used, and may be less of an issue after amendments in 2004, which allow for the preliminary inquiry judge to permit the accused to be absent for part or all of the inquiry (section 537(1)(j.1)). Under such circumstances, the judge is required to inform the accused that evidence taken in their absence could be admissible under section 715 (section 537(1.01)).

Uses of Preliminary Inquiry

Evidence given at the preliminary inquiry has several uses other than just for the judge to decide whether to order an accused to stand trial. The preliminary inquiry preserves evidence in case a witness dies, becomes ill or insane, is unavailable, or refuses to testify at the trial of the accused. In such circumstances, the evidence taken at the preliminary inquiry can be read into evidence at the trial of the accused, pursuant to section 715 of the *Criminal Code.* This is sometimes referred to as **"perpetuated evidence."**

The Supreme Court of Canada dealt with section 715 in *Potvin.* Potvin was convicted at trial of second-degree murder. Two men and one woman were alleged to have planned to rob a woman, who was severely beaten and who subsequently died. The Crown intended to proceed with charges against Potvin, and to call the other two as Crown witnesses. When one witness refused to testify at the trial, the trial judge allowed the witness's evidence from the preliminary inquiry to be read in at Potvin's trial. The accused's appeal to the Quebec Court of Appeal was dismissed.

On further appeal by Potvin, the Supreme Court of Canada ordered a new trial. The Court held that section 715 did not violate an accused's rights under sections 7 and 11(d) of the *Charter*. The Court explained that courts in Canada (prior to the *Charter*), England, and the United States have all established a practice of admitting statements that were taken under oath, provided that the accused had had an opportunity to cross-examine the witness at the time they were made (541). "In the absence of circumstances which negated or minimized the accused's opportunity to cross-examine the witness when the previous statement was given," the accused's rights under section 7 are not violated (542).

The Court in *Potvin* stressed that section 715 also allows the trial judge the discretion to exclude perpetuated evidence where its admission would be unfair to the accused, although such circumstances would be "relatively rare" (548). Situations in which it would be unfair to admit such evidence could arise from "the manner in which the evidence was obtained," or from the fact that it was "highly prejudicial to the accused and of only modest probative value" (552). In exercising this discretion, the court must weigh "two competing and frequently conflicting concerns" in the administration of justice: "fair treatment of the accused and society's interest in the admission of probative evidence in order to get at the truth of the matter in issue" (553). In this case, the Court held, the trial judge had stressed the probative value of the evidence without giving adequate consideration to the possible sources of unfairness to the accused. The Court ordered a new trial so that the trial judge could properly exercise his or her discretion under section 715. The Court added that it was inappropriate for the trial judge to refer to the evidence from the preliminary inquiry as "evidence like any other testimony given in the course of the trial" when instructing the jury. The jury should have been told that it had not had "the benefit of observing the witness giving the testimony" (555).

In addition to using the evidence from a preliminary inquiry at trial under section 715, transcripts of evidence at the preliminary inquiry can be used by the defence to cross-examine Crown witnesses (and by the Crown to cross-examine defence witnesses, if any testify at the preliminary inquiry).

A preliminary inquiry also provides an opportunity for the defence to see the Crown witnesses and assess them, and for the Crown to find out how its witnesses will stand up. After a preliminary inquiry, the accused may decide that the evidence is so overwhelming that a guilty plea is appropriate. The Crown may see sufficient weaknesses in its case that it decides to drop it. A day in court may satisfy the victim that a case is too weak for the Crown to proceed.

The Abolition of the Preliminary Inquiry

In 1992, the Chief Justice of the Supreme Court of Canada suggested that the preliminary inquiry be abolished (*Vancouver Sun* A11). In February of 1994, the Department of Justice released a discussion paper dealing with the possibility of eliminating preliminary inquiries.

Arguments in favour of abolishing the preliminary inquiry included the following:

1. When deciding whether or not to proceed with a case, prosecutors now apply a tougher test than a justice must use at a preliminary inquiry.
2. The Supreme Court of Canada's decisions on an accused's rights under the Charter require the prosecution to disclose its evidence to the defence before a preliminary inquiry, making the preliminary inquiry superfluous with respect to the disclosure of evidence.
3. Victims of crime and witnesses may have to give evidence at both the preliminary inquiry and the trial, which is an unfair burden to them.
4. The preliminary inquiry has no judicial purpose and is now being used by the prosecution and defence only to assist with trial preparation.
5. A preliminary inquiry creates another layer in the criminal justice process, delaying the resolution of a case and increasing the difficulties for victims of crime and witnesses.

6. A preliminary inquiry is an additional and unnecessary expense.

7. The preliminary inquiry is a weak link in the justice process, with some inconsistencies and anomalous elements (6–8; also see Epp 1996).

In defence of the preliminary inquiry, the Department of Justice stated:

1. The preliminary inquiry is the best way for the defence to properly prepare for the trial.

2. The "dress rehearsal" role of a preliminary inquiry helps all parties to prepare for the trial, to everyone's benefit.

3. Once an accused has had a chance to hear the evidence proving his or her guilt at the preliminary inquiry, he or she is more likely to plead guilty before the trial.

4. The preliminary inquiry provides an excellent forum for the defence and the prosecution to discuss guilty pleas, reduced charges and appropriate punishment.

5. After the preliminary inquiry, the defence can decide to accept some of the evidence and limit the issues that it will argue at trial.

6. Preliminary inquiries do not delay trials and may, in fact, contribute to earlier trial dates (9–10).

According to the Department of Justice (12), any changes to the preliminary inquiry will have to take into account an accused's right to make full answer and defence under section 7 of the *Charter*, the right to a trial within a reasonable time (section 11(b)), and the right to equality under the law (section 15). If the federal government follows through with a plan to increase the number of hybrid offences (by decreasing the number of strictly indictable offences), Crown counsel would be allowed to bypass more preliminary inquiries simply by electing to proceed by way of summary conviction (Skelton and Ayed 1999, A1). Some of the amendments made historically and in 2004 (referred to above) may have an impact on the number and nature of preliminary inquiries (see Paciocco 2004).

SUMMARY

The Supreme Court of Canada, in the context of the right to make full answer and defence founded in section 7 of the *Charter,* has placed a heavy onus on the Crown to provide complete pre-trial disclosure to the defence in criminal prosecutions. Pre-trial disclosure is, in a general sense, a form of informal particulars (circumstances). The purpose of providing pre-trial disclosure is to permit the accused to know the case to be met at trial. Formal particulars, however, are court-ordered and form part of the specific allegation against the accused that must be proved to obtain a conviction.

Defence disclosure, in which the defence has a duty to advise the Crown of certain aspects of its case before a trial, exists in the United States and England. Many jurisdictions in Canada have implemented forms of disclosure court, which seek to ensure that the defence has obtained timely disclosure of the Crown's case and has considered possible admissions by the accused prior to the matter being set down for trial. Some of the 2004 amendments surrounding preliminary inquiries require defence disclosure.

The preliminary inquiry is designed to ensure that there is a case to meet before an accused is required to face trial on a more serious indictable matter. It has also evolved into a vehicle for pre-trial discovery by the defence of the prosecution's case. At the end of the preliminary inquiry, the court will consider whether the Crown has led some evidence on each essential element of the charge; the court cannot weigh the evidence. If a *prima facie* case exists, the accused will be committed to stand trial; if not, the accused will be discharged. An accused may consent to being committed to stand trial at any stage of the proceedings. The higher courts will not normally interfere with a provincial court judge's decision respecting committal.

A provincial court hearing a preliminary inquiry is not a "court of competent jurisdiction" within the meaning of section 24 of the *Charter,* and therefore cannot exclude evidence because of alleged breaches of the accused's *Charter* rights. The Crown can avoid a preliminary inquiry by preferring a direct indictment, as discussed in Chapter 3. The public can be excluded from a preliminary inquiry, and a ban on the publication of proceedings at the preliminary inquiry must be made if requested by the defence.

Besides ensuring that the charges have merit and providing the defence with discovery, the preliminary inquiry also serves the purpose of perpetuating evidence. If a witness is unable to or refuses to testify at trial, her or his testimony from the preliminary may be read into evidence at trial.

QUESTIONS TO CONSIDER

(1)   "Should Crown counsel accept a plea of guilty where he or she knows that the prosecution cannot prove a material element in its case (for example due to an unavailable witness)? Does Crown counsel have a duty to disclose to the accused or defence counsel that the prosecution is unable to prove a material element of its case?" (Questions raised in the Martin Report.)

(2)   What are some of the problems in our criminal justice system that have led to calls for greater disclosure by the Crown?

(3)   Should Crown disclosure rules be tailored to local conditions, or should there be standard rules that apply across Canada?

(4)   Should the federal government introduce defence disclosure in Canada? What are the advantages and disadvantages of such a system? Would it change the nature of the adversarial system? Would it violate an accused's rights under the *Charter*?

(5)   What changes to the law have moved Canada toward more defence disclosure?

(6)   What is the purpose of a focus hearing?

(7)   What are the pros and cons of a preliminary inquiry in criminal cases? Would you recommend that it continue or be abolished?

(8)   What is the test applied at the preliminary hearing?

(9)   Compare the test used at a preliminary inquiry with the test used by prosecutors in deciding whether to proceed with a case. What impact does this have on your recommendation in question 7?

(10)   On what basis can the public be excluded from a preliminary inquiry?

BIBLIOGRAPHY

Baron, Ethan. "Pickton Pleads Not Guilty to 26 Murder Counts." (31 January 2006) *Vancouver Province.*

Brauti, Peter M. and Gena Argitis. "Possession of Evidence by Counsel: Ontario's Proposed Solution." (2003) 47 *Criminal Law Quarterly* 211.

Cooper, Austin M. "The Ken Murray Case: Defence Counsel's Dilemma." (2003) 47 *Criminal Law Quarterly* 141.

Costom, S. "Disclosure by the Defence: Why Should I Tell You?" (1996) 1 *Canadian Criminal Law Review* 73.

Davison, Charles B. "Putting Ghosts to Rest: A Reply to the 'Modest Proposal' for Defence Disclosure of Tanovich and Crocker." (1996) 43 *Criminal Reports* (4th) 105.

Davison, Charles B. "Disclosure, Due Diligence and Defence Counsel—Increasing the Burden and Raising the Standards." (1998) 13 *Criminal Reports* (5th) 269.

CHAPTER 4: *Crown Disclosure and the Preliminary Inquiry*

Department of Justice. *Do We Still Need Preliminary Inquiries?* Ottawa: Department of Justice, 1994.
   This is a consultation paper on the options for changing the *Criminal Code* regarding preliminary inquiries. The issue is examined in light of the principles of the criminal justice system and the purpose of the preliminary inquiry. The paper considers options: keep the preliminary inquiry as it is, change it, or abolish it.

Department of Justice. *Disclosure Reform–Consultation Paper* (November 2004; http://www.justice.gc.ca/en/cons/disc-ref/index.html; accessed September 18, 2005).
   The federal Minister of Justice called for comments on amendments to five areas: 1) the use of electronic disclosure; 2) clarifying the core materials to be disclosed to alleviate the burden of disclosure; 3) setting up specialized court proceedings to expedite disclosure; 4) establishing disclosure-management procedures and timelines; and 5) addressing the improper use of disclosed materials.

Devlin, Patrick. *The Criminal Prosecution in England.* New Haven: Yale University Press, 1958.

Epp, John Arnold. "Abolishing Preliminary Inquiries in Canada." (1996) 38 *Criminal Law Quarterly* 495.

Ferguson, Gerry. "Judicial Reform of Crown Disclosure." (1992) 8 *Criminal Reports* (4th) 295.

Glynn, Joanna. "Disclosure." [1993] *Criminal Law Review* 841.

Kaiser, Archibald. 2009. "*McNeil*: A Welcome Clarification and Extension of Disclosure Principles: 'the adversary system has lingered on.'" (2009) 62 *Criminal Reports* (6th) 36.

Law Reform Commission of Canada. *Controlling Criminal Prosecutions: The Attorney General and the Crown Prosecutor.* Working Paper #62. Ottawa, 1990.

Law Reform Commission of Canada. *Disclosure by the Prosecution.* Report #22. Ottawa, 1984.

Law Reform Commission of Canada. *Discovery in Criminal Cases.* Ottawa, 1974.

Martin, Arthur G. *Charge Screening, Disclosure, and Resolution Discussion.* Toronto: Report of the Attorney-General's Advisory, 1993. Chapter III Disclosure.
   The Report summarizes 15 propositions arising from the Supreme Court of Canada's decision in *Stinchcombe* (based on the right to make full answer and defence under section 7 of the *Charter*), and recommends that all information be disclosed unless the public interest (to protect informants, ongoing investigations, or the integrity of investigative techniques) requires non-disclosure. It also makes recommendations with respect to the importance of proper note-taking by the police, and the preference for video-taped statements where feasible.

Maude, Brian Edward. "Reciprocal Disclosure in Criminal Trials: Stacking the Deck Against the Accused, or Calling Defence Counsel's Bluff?" (1999) 37 *Alberta Law Review* 715.

McKinnon, G.D. "Accelerating Defence Disclosure: A Time For Change." (1995) 53(1) *The Advocate* 25.

Mitchell, Graeme G. "*R. v. Dixon*: The Right to Crown Disclosure—A Road Map For the Future?" (1998) 13 *Criminal Reports* (5th) 260.

Paciocco, David M. "A Voyage of Discovery: Examining the Precarious Condition of the Preliminary Inquiry." (2004) 48 *Criminal Law Quarterly* 151.

Paciocco, David M. "Stinchcombe on Steroids: The Surprising Legacy of *McNeil*." (2009) 62 *Criminal Reports* (6th) 26.

Pomerant, David, and Glenn Gilmour. *A Survey of the Preliminary Inquiry in Canada.* WD1993-10e. Ottawa: Department of Justice, April 1993.

CHAPTER 4: *Crown Disclosure and the Preliminary Inquiry*

Quigley, Tim. *Procedure in Canadian Criminal Law.* Toronto: Carswell, 1997 (the second edition, 2005, is available in looseleaf).
> Chapter 14 examines preliminary inquiries, and Chapter 16 discusses prosecutorial powers.

Roach, Kent, Gary Trotter and Patrick Healy. *Criminal Law and Procedure: Cases and Materials*, 9th ed. Toronto: Emond Montgomery Publications Limited, 2004.
> Chapter 4 covers the role of the prosecutor and defence counsel and disclosure.

Renke, Wayne N. "Real Evidence, Disclosure and the Plight of Counsel." (2003) 47 *Criminal Law Quarterly* 174.

Salhany, Roger E. *Canadian Criminal Procedure,* 6th ed. Toronto: Canada Law Book Inc., (available online on Criminal Spectrum) .
> Chapter 5 discusses preliminary inquiries, the proceedings, Crown disclosure, committal, and discharge.

Scott, Ian D. "Can Documents Smoke? The *R. v. Murray* Decision and Documents Characterized as Evidence of a Crime." (2003) 47 *Criminal Law Quarterly* 157.

Skelton, Chad, with Nahlad Ayed. "Fast Tracking for Trials Wins Support." *Vancouver Sun,* (16 August 1999) A1, A2.

Slobogin, Christopher. "Discovery by the Prosecution in the United States: A Balancing Perspective." (1994) 36 *Criminal Law Quarterly* 422.

Stalker, M. Anne. "Charter Roadblocks to Defence Disclosure." (2002) 40 *Alberta Law Review* 701.

Stuart, Don, Ron Delisle, and Tim Quigley. *Learning Canadian Criminal Procedure*, 9th ed. Toronto: Carswell, 2008.
> Chapter 8 talks about disclosure and discovery, and Chapter 9 covers the preliminary inquiry.

Tanovich, David M., and Lawrence Crocker. "Dancing with Stinchcombe's Ghost: A Modest Proposal for Reciprocal Defence Disclosure" (1994) 26 *Criminal Reports* (4th) 333.

Tanovich, David M. "When Does *Stinchcombe* Demand That the Crown Reveal the Identity of a Police Informer?" (1995) 38 *Criminal Reports* (4th) 202.

Tomljanovic, Goran. "Defence Disclosure: Is the Right to 'Full Answer' the Right to Ambush?" (2002) 40 *Alberta Law Review* 689.

Tochor, Michael D. and Keith D. Kilback. "Defence Disclosure: Is it Written in Stone?" (2000) 43 *Criminal Law Quarterly* 393.

Vancouver Sun. "Abolish Preliminary Hearings, Supreme Court Chief Justice Says," June 12, 1992, A11.

CASES

R. v. Arcuri, [2001] 2 S.C.R. 828. The Court reaffirmed a "well-settled rule that a preliminary inquiry judge must determine whether there is sufficient evidence to permit a properly instructed jury, acting reasonably, to convict, and the corollary that the judge must weigh the evidence in the limited sense of assessing whether it is capable of supporting the inferences the Crown asks the jury to draw. As this Court has consistently held, this task does not require the preliminary judge to draw inferences from the facts or to assess credibility. Rather, the preliminary inquiry judge must, while giving full recognition to the right of the jury to draw justifiable inferences of fact and assess credibility, consider whether the evidence taken as a whole could reasonably support a verdict of guilty" (para. 1).

R. v. Bjelland, 2009 SCC 38. Exclusion of evidence under section 24(1) as a remedy for late disclosure will occur only in exceptional circumstances.

Boucher v. The Queen, [1955] S.C.R. 16.

R. v. C. (M.H.) (1991), 63 C.C.C. (3d) 385 (S.C.C.). There is a common law duty on the Crown to disclose all material evidence, even favourable to the defence.

R. v. Dixon, [1998] 1 S.C.R. 244. Duty of the Crown to disclose, and the effect of defence counsel's lack of due diligence.

R. v. Egger, [1993] 2 S.C.R. 451.

R. v. Gill, [2006] B.C.J. No. 1378 (BCSC).

R. v. Hynes, [2001] 3 S.C.R. 623. Although the preliminary inquiry judge may consider the admissibility of the accused's statement based on voluntariness, the judge cannot consider the admissibility of evidence based on a breach of the accused's rights under the *Charter*.

R. v. Jewitt (1985), 21 C.C.C.(3d) 7 (S.C.C.). A judicial stay of proceedings may be entered in the clearest of circumstances for an abuse of process. Such a stay constitutes a "judgment or verdict of acquittal," against which the Crown may appeal.

R. v. McNeil, [2009] 1 S.C.R. 66. The Court elaborates on the requirements of Crown disclosure under *Stinchcombe* and clarifies the process for applications for third party records under *O'Connor*.

R. v. Mills, [1986] 1 S.C.R. 863. A provincial court judge conducting a preliminary inquiry is not a "court of competent jurisdiction," capable of considering *Charter* breaches and ordering remedies under s. 24(1), nor can such a court order the exclusion of evidence under s. 24(2). Appeals of *Charter* decisions may be taken following trial in the usual manner; there is no independent basis for appeals from interlocutory *Charter* motions.

R. v. Moore, [1992] 1 S.C.R. 619. A provincial court judge conducting a preliminary inquiry does not have jurisdiction to consider the constitutionality of legislation under the *Charter*.

R. v. Murray (2000), 144 C.C.C. (3d) 289 (Ont. S.C.J.). The acquittal of Kenneth Murray, Paul Bernardo's lawyer, on charges of attempting to obstruct justice by concealing videotapes of the sexual assaults of Bernardo's victims.

R. v. O'Connor, [1995] 4 S.C.R. 411

R. v. Potvin, [1989] 1 S.C.R. 525. Section 715 does not violate an accused's rights under section 7 or 11(d). The court sets out requirements for using evidence from a preliminary hearing at trial when a witness is unavailable.

R. v. S.J.L., [2009] S.C.J. No. 14.

Skogman v. The Queen, [1984] 2 S.C.R. 93.

Re S.P.I. and The Queen (2005), 193 C.C.C. (3d) 240 (Nunavut C. J.).

Stinchcombe v. The Queen (1992), 68 C.C.C. (3d) 1 (S.C.C.). Details the requirements of disclosure by the Crown under section 7 of the *Charter*.

U.S.A. v. Sheppard (1976), 30 C.C.C. (2d) 424 (S.C.C.). The test on extradition is the same as that for directed verdicts at trial (and committals for trial after preliminary inquiry!): "Is there admissible direct evidence which, if believed, could result in a conviction?" The judge is not to assess credibility or otherwise weigh the evidence in answering that question.

R. v. Taillefer; R. v. Duguay, [2003] 3 S.C.R. 307

Thatcher v. The Queen, [1987] 1 S.C.R. 652.

R. v. Trotta, [2007] 3 S.C.R. 453.

CHAPTER 5: *Juries and Procedure at Trial*

CHAPTER OBJECTIVES

In studying this chapter, you should develop an understanding of the following topics and concepts:

- what rights an accused has to a trial by jury, and how those rights can be lost
- who may or may not serve as a juror
- the process by which a jury is chosen
- the types of challenges available, and when and how they may be exercised
- how the jury selection system accommodates concerns of bias or racism
- what happens when the panel is exhausted before a jury is selected
- the conditions under which a juror may be discharged during the course of the trial, and the effect of such a discharge on the trial
- the requirements of unanimity and secrecy
- the impact of social science research on jury trials, and on the assumptions underlying them
- the extent to which juries may disregard the law
- the mechanics of *voir dires* occurring within the context of a jury trial
- the extent to which the publication of details of proceedings at trial may be prohibited
- the general order in which things happen at trial

JURY TRIALS

The Right to Trial by Judge and Jury

Under section 11(f) of the *Charter*, every accused (except those charged under military law) has a right to a jury trial if charged with an offence for which the maximum punishment is five years or more. This section does not entitle an accused to be tried *without* a jury where one is required under the *Criminal Code* (*Turpin*). Everyone charged with an offence listed in section 469 will be tried by judge and jury, unless both the Crown and the accused agree to have a trial before a judge of a superior court of criminal jurisdiction without a jury, pursuant to section 473. An accused can elect to be tried by a judge and jury for most other indictable offences (see Chapter 1, "Elections"). If the Crown proceeds by way of direct indictment, the accused is deemed to have elected trial by judge and jury, but may re-elect to be tried without a jury (section 565(2)). The necessity of gaining the prosecutor's consent to a trial without a jury in these circumstances was removed in 2008. Under section 568 (section 569 re: Nunavut), the Crown can veto an election or re-election (e.g., under 565(2)) of trial without a jury if the offence is one punishable by more than five years in prison, and thereby force the accused to have a jury trial. See Berger (2004) for an argument that forcing an accused to have a jury trial violates the accused's rights to a fair trial, an impartial tribunal, and to make full answer and defence.

Accused persons risk losing the right to a jury trial if they fail to appear for trial at the scheduled time. Section 598 states that, having failed to appear for trial, an accused "shall not be tried by a court composed of judge and jury unless (a) he establishes to the satisfaction of a judge…that there was a legitimate excuse for his failure to appear or remain in attendance for his trial or (b) the Attorney General requires [one] pursuant to section 568 or 569." Section 598(1)(a) was challenged under section 11(f) of the *Charter*, and the Supreme Court of Canada decided that although the section violates the *Charter*, it is "demonstrably justified in a free and democratic society" under section 1 (*Lee*).

Qualification and Selection of Jurors: The Out-of-Court Process

Madame Justice L'Heureux-Dubé, in *Sherratt*, explained that the division between "out-of-court" selection of jurors (governed by provincial legislation) and "in-court" selection (governed by federal legislation) is the result of sections 92(14) and 91(27) of the *Constitution Act*. Provincial governments have jurisdiction over "The Administration of Criminal Justice in the Province," which includes composition of jury panels, and the federal government has jurisdiction over the "Criminal Law," including "the Procedure in Criminal Matters" (521), such as how a jury is selected.

The first step in selecting a jury is the out-of-court determination of who is eligible to sit on a jury, and the establishment of the initial population from which the **array** or **panel** is chosen. Jurisdictional conflicts between the federal and provincial governments are resolved by section 626(1) of the *Criminal Code,* which delegates this task to the provincial governments, except that a person cannot be "disqualified, exempted or excused from serving as a juror in criminal proceedings on the grounds of his or her sex." This section was enacted in 1972 to override legislation in some provinces that excluded women from serving on juries.

Although all provinces disqualify people who work in the criminal justice system from serving on juries (for example, police officers, lawyers, prison guards), other disqualifications vary from province to province. Section 3(2) of the British Columbia *Jury Act*, R.S.B.C. 1996, c.242, allows chiropractors, dentists, naturopaths, and persons "regularly employed in the collection, management or accounting of revenue under the *Financial Administration Act*" to exempt themselves from jury duty, if they so desire. The Ontario *Juries Act*, R.S.O. 1990, c.J.3 section 3(1) makes "every legally qualified medical practitioner and veterinary surgeon who is actively engaged in practice and every coroner" ineligible to serve as jurors. The Ontario Court (General Division) suggested that these professionals were likely ineligible "because the uninterrupted performance of their work was considered to be in the public interest" (*Church of Scientology et al. (No. 1)* 339).

Most provinces disqualify people who are not Canadian citizens from sitting on a jury. The Nova Scotia Law Reform Commission recommended that landed immigrants who are not Canadian citizens be allowed to serve on juries (Conrod 1993, 2); however this recommendation was never implemented. The restriction of jury eligibility to Canadian citizens has been unsuccessfully challenged under the *Charter* (*Laws*).

Some provincial legislation gives the Sheriff's office the discretion to exempt a potential juror on the basis that "serving as a juror may cause serious hardship or loss to the person or to others" (section 6 of the *Jury Act* R.S.B.C.). On October 3, 1994, the Sheriff's office in Vancouver stated to the authors that "serious hardship" provides the Sheriff with discretion to excuse, for example, "single moms [and, presumably, single dads], a woman with young children or one who is breastfeeding, persons running their own business if the business cannot carry on without them, and full-time university students who cannot take the time off."

The array or panel is drawn from the list of eligible jurors, by a procedure that varies from area to area. Computers are often used to generate random samples from the list of eligible jurors in a particular location. The initial array may consist of 100 to 300 or more persons, depending on the location of the trial, the number of jury trials for which juries are being selected, the concern over the number of jurors who might be disqualified, and so forth. In the unusually high profile case of Paul Bernardo, 1500 potential jurors were subpoenaed.

Historically, women, various minority groups, and Aboriginals were excluded from voting, and through this exclusion, from serving on juries. Even after the franchise was granted to everyone, other means of establishing the initial population from which jurors were chosen resulted in a disproportionate number of white, wealthy, male jurors. For example, the Donald Marshall Inquiry in Nova Scotia in 1989 concluded that no natives had ever sat on a jury in Nova Scotia (Canadian Press 1994, A8). A study by the Nova Scotia Law Reform Commission in 1992–93 found that up until 1985, jurors were selected from property assessment rolls. This resulted in 85 percent of jurors being men (Conrod 1993, 2). In 1985, Nova Scotia switched to the electoral list for the initial out-of-court selection of the jury pool.

Use of the voters' list is not a perfect solution. Some people (especially those who are unemployed, underemployed, or renting) move more often than others, and are less likely to be found at their last known address. The Nova Scotia Law Reform Commission recommended broadening the jury pool to include names selected from motor vehicle registration, telephone directories, and health insurance records. Nova Scotia now uses the health registration list. For a more in-depth commentary on this out-of-court process, see Israel (2003).

Juries are supposed to be representative of the community. The Supreme Court of Canada concluded in *Sherratt* that "Provincial legislation guarantees representativeness, at least in the initial array," and that "Canadian laws by and large have long met the standard" of selecting a "representative cross-section of the community" (526). These conclusions are doubtful, however, given some of the studies on jury representation and some of the cases. For example, in 1984 it came to light that sheriffs in Vancouver routinely and purposely excluded Aboriginals from the initial panel or array (*Butler*; see Israel 2003 for other examples).

Challenging the Array

It is possible for either the accused or the Crown to **challenge the array** under section 629. Such challenges are directed at removing the entire panel of jurors and are limited to grounds of "partiality, fraud or wilful misconduct on the part of the sheriff or other officer" who put the array together. Form 40 in the *Code* provides a format for such a challenge.

Historically, there have been few challenges at this stage, and they were rarely successful. There is a presumption in law that when people act under statutory duty (here, sheriffs), they act properly, unless there is evidence to the contrary. More recently, greater concern has been expressed about how the array is chosen, and about the groups that are excluded from the out-of-court selection of potential jurors. Challenges have focused on who does not make it onto the initial list of potential jurors.

Some efforts by sheriffs to broaden the diversity of the initial panel have gone unchallenged. However, in *Born with a Tooth*, such efforts were rejected by the Alberta Court of Queen's Bench because they did not meet the requirement of randomness. The sheriff, in an effort to find a more diverse group of potential jurors (in accordance with instructions he received), added 52 names of residents of three near-by Indian Reserves, selected from the customer list of a utility company, to the 200 names drawn randomly from the voters' list. Even the latter was not truly random, because the sheriffs had started the practice of alternating the selection of women and men in order to ensure that every panel was composed of an equal number of women and men. The *Jury Act* in Alberta required the names to be selected randomly, although the attempt to ensure an equal number of women and men was not challenged. The judge concluded that the panel was partial and instructed the sheriff to select a new panel. In doing so, the judge quoted former Dean Lynn Smith, of the University of British Columbia law school, as saying that if a jury is to be representative of the community, the array must be "drawn from the community as a whole at random, promiscuously and indiscriminately, otherwise juries in individual cases are not truly representative of the community" (397). How can the sheriff ensure that the sample is indeed representative of the community?

Selecting the Jurors: In-Court Process

Prior to codification in 1992, the Supreme Court of Canada in *Sherratt* stated that the common law practice that had developed, permitting the judge to pre-screen and excuse potential jurors before the more formal jury selection process commenced, was limited to obvious cases, and should not usurp the function of **triers**. Section 631 now codifies the practice. In addition to personal interest in the matter before the court or connections to the players involved, the section allows the judge to excuse potential jurors for "personal hardship or any other reasonable cause that, in the opinion of the judge, warrants that the juror be excused" (632(c)). Judicial pre-screening of jurors must be done in the presence of the accused (*Barrow*). A journalist for *The Globe and Mail* attended a jury selection in Toronto in 2003 and found that there was a "rush of people" who came forward to try and get excused from jury duty for numerous and "often so patently transparent" reasons (Blatchford 2003).

The judge may also order that potential jurors, who want to be excused for personal hardship or other reasonable cause, to **"stand by"** under section 633 until the end of the jury selection. To stand by

---

Box 5.1 Pre-Vetting Prospective Jurors

The Latimer case from Saskatchewan involved an alleged "mercy killing" by the accused of his severely disabled daughter. Prior to the accused's first trial, the prosecutor and the RCMP prepared a questionnaire that was administered to 30 of 198 prospective jurors, covering such subjects as attitudes toward abortion, euthanasia, religion, and so on. Five of those 30 people actually ended up serving as jurors at trial. Lamer, C.J., for the court, expressed unqualified disapproval:

> The actions of Crown counsel at trial…were nothing short of a flagrant abuse of process and interference with the administration of justice. The question of whether the interference actually influenced the deliberations of the jury is quite beside the point. The interference contravened a fundamental tenet of the criminal justice system, which Lord Heward C.J. put felicitously as "justice should not only be done, but should manifestly and undoubtedly be seen to be done" (citations omitted; *Latimer* 1997 para. 43).

The prosecutor was later acquitted of attempting to obstruct justice (*Kirkham*).

Despite these scathing comments from the Supreme Court of Canada, jury vetting by Crown and police was recently uncovered in three Ontario cities. The Crown was using notations by the police on jury lists (such as "calls a lot for minor complaints;" "dislikes police;" "family issues;" "criminal associates;" "dad is a drinker") to screen potential jurors (Tanovich 2009, A6).

means that the juror may still be called later, because section 641 requires that once all the potential jurors have been called, if a full jury has still not been selected, the "stand asides" have to be called again.

The twelve jurors are selected from the array, following the procedure in section 631. Each juror's name is recorded on a card, and the clerk randomly draws these cards and calls the names and numbers of potential jurors. Since 2002, the court, upon application or its own motion, may order the clerk to call out only the number of the juror if "satisfied that it is in the best interests of the administration of justice to do so, including in order to protect the privacy or safety of the members of the jury" (section 631(3.1)). Once a juror is called, challenges to the juror may be made by the Crown and defence counsel. The *Code* was amended in 1992 to give the Crown and the accused an equal number of peremptory challenges (20, 12, or 4, depending on the offence–discussed below). The defence and Crown take turns being the first to indicate whether they are challenging a particular juror. To illustrate the process, a potential juror (#1) will be called, defence counsel will indicate whether the accused is content with or challenges the juror, and if content, the Crown will then indicate whether it is content or challenges. If both are content, the person is sworn in as a juror for the trial. Next, potential juror #2 will be called, and the Crown will first indicate whether it is content with or challenges that person, and so on. Since 2002, one or two alternate jurors may be selected if the judge "considers it advisable" (section 631(2.1)). If there is not a full jury at the commencement of the trial, an alternate juror is substituted. If there is a full jury, the alternate jurors are excused (section 642.1).

Challenges are either peremptory or for cause. A **peremptory challenge** (under section 634) is a challenge for which the accused (or Crown) does not have to give any reason. It is a challenge as of right, and there is no debate over it. The number of peremptory challenges depends on the type of offence. Section 634(1) stipulates that the prosecutor and the accused are each entitled to 20 challenges in respect of high treason or first-degree murder, to 12 challenges in respect of any other offence for which the accused may be sentenced to more than five years imprisonment, and to 4 challenges in respect of all other offences. Where several offences are included on a single indictment, each party is entitled to the number of peremptory challenges available in respect of the most serious count only (s. 634(3)). Where several accused are jointly charged on a single indictment, the Crown is entitled to the same number of peremptory challenges as the total of all of the challenges available to the accused (s. 634(4)).

A **challenge for cause** under section 638 requires the Crown or the accused to show cause why the person should not be a juror, on the ground that:

    (a) the name of a juror does not appear on the panel...;
    (b) a juror is not indifferent between the Queen and the accused;
    (c) a juror has been convicted of an offence for which he was sentenced to death or to a term of imprisonment exceeding twelve months;
    (d) a juror is an alien;
    (e) a juror, even with the aid of technical, personal, interpretative or other support services provided to the juror under section 627, is physically unable to perform properly the duties of a juror; or
    (f) a juror does not speak the official language of Canada that is the language of the accused or the official language of Canada in which the accused can best give testimony....

Note that "no challenge for cause shall be allowed on a ground not mentioned in subsection (1)" (section 638(2)). Both sides are permitted an unlimited number of challenges for cause. Challenges for cause are questions of fact, and accordingly are determined not by the judge but by the jury itself (except if the challenge is under (a) set out above). Section 640(2) specifies the procedure whereby two jurors or potential jurors ("triers") are selected to rule on the challenge. In 2008, the *Code* was

---

Box 5.2 Challenging Potential Jurors

In Canada, we do not require potential jurors to fill in lengthy questionnaires, with questions like "Are you divorced?" "Did you initiate the divorce?" or "Have you ever felt sufficiently frustrated within a domestic situation that you considered violence?" Potential jurors for the trial of O.J. Simpson, who was accused of killing his wife, were required to fill in 75-page questionnaires containing such questions (*Vancouver Sun* 1994, A9).

---

amended so that static triers (the same triers act throughout the challenge for cause process) replaced the rotating method (the two most recently sworn jurors tried the challenge for cause).

A challenge for cause (for example, that "a juror is not indifferent between the Queen and the accused") requires that the challenger show the existence of the alleged ground (such as a pre-existing attitude on the part of the juror). The courts have made it very clear that this type of challenge is not to be a fishing expedition. The challenge for cause must have some substance to it, and the foundation for the challenge has to be found in material filed with the court, or presented through witnesses (*Hubbert, Makow,* and *Sherratt*). We do not follow the American method of interrogating prospective jurors. In commenting on this aspect of the Canadian way, Mr. Justice Seaton of the British Columbia Court of Appeal wrote: "in our jurisdiction an accused is entitled to an indifferent jury not a favourable one" (*Makow*, 519).

Canadian law presumes that "jurors are capable of setting aside their views and prejudices and acting impartially between the prosecution and the accused upon proper instruction by the trial judge on their duties" (*Find* para. 26). According to the Supreme Court of Canada, "This presumption is displaced only where potential bias is either clear and obvious (addressed by judicial pre-screening), or where the accused or prosecution shows reason to suspect that members of the jury array may possess biases that cannot be set aside (addressed by the challenge for cause process)" (*Find* para. 26). Reasons why the court might allow specific questions to be asked of each potential juror include pre-trial publicity and notoriety, and more recently possible prejudice or discrimination. The Supreme Court of Canada has so far rejected expanding this challenge to include type of offence.

The Effect of Pre-trial Publicity and Notoriety on the Selection of Jurors

Defence counsel are sometimes given leeway to challenge potential jurors on the ground of partiality in cases where there has been a great deal of pre-trial publicity, or where the case has become notorious in the community. Historically, these cases were quite rare. Examples include the "Squamish Five" case in British Columbia (activating explosives and arson for political purposes), the *Zundel* case in Ontario (spreading false news through anti-Semitic publications), and the *Keegstra* case in Alberta (promoting hatred of Jews during social studies class).

More recently, defence counsel have been more successful in challenging jurors on the basis of publicity; however, they have not been successful when asking the courts to stay charges under section 24(1) of the *Charter* because the publicity has made it impossible for the accused to have a fair hearing. In *Kenny*, the accused was charged in 1989 with sexual and physical abuse of boys at the Mount Cashel Boys' Home and Training School in Newfoundland. Two weeks before Kenny was charged, the Newfoundland government appointed Mr. Justice Samuel Hughes to conduct an inquiry into the alleged cover-up of the complaints by the police in 1975 (42). Most of the evidence at the Hughes Inquiry was carried live on cable TV and replayed in the evening. Portions of the evidence were also reported in the print media. More than 525 news stories, and Michael Harris's book, *Unholy Orders* (see Appendix C), related many of the allegations against Kenny.

Kenny argued that the effect of the prejudicial pre-trial publicity, along with the inquiry, the delay, and statements by Crown officials, amounted to an abuse of process, and that it would be impossible for Kenny to have a fair trial. The defence asked for a stay of proceedings. In support of their application, the defence submitted copies of the pre-trial publicity, as well as the expert testimony of Dr. James Ogloff and Dr. Neil Vidmar (see Ogloff and Vidmar, 1994, for some of the research they did in support of their testimony). In rejecting the defence application, the Newfoundland trial court stated that the "risk of a biased jury [could] be neutralized by proper jury selection procedures and by proper judicial instructions" (60). Kenny was later convicted in a trial by judge alone.

Following charges of manslaughter and criminal negligence causing death in the Westray Mine explosions in Nova Scotia, two accused applied for a declaration that the Commission of Inquiry, established to investigate the disaster, would (amongst other things) violate their right to a fair jury trial. Although the majority in the Supreme Court of Canada judgment refused to address this issue, because the accused had in the meantime elected trial by judge along, a minority thought some guidance would be useful. Mr. Justice Cory, with Iacobucci and Major JJ. concurring, wrote: "Negative publicity does not, in itself, preclude a fair trial. The nexus between publicity and its lasting effects may not be susceptible of scientific proof, but the focus must be upon that link and not upon the mere existence of publicity" (*Phillips* para. 129). Furthermore, the effects of publicity have to be examined "in the context of the highly developed system of safeguards which have evolved in order to prevent just such a problem" (para. 130). See the comment by Gorman (2000), supporting a number of decisions by the Supreme Court of Canada which rejected judicial stays on the basis of pre-trial publicity.

The Impact of Systemic Racism on the Selection of Jurors

In *Parks*, the Ontario Court of Appeal ruled that it was permissible for defence to ask potential jurors the following question after a preamble:

> As the judge will tell you, in deciding whether or not the prosecution has proven the charge against an accused a juror must judge the evidence of the witnesses without bias, prejudice or partiality:
>
>> Would your ability to judge the evidence in the case without bias, prejudice or partiality be affected by the fact that the person charged is black and the deceased is a white man?

In reaching this conclusion, the Ontario Court of Appeal used the threshold test developed by the Supreme Court of Canada in *Sherratt*: "is there a realistic potential for the existence of partiality?" With regard to the existence of racism (that can exist as part of a personal credo, subconsciously, or systemically within our institutions), the Court examined a number of reports in Canada that established the existence of racism, and considered government initiatives to combat racism (paras. 44-53). They specifically considered a report by Stephen Lewis that pointed out the anti-Black racism that was particularly strong in Toronto. The Court noted that, given the government initiatives in Ontario surrounding racism, it was ironic that the Crown was objecting to the questions (para. 51). After reaching the conclusion that racism did exist, the Court considered whether it could rely on the presumption that sworn jurors would do their duty and whether the other safeguards in the system, such as jurors' oath, the solemnity of the task, the "diffused impartiality" created by 12 jurors, the dynamics of focusing on the evidence, and the warnings provided to the jurors by the trial judge would support the presumption (paras. 56-7). In this particular case, the trial judge had given the jury a strong warning against allowing prejudices to interfere with the deliberation process (para. 57). Despite these safeguards, the Court of Appeal thought that there was a realistic possibility of racial partiality, based on the extent of racial bias and on the results of some juror simulation tests (although the evidence was not conclusive).

The Court of Appeal found that the process of asking the question would not take long, and would have three benefits: (1) some jurors would screen themselves out; (2) the jurors would be sensitized to the need to ensure that racial bias not enter the deliberations; and (3) the question "enhances the appearance of fairness in the mind of the accused" (para. 92). An application by the Crown for leave to appeal to the Supreme Court of Canada was refused.

In British Columbia, the Court of Appeal's decisions had tended to disagree with the Ontario courts. In *Williams*, Chief Justice Esson dismissed an application by the defence to challenge jurors for cause. The defence wanted to ask potential jurors whether their ability to be impartial would be affected by the fact that the person charged was an Indian and the complainant was white. The case involved an armed robbery at a pizza parlour in Victoria. There were no obvious racial overtones to the robbery. The Crown did not dispute the existence of widespread prejudice against Aboriginals, but took the position that widespread prejudice was not enough to displace the presumption that jurors could be relied on to do their duty in an impartial manner, relying on other safeguards within the criminal justice system, as listed above by the Ontario Court of Appeal in *Parks*. The British Columbia Court of Appeal dismissed the accused's appeal, distinguishing *Parks* on the basis of the "peculiar conditions existing in metropolitan Toronto, in particular the attitudes held by many regarding the link between black persons and serious crime" (207).

The Supreme Court of Canada overturned the British Columbia Court of Appeal, and explained the two stages under section 638. At the first stage, where the judge determines whether a challenge for cause should be allowed, the test is "whether there is a realistic potential or *possibility* for partiality." That is, is there "reason to suppose that the jury pool *may* contain people who are prejudiced and whose prejudice might not be capable of being set aside on directions from the judge" (para. 32; emphasis in original). At the second stage, the actual challenge for cause, the question "is whether the candidate in question *will* be able to act impartially" (para. 33; emphasis in original). The Court held that although there is no automatic right to challenge for cause based on the accused's membership in a disadvantaged or minority group, "absent evidence to the contrary, where widespread prejudice against people of the accused's race is demonstrated at a national or provincial level, it will often be reasonable to infer that such prejudice is replicated at the community level" (para. 41). Thus, if widespread racial prejudice of a given type is established in one case, judges in subsequent cases may take judicial notice of the fact. The Court referred to numerous reports that established the existence of racism against Aboriginals in Canada and British Columbia.

Following *Parks*, the Ontario Court of Appeal decided that the existence of anti-black attitudes existed outside of Metropolitan Toronto (*Wilson*), and that there was a realistic potential or possibility for partiality against persons of Asian origin, but rejected a challenge question that included geographical origins (*Koh*). In *Campbell*, the Ontario Court of Appeal found that the trial judge was in error when it allowed a *Parks* type question but refused to allow the victim to be identified as white. According to the Court, the interracial nature of sexual assault by a black man on a white woman increased the possibility of partiality (262).

In *Spence*, a black accused was charged with robbing two pizza delivery men, one white and the other East Indian. The trial judge allowed the following question: "Would your ability to judge the evidence in this case without bias, prejudice or partiality be affected by the fact that the accused person is a black man?" But, he refused to allow the question to include the race of the complainants. The trials were then severed, and the accused appealed his conviction of robbing the East Indian complainant on the basis that the trial judge should have allowed the question to include the complainant's race. The majority of the Ontario Court of Appeal overturned the conviction and found that the trial judge should have allowed the race of the complainant to be included in the question. The Supreme Court of Canada allowed the further appeal, and restored the conviction. The Court rejected defence counsel's argument that East Indian members of the jury may be more sympathetic or empathetic to a victim of the same race. No evidence was presented for such a proposition, and possible racial sympathy was quite different from the potential racial hostility that might exist against an Aboriginal or black accused charged with a crime against a white complainant (para. 7).

Broadening the Challenge for Cause

A number of lower courts allowed challenges for cause based on the attitudes of prospective jurors toward particular types of offences (Rose and Ogloff 2002, 218). However, the Supreme Court of Canada in *Find* decided that no basis had been presented to support such a challenge in that case. Find was charged with 21 counts of sexual assault involving three complainants, aged 6 to 12 at the time of the alleged offences. The defence failed to "establish widespread bias arising from charges relating to sexual abuse of children" (para. 88). The most the evidence established was that "the crime of sexual assault, like many serious crimes, frequently elicits strong attitudes and emotions" (para. 88). "Out of an abundance of caution," the Court addressed the second question: "Is it reasonable to infer that some jurors may be incapable of setting aside their biases despite trial safeguards?" (para. 89). In rejecting this proposition the Court wrote, "Absent proof, we cannot simply assume that strong beliefs and emotions translate into a realistic potential for partiality, grounding a right to challenge for cause" (para. 109).

According to Coughlan (2001, 34) this was the only decision that the court could have arrived at while keeping the presumption of innocence intact in our criminal justice system. Plaxton (2001, 303) disagreed, arguing that some jurors will be incapable of setting aside their biases, and that these potential jurors should be screened out. For an argument against offence-based challenges for cause, see Dufraimont (2000).

Talesmen—Running out of Potential Jurors

A curious phenomenon, that is uncommon but not rare, occurs when the panel is exhausted (the court has "run out" of jurors) before selecting the 12 needed to form a jury. Under section 642, the court may order the sheriff "forthwith to summon as many persons, whether qualified jurors or not, as the court directs for the purpose of providing a full jury and alternate jurors." These people, called **talesmen,** are then treated as if they were part of the panel. Typically, the sheriffs grab people off the street in the vicinity of the courthouse, and haul their disconcerted catch into the courtroom, to be challenged or accepted for service on the jury being empanelled.

Discharging a Juror

Under section 644, a judge may **discharge** a juror "in the course of a trial," for "reason of illness or other reasonable cause." The trial can continue, so long as the number of jurors does not fall below 10 (section 644(2)). If the trial has not begun when the juror is discharged, another juror must be selected to fill the former juror's place (section 644(1.1). The trial begins when the accused is **placed in the charge of the jury,** marked by the clerk addressing the accused in a prescribed manner after a full jury is sworn or affirmed.

Defects in Selecting a Jury

A number of sections of the *Criminal Code* deal with defects in the selection of jurors. Section 643(3) provides that irregularities in selecting a jury under sections 631, 635, and 641 do not invalidate the proceedings. Section 670 states that verdicts shall not be overturned "by reason of any irregularity in the summoning or empanelling of the jury," or "for reason that a person who served on a jury was not returned as a juror by a sheriff or other officer." Section 671 is similar to 670, in that it specifies that a verdict shall not be quashed because of specific omissions and irregularities in the jury selection process.

Jury Unanimity

Jurors must be unanimous in their verdict to render a legally effective verdict. This requirement does not mean that, to convict, the jurors must reach a unanimous decision as to *how* the offence was committed. For example, in *Thatcher* the Supreme Court of Canada decided that the jury could be left with two options (he killed his wife or he had someone else kill her). The jury could be unanimous on the outcome (i.e., guilty), without agreeing on the method. Similarly, the Quebec Court of Appeal in *Pearson* decided that the jury need not be unanimous on how the accused trafficked in drugs (whether the accused sold, delivered, transported, or offered to do so), so long as they were unanimous that the accused committed the offence of trafficking (565).

If jurors cannot reach a unanimous verdict (resulting in a "hung" jury), a **mistrial** is declared and the Crown has the option of conducting another trial. A new trial was ordered in a case where the trial judge told the jurors they had to be unanimous one way or the other, leaving them with the impression they could not disagree (*Naglik*). There are arguments for and against the requirement of unanimity, rather than a simple majority, a 75 percent majority, or some other formula. The Law Reform Commission of Canada (1980) examined the requirement of unanimity and concluded that it should continue.

Directed Verdicts

The common law rule, requiring judges to instruct a jury to enter a not guilty verdict in cases where the judge was of the view that there was no evidence on which to convict an accused, ended in June of 1994 with the Supreme Court of Canada's decisions in *Rowbotham* and *Roblin*. The Toronto trial judge instructed the jury, after a four-week trial, to enter a verdict of not guilty against two men charged with conspiring in Canada to sell drugs in Texas. The judge decided that the men could not be convicted in Canada for the offence (because the Crown had failed to show that selling marijuana was illegal in Texas). The jurors objected to the judge's instruction, and one juror told the judge, "Sorry, some of us still believe a guilty verdict should go through." They finally acceded to the judge's instructions and entered a not guilty verdict. On appeal to the Supreme Court of Canada, the court held that the common law rule governing **directed verdicts** made a mockery of the jury system, and if there is no evidence for a conviction, the judge should enter the not guilty verdict, rather than instructing the jury to do so. The Court decided that the trial judge was wrong in deciding there was no evidence to go to the jury, and ordered a new trial (*Rowbotham and Roblin*, discussed in Bindman 1994 and Schmitz 1994).

Secrecy of Jury Proceedings

The rationales for jury secrecy are to foster free and frank debate among the jurors, to preserve the finality of the verdict, and to protect jurors from post-verdict harassment (*Pan* para. 50-53). Until the amendments to the *Criminal Code* in 1972, the secrecy of the jury room was enforced by common law, and jurors who talked about what went on in the jury room were subject to contempt charges (Law Reform Commission of Canada 1980, 143). The law in Canada differs from what most Canadians expect, based on television programs from the United States showing jurors being questioned about their decisions. Section 649 of the *Criminal Code* makes it an offence to "disclose any information relating to the proceedings of the jury when it was absent from the courtroom that was not subsequently disclosed in open court." Members of the jury cannot discuss what went on in the jury room during their deliberations. The only exception to this secrecy is that disclosure may be made for the purposes of investigating or giving evidence of an offence by a juror of wilfully attempting to "obstruct, pervert or defeat the course of justice" under section 139(2)).

Still, there are those in Canada who have interviewed jurors. Pron and Donovan (1992; see Appendix C) managed to talk to two jurors, as discussed in their account of the trials against Rui-Wen Pan. One, who refused to go along with the acquittal at Pan's second trial, described a terrifying time during jury deliberations that included being assaulted and "manhandled" by one of the other jurors. She said, "one of the men told me I would be locked in that jury room until I reached the decision they wanted..." (338). At Pan's trial, the juror sent the judge a note making reference to what she had "been put through in the jury room," and asking that the jurors be polled for their verdicts. After reviewing the note, both counsel opposed conducting an inquiry of the juror, and the judge declared a mistrial. Prior to his third trial, Pan brought an application to have the proceedings stayed, and asked the judge to admit evidence from the jurors at the second trial as to what transpired during their deliberations. The judge denied his application, and Pan's appeal to the Ontario Court of Appeal was dismissed. On further appeal, the Supreme Court of Canada dismissed Pan's appeal, upholding the constitutionality of the common law and section 649 of the *Code*. The Court summarized the common law: "statements made, opinions expressed, arguments advanced and votes cast by members of a jury in the course of their deliberations are inadmissible in any legal proceedings. In particular, jurors may not testify about the effect of anything on their or other jurors' minds, emotions or ultimate decision" (para. 77). However, this does not prevent the admission of "evidence of facts, statements or events extrinsic to the deliberation process, whether originating from a juror or from a third party, that may have tainted the verdict" (para. 77).

Social Science and Juries

In 1980, the Law Reform Commission of Canada recommended that the Chief Justice of a province be permitted to make exceptions under section 649 for the purpose of research, or if the verdict is being impeached upon application to the Minister of Justice. The Commission was of the view that research findings could be important for law reform, and as an evaluation of the present system (1980, 143). In 1990, *The Globe and Mail* (June 15) published an editorial, endorsing the Law Reform Commission's suggestion that jurors be allowed to talk about their deliberations to those authorized to collect such information for scientific research, concluding that the *Criminal Code* should permit officially sanctioned research and analysis of jury deliberations (Lawrie 1990, A4). The Supreme Court of Canada in *Pan* stated that it would welcome more research in this area, as it would "add to

---

Box 5.3 Jurors Testify

Gillian Guess was a juror on a high-profile, eight-month murder trial in Vancouver, that ended in October 1995 with an acquittal. Guess was subsequently charged with obstruction of justice, on allegations that she had had a sexual affair with one of the accused while the trial was going on. At Guess's trial, six jurors from the original murder case testified about Guess's conduct and attitudes during the course of the trial and deliberations. This was said to be the first time that jurors have testified in Canada under the exception in section 649. Guess was ultimately convicted by a jury and sentenced to 18 months in prison, followed by a year of probation. Her appeal to the British Columbia Court of Appeal was dismissed, as was her application for leave to appeal to the Supreme Court of Canada (*Guess*). On October 20, 1999, the British Columbia Court of Appeal ruled that Guess could not be called as a witness in the appeal of an accused whose jury she sat on, because she was exercising a judicial function as a juror in his trial (*Budai*).

---

the legitimacy of the existing rules and, if need be, would trigger judicial or legislative modifications" (para. 107). However, it was left to Parliament to change the law and set the parameters surrounding research and jury secrecy (para. 107).

Defence counsel might use social science research in a number of ways in, or when preparing for, jury trials. Surveys might be used to test the degree and effect of pre-trial publicity on the accused's right to a fair trial. For instance, the lawyer for Colin McGregor, who killed Patricia Allen (his estranged wife) with a cross-bow in Ottawa, used a public opinion poll to convince a judge that his client would not get a fair trial by a jury. Fifty-nine percent of the respondents in the poll said they would find unacceptable a verdict of not guilty by reason of insanity (*Lawyers Weekly*, January 20, 1993).

Pre-trial surveys can be used to convince the judge that massive pre-trial publicity has created sufficient question about possible prejudice that the judge should allow flexibility in questioning jurors prior to a challenge for cause. Survey results may also suggest the type of juror counsel might wish to select for the jury. In the United States, jury selection is a major industry, as hundreds of consulting firms "rake in several hundred million dollars a year" (Hutson 2007 online). In Canada, lawyers view jury selection as a "crap shoot," and typically "listen to their gut" (Spencer 1994, 5; also see Tanovich, Paciocco, and Skurka 1997 Chapter 10). However, surveys and commission reports are used to support a challenge for cause on, for example, racial bias (discussed above). In addition, social scientists may testify about basic psychological processes, etc., or specific research testing assumptions the law makes about human behaviour (Rose and Ogloff, 2002).

Some defence counsel might have experts observe the jury selection process, in an attempt to predict the juror's attitude toward the defence through appearance and demeanour. In cases where the stakes are high and the accused has a lot of money, a shadow jury might be used, to observe the trial in progress and give defence counsel insight into the jury's reaction to the evidence. Jury simulations may be used in advance to test defence lines of argument, styles of presentation, and so on.

Box 5.4 Jury Secrecy and Jury Research

Perhaps the biggest problem caused by the requirement of jury secrecy is the impediment it raises to scientific research. While we may *believe* we know how juries operate, based on anecdotal information or introspection, we do not *know* how they operate. The laboratory research done on simulated juries suggests that many of our long-cherished beliefs about juries may be wildly inaccurate, but the applicability of such analogue research to real-world trials is questionable, and is certainly questioned by the courts (see, for example, *Kenny* at 51ff). Real-world evidence, however questionable it may be methodologically, also raises serious concerns about our misunderstanding of juries. A disturbing example arises from the wrongful conviction of Morin.In an episode of the CBC's program *The Fifth Estate,* entitled "Innocent—beyond a shadow of a doubt," one of the jurors who convicted Morin on his second trial explained that it was an aspect of the testimony of Morin himself that convinced her of his guilt: "the lack of eye contact with us. He never looked at the Jury once while he was on the stand." Given this sort of shocking but rare insight into the workings of juries, can the legal system continue to justify not studying juries directly? Quinlan suggests it is only by lifting the veil of secrecy that errors made by jurors could be uncovered. In his view, precluding the effective review of verdicts promotes injustice (134). Chopra and Ogloff (2000) argue that jury secrecy should be lifted for evaluative research using former jurors, and to allow jurors to discuss their experiences to relieve stress. For a commentary on how to deal with jurors' stress see Anand and Manweiller (2005).

Jurors' Comprehension of Evidence and Instructions

As stated at the beginning of this chapter, Berger (2004) suggests that forcing an accused to have a jury trial could violate the accused's right to a fair trial under the *Charter*. He uses some of the social science research on jurors' comprehension of evidence and their inability to follow instructions to ignore certain evidence to support his arguments.

The Role of the Jurors

Two of the functions of the jury are to protect us from oppressive laws and to ensure fairness in criminal justice proceedings (Law Reform Commission of Canada 1980, 11). The Law Reform Commission of Canada has taken the view that in the rare case where a jury does not apply the law, "the social good caused by such a 'revolt against the law' outweighs whatever dangers may arise" (1980, 13).

Jury Nullification

The Supreme Court of Canada defined jury nullification as referring to "that rare situation where a jury knowingly chooses not to apply the law and acquits a defendant regardless of the strength of the evidence against him" (*Latimer* 2001 para. 57). Should juries be allowed to refuse to apply laws they disagree with? Dr. Henry Morgantaler was repeatedly acquitted by juries of charges of illegally performing abortions in the face of overwhelming evidence of his guilt. This result has prompted the government to change the abortion laws. In another example, some people doubt that the Crown could ever get a jury conviction of any doctor who assisted a terminally ill patient to commit suicide. If the jury can ignore the law in order to acquit, should they be allowed to ignore the law in order to convict? (See the debate the jury had with the judge under "Directed Verdicts" above).

The Supreme Court of Canada in *Krieger*, stated that jurors have the power, but not the right, to disregard the law or its application "when their conscience permits no other course" (para. 27). A trial judge is not allowed to direct a jury to enter a guilty verdict as such an order deprives accused of their constitutional right to a jury trial (para. 2). Is there a duty to inform the jury that they have the power to refuse to apply the law? Judges are not allowed to instruct a jury that they have the power to disregard the law if they find it oppressive or unfairly applied in the circumstances. In fact, judges are obliged to "take steps to ensure that the jury will apply the law properly" (*Latimer* 2001, para. 70). Defence counsel cannot encourage a jury to refuse to apply the law (para. 68). See Dufraimont (2996) and Nowlin (2008) for further discussion of jury nullification.

Box 5.5 Do Jurors Understand Legal Instructions?

Dr. James Ogloff has been studying jurors' comprehension of judges' instructions since the 1980s. His research has found that "people clearly have a very difficult time comprehending jury instructions. Our research shows that people may render their verdicts in ignorance of the law rather than in light of it."

Ogloff also suggests that the difficulty jurors have understanding the judge's instructions may be the result of the very task they are being asked to complete. "After hearing days, even weeks, of testimony, jurors then receive the judge's instructions. The instructions alone can last for two days. The law expects that jurors can then apply the instructions to the evidence they received at trial and render a verdict."

Ogloff notes that, unfortunately, rather than focusing on the legal instructions and using them to assess the evidence, jurors often launch into their deliberations without paying much attention to the instructions. In fact, "we found in one study that participants spent on average just over 70 seconds considering the instruction." Even then, the participants were correct in defining those instructions they did mention only 60 percent of the time

Finally, Ogloff has found that none of the commonly thought-of remedies to increase to increase jurors' comprehension of instructions are successful (e.g., jurors taking notes, giving instructions at the beginning of the trial or at both the beginning and the end, giving jurors written copies of the instructions, using simplified instructions, using a "decision tree" to help with structured decision-making) (interview by authors with Dr. Ogloff, February 12, 2000).

For a comprehensive review of the literature on juror understanding of judicial instructions, see Ogloff and Rose, 2005. For the difficulty that jurors have in understanding DNA evidence, see Holmgren 2005a and 2005b.

The Fully Informed Jury Association (FIJA), located in Helmville, Montana (www.fija.org/), is an American grassroots organization whose purpose is to re-educate U.S. citizens about their inherent right as jurors to reach a verdict based on their own sense of justice. They are lobbying to have the law changed, such that jurors would have to be told that they have this option. The group is of the view that this would re-establish the jury as an authority on the law in addition to their role as finders of fact. The group also educates potential jurors about their rights and role in the trial process.

Should the Jury System Be Abolished?

The Law Reform Commission of Canada (1980), relying on a number of opinion polls of judges, former jurors, and other members of the public, supported the retention of jury trials. The Commission approved of the jury as being:

- a good fact finder
- the conscience of the community (a concern with fairness enters their decision)
- a protector from oppressive laws (this casts the jurors in the political role of lawmaking)
- a tool for education
- a means of legitimizing the criminal justice system.

The jury system also:

- relieves judges of the constant burden of making decisions
- deflects criticisms from judges
- allows for a fresh approach to each case
- prevents fact finders from considering inadmissible evidence
- decentralizes authority

Box 5.6 James Tyhurst

"Fuck you, you bastards, I hope you have a shitty Christmas," a former patient of James Tyhurst screamed at the jury that had just acquitted Tyhurst of sexual assault. The prosecutor was quoted as having described it as, "the most bizarre verdict I have heard. The evidence against [Tyhurst] was overwhelming" (Todd 1993, D10). Cases such as these leave the public, and some members of the legal profession, questioning whether the jury trial should be abolished.

VOIR DIRES

A *voir dire,* often referred to as a "trial within a trial," is held (usually) to determine the admissibility of evidence, such as whether a confession is voluntary and admissible, whether evidence should be excluded under section 24(2) of the *Charter*, and so on. A *voir dire* is usually held in the absence of the jury; that is, the jury leaves the courtroom and waits for the completion of the *voir dire*. To avoid inconveniencing the jurors when there are lengthy *voir dires*, or a number of them on anticipated issues, *voir dires* will often be held at the beginning of the trial, without the jury waiting. This avoids having to send the jury in and out of the courtroom over the course of the trial. A *voir dire* to determine whether a witness is qualified to give expert testimony is normally conducted in the presence of the jury, to enable the jurors to decide how much weight to give to the expert's opinion (discussed further in Chapter 13).

The only issue on a *voir dire* concerning the admissibility of evidence is whether the evidence is admissible; generally, its probative value or weight is irrelevant. The usual procedure is that if the Crown or accused seeks to introduce evidence of questionable admissibility, or one of them objects to the admissibility of certain evidence, a *voir dire* will be requested, and the judge will declare a *voir dire*. If the Crown requested the *voir dire*, it will lead evidence on the issue of the admissibility of the evidence it wants to tender, and the defence cross-examines the Crown witnesses. The defence may call its own witnesses on the issue. If an accused testifies at a *voir dire* within his or her trial, this evidence cannot be used against the accused during the trial (see Chapter 10). Both Crown and defence counsel will present arguments, and the judge will decide whether the evidence is admissible. If admissible, and if it is a jury trial, the evidence is then repeated in front of the jury. In a non-jury trial, the defence and Crown will often agree to have the evidence heard on the *voir dire* become part of the evidence on the trial proper, so there will be no need to repeat it.

EXCLUSION OF THE PUBLIC AND PUBLICATION BANS

Section 486(1) of the *Code* states that criminal proceedings are to be in open court, except that the judge may exclude all or any members of the public "for all or part of the proceedings if the judge . . . is of the opinion that such an order is in the interest of public morals, the maintenance of order or the proper administration of justice, or is necessary to prevent injury to international relations or national defence or national security." Although an earlier version of this section was found to violate section 2(b) of the *Charter*, it was demonstrably justified under section 1 of the *Charter (Canadian Broadcasting Corp)*. The proper administration of justice includes "ensuring that the interests of witnesses under the age of eighteen years are safeguarded in all proceedings" and that "justice system participants who are involved in the proceedings are protected" (section 486(2)). Judges who refuse to make such orders when requested by the Crown or defence for specified sexual assault offences must provide reasons (section 486(3)).

Several provisions in the *Code* (relating both to pre-trial and trial proceedings) govern restrictions on media and public access to documents, information, and testimony in the criminal justice system. For example, section 486.4(1) allows a judge to order a ban on the publication of the identity of a complainant or witness, or of information that might identify the complainant or witness in trials of a variety of sex- and exploitation-related offences. Where such a ban is available, section 486.4(2) requires a trial judge to advise any complainant or witness (under the age of 18) of their right to apply for a ban, and to grant such a ban if it is requested. Although this section violates section 2(b) of the *Charter*, the Supreme Court of Canada has held that it was a reasonable limit under section 1 (*Canadian Newspapers*).

If section 486.4 is not applicable, section 486.5 can be used to apply for a publication ban on "any information that could identify the victim or witness," if the judge, after considering the factors in section 486.5(7), determines that publication ban is "necessary for the proper administration of justice." The section sets out the procedure to be followed when applying for a publication ban.

Section 486.6 creates a summary conviction offence for the violation of a publication ban under sections 486.4 or 486.5.

The Supreme Court of Canada in *Dagenais* and *Mentuck* modified the common law rule, allowing judges in their discretion to order bans on the publication of information related to a criminal trial. In *Dagenais*, the Canadian Broadcasting Corporation advertised their mini-series, "The Boys of St. Vincent," a fictional story about the physical and sexual abuse of boys in a Catholic institution. The program was to run during the trial of some of the accused, and before others went to trial, on charges of physical and sexual abuse of boys in a Catholic training school in Ontario. The accused applied for an injunction to prohibit the CBC from broadcasting the program, and an injunction was granted for the duration of the trials. The Ontario Court of Appeal narrowed the order to prohibit the broadcast in Ontario, and by one Montreal station.

Although at common law the right to a fair trial took priority over freedom of expression, according to Chief Justice Lamer (for the majority in *Dagenais*) it is inappropriate for the courts to favour one *Charter* right (section 11(d)) over another (section 2(b)). When two *Charter* rights come into conflict, the courts have to balance both sets of rights. Lamer, C.J.C. was of the view that there were many alternatives to a publication ban—"adjourning trials, changing venues, sequestering jurors, allowing challenges for cause and voir dires during jury selection, and providing strong judicial direction to the jury" (881), and that therefore the publication ban did not meet the common law requirements. The Chief Justice expressed doubts about the efficacy of bans. He opined that jurors are capable of following instructions and that modern technology (satellite dishes, computer networks, and so forth) has reduced the effectiveness of publication bans. If the benefits of the ban are limited, its negative impact on freedom of expression might outweigh the ban's usefulness (886).

In *Mentuck*, the Crown wanted publication bans to protect the identities of undercover officers, and their "Mr. Big" operational methods (discussed in Chapter 10). The trial judge imposed a one-year ban on the officers' identities, but refused to impose a ban on their operational methods. The Supreme Court of Canada upheld the trial judge's decision. The test for publication bans (whether at common law, or under section 486.4 or 486.5, now referred to as the Dagenais/Mentuk test), is summarized by the Supreme Court of Canada in *Re Vancouver Sun*. Publication bans should be limited to situations where:

> (a) such an order is necessary in order to prevent a serious risk to the proper administration of justice because reasonably alternative measures will not prevent the risk; and
> (b) the salutary effects of the publication ban outweigh the deleterious effects on the rights and interests of the parties and the public, including the effects on the right to free expression, the right of the accused to a fair and public trial, and the efficacy of the administration of justice (*Re Vancouver Sun* para. 29 quoting *Mentuck*, para. 32).

Iacobucci and Arbour, JJ for the majority in *Vancouver Sun*, elaborated:

> The first part of the Dagenais/Mentuck test reflects the minimal impairment requirement of the Oakes test, and the second part of the Dagenais/Mentuck test reflects the proportionality requirement. The judge is required to consider not only "whether reasonable alternatives are available, but also to restrict the order as far as possible without sacrificing the prevention of the risk"(*Re Vancouver Sun* para. 30 quoting *Mentuck*, para. 36).

The Supreme Court of Canada in *Vancouver Sun*, and its companion case *Application under s. 83.28 of the Criminal Code,* used the Dagenais/Mentuck test to determine if investigative hearings under section 83.28 (part of the *Anti-terrorism Act* of 2001) should be held *in camera*. The presumption is that such hearings will be in public, and the onus is on the party who wants to exclude the public.

The *Code* also contains a number of automatic publication bans; see sections 276.3, 278.9, 517, 542(2), 648(1), and 672.5(11).

PROCEDURE AT TRIAL

The trial begins with the Crown's opening address. After its opening address, the Crown calls witnesses to testify. Each Crown witness presents his or her direct evidence, or evidence-in-chief, and is then cross-examined by the accused or by defence counsel. If any new matter arises in cross-examination, or if there is some uncertainty in what was said, the Crown may re-examine its witness.   After the Crown's case is in (all of the Crown's witnesses have given their evidence-in-chief and been cross-examined), and before calling evidence, the defence can make a **no evidence motion** to have the charges dismissed on the basis that there is no case to go to the jury (i.e., that there is no evidence on some essential element of the offence). If the no evidence motion is dismissed, the defence still has the option of calling evidence.

If the defence plans to call evidence, defence counsel may give an opening address and call witnesses (section 651(2)). Some courts have allowed defence counsel to give their opening address immediately after the Crown's opening address (*Morgan*), although other courts have limited this deviation from section 651(2) to special circumstances (*D. (A.)*). Some argue that this sequence gives the Crown an unfair advantage, because of the persuasiveness of first impressions.

If defence witnesses are called, they give evidence-in-chief, and may then be cross-examined by the Crown, and re-examined by defence about anything that arises from the cross-examination. After the evidence is all in, Crown and defence counsel address the jury. Section 651 provides that if the defence calls no witnesses, the Crown will address the jury first, and if the defence calls witnesses, the Crown will address the jury last. The Supreme Court has ruled that requiring the accused to address the jury first if he or she calls evidence does not violate sections 7 and 11(d) of the *Charter* (*Rose*). Again, the issue revolves around social science evidence relating to the "primacy" and "recency" effects, and the belief that it is advantageous to address the jury last. For a discussion of this issue as raised by the *Rose* case, see Manson (1999), and Sankoff and Hendel (1999).

There are limits to what lawyers can say to the jury in their addresses. They may not refer to any facts other than evidence that was presented to the jury, and they should not provide their own opinion as to the guilt or innocence of an accused or the credibility of a witness. Crown counsel must refrain from comments that may inflame the jury against the accused (for further discussion see Granger 1996, 226–42). Inappropriate comments may result in the judge declaring a mistrial. Following the closing addresses, the judge charges the jury, summarizing the case and instructing the jury on the law they are to apply (for examples of model jury instructions, see the Canadian Judicial Council, online, and Ferguson and Bouck 2004). The jury then retires to deliberate on its verdict.

REASONS FOR JUDGMENT

Juries do not give reasons for their decisions; however, when a trial is by judge alone, the judge is required to give reasons in order to inform the accused why a conviction has been entered. The requirement also exists when an accused is acquitted, so that the Crown has adequate grounds to consider an appeal (*Walker* para. 2). The use of "boiler plates" reasons, such as those used in *Sheppard,* are inadequate. In that case, the trial judge, without addressing the issues raised in the evidence, stated: "Having considered all the testimony in this case and reminding myself of the burden on the Crown and the credibility of witnesses, and how this is to be assessed, I find the defendant guilty as charged" (para. 2).

The functional test for the adequacy of judicial reasons includes whether an appellate court could meaningfully review the correctness of the decision (*Braich*). Boyle and MacCrimmon argue that the Supreme Court of Canada did not go far enough in *Braich*: "judges can make their decisions appeal-proof without providing much insight into their reasoning process and, in particular, without revealing their approach to issues of social context" (2002, 57). Also see Plaxton (2002) and Quigley (2002). The Supreme Court of Canada recently addressed the adequacy of trial judges reasons in cases where credibility was in issue (*R.E.M.*; *Dinardo* and *H.S.B.*).

SUMMARY

The *Charter* guarantees every accused the right to a jury trial in the case of all offences for which the maximum penalty is five years or more. The accused can typically elect to have a jury trial for most other indictable offences. The right to, or election of, trial by jury can be lost if an accused fails to appear for trial as scheduled. So far the courts have not recognized a right to *not* have a jury trial.

Who is qualified to sit as a juror is generally a matter for the provinces to decide, although the *Code* prohibits discrimination in this regard on the basis of sex. Generally, non-citizens, medical personnel, and people who work in the legal or justice system may be excluded or exempted from sitting on juries by various provincial legislation.

The first step in selecting a jury is for the Sheriff's office to draw an array of eligible jurors. The composition of the array may be challenged on the basis of partiality, fraud, or misconduct by the sheriff who constructed it. Once an array is called, the judge may excuse or direct to stand by any potential juror who claims an interest in the matter to be tried or professes personal hardship.

Prospective jurors are then called, one at a time. Counsel for the Crown and defence take turns indicating whether they are challenging the potential juror or are content. If both are content, the person is sworn as a member of the jury. Counsel may exercise either of two types of challenges. Peremptory challenges require no reason to be given, but are limited in number (depending upon the nature of the charge). Challenges for cause are of unlimited number, but require the party challenging the juror to show cause, on specified grounds, such as partiality. Should the array be exhausted before a full jury is selected, the Sheriffs may be authorized to go out and immediately subpoena talesmen off the street.

Jurors are not generally questioned by counsel or the court. In limited circumstances, based on factors such as pre-trial publicity or a demonstrated prospect of systemic racism, limited questioning may be permitted to support the exercise of challenges for cause.

The trial, which technically starts when the accused is called to enter a plea prior to the selection of the jury, formally commences after the jury is selected. At this time, the indictment is read to the accused, and he or she is placed in the charge of the jury.

During the course of a trial, the judge may discharge a juror for illness or similar reason. The trial will continue, and the jury will be validly constituted, so long as at least 10 jurors remain. Where the judge is of the view at the close of the case that there is no evidence to support a conviction, the judge should enter an acquittal, rather than directing the jury to do so.

Following the presentation of evidence, addresses, and the judge's charge, the jury retires to consider its verdict. To be effective, the verdict must be unanimous; otherwise, a mistrial will be declared, and the accused may be retried by a new jury. Jury deliberations are secret, and it is unlawful for their deliberations to be divulged to anyone. Jurors may not be instructed that they may choose to disregard the law, although juries plainly do so from time to time, and the Supreme Court of Canada has recognized their power to do so.

A *voir dire* is a hearing during the course of the trial, in the absence of the jury, to determine an issue relating to the admissibility of evidence.

The general rule is that proceedings at trial are public and may be published. There are provisions for excluding the public, and granting bans on the publication of trial proceedings, but such restrictions should be imposed only rarely, to ensure a fair trial.

At trial, after selection of the jury, the Crown will address the jury by way of opening its case. Next, the Crown will call its witnesses, who will give evidence-in-chief, and be subject to cross-examination by the defence, with a limited possibility of re-examination by the Crown. Following the Crown's evidence, the defence may make a no evidence motion. If the court concludes that there is evidence capable of supporting a jury verdict of guilty, the defence may elect to give an opening statement to the jury, and to call evidence in the same manner as the Crown. Following the completion of evidence, counsel will address their arguments to the jury; if the defence called evidence, the Crown will address last, otherwise the defence will address last. The judge will then summarize the evidence and instruct the jury on the law they must apply to whatever facts they find, and the jury retires to consider its verdict.

QUESTIONS TO CONSIDER

(1) On what grounds can an array be challenged?

(2) How can a 100 randomly drawn names for a jury trial not be representative of the community? What are the benefits of having a jury that is representative of the community? Are there any drawbacks to a jury that is representative of the community?

(3) What are talesmen?

(4) How many peremptory challenges will an accused have if charged with being unlawfully in a dwelling house under section 349? How many challenges for cause will an accused have if charged with being unlawfully in a dwelling house under section 349?

(5) When a panel member is challenged for cause, who decides whether the potential jurors are impartial or not?

(6) On what bases can the Crown or the defence challenge a juror for cause? Why does each side have an unlimited number of challenges for cause?

(7) Under what conditions will defence counsel be allowed to challenge prospective jurors for racial bias?

(8) What safeguards are built into the jury system against having racism affect a jury's decision?

(9) Should jury trials be abolished? Why or why not?

(10) Should juries be allowed to find an accused not guilty on the basis that the offence charged is one contrary to a law that does not have the support of the community or the application of the law is unfair in the circumstances? Should a judge be required to inform jurors of this option?

(11) What is the impact on the jury nullification issue of section 11(d) of the *Charter*, guaranteeing the right to be "presumed innocent until proven guilty *according to law*" (emphasis added)?

(12) What test does the court apply when determining whether to impose a publication ban?

(13) What is a voir dire?

(14) What is a no evidence motion and when does it take place?

BIBLIOGRAPHY

Anand, Sanjeev and Heather Manweiller. "Stress and the Canadian Criminal Jury Trial: A Critical Review of the Literature and the Options for Dealing with Stress." (2005) 50 *Criminal Law Quarterly* 403.

Berger, Benjamin L. "*Peine Forte et Dure*: Compelled Jury Trials and Legal Rights in Canada." (2004) 48 *Criminal Law Quarterly* 205.

Bindman, Stephen. "Judges Ordered to Enter Acquittals Rather Than Direct Juries to Do So." (25 June 1994) *Vancouver Sun* A11.

Blatchford, Christie. "Watching a Jury Being Chosen is a Lesson in Human Nature and the First Step to a New Trial" *The Globe and Mail*, September 12, 2003 (web version).

Boyle, Christine and MacCrimmon, Marilyn T. "Reasons for Judgement: A Comment on *R. v. Sheppard* and *R. v. Braich"* (2002) 47 *Criminal Law Quarterly* 39.

Canadian Judicial Council. *Model Jury Instructions in Criminal Matters*. www.cjc-ccm.gc.ca/article.asp?id=2337; accessed December 31, 2005.

Canadian Press. "Jury System Unfair to Minorities, Report Says."(21 July 1994) *Vancouver Sun* A8.

Chopra, Sonia R. and James R. P. Ogloff. "Evaluating Jury Secrecy: Implications for Academic Research and Juror Stress." (2001) 44 *Criminal Law Quarterly* 190.

Conrod, M. "NS Law Reform Commission Discussion Paper Proposes Changes to Province's Jury System." (16 July 1993) *Lawyers' Weekly* 2.

Coughlan, Steve. "*R. v. Find*: Preserving the Presumption of Innocence." (2001) 42 *Criminal Reports* (5th) 31.

Der, Balfour. *The Jury: A Handbook of Law and Procedure.* LexisNexis Canada, looseleaf.
   The book addresses empanelling a jury, discharge of a juror, mistrials, *voir dires*, jury deliberation, questions by jurors, sequestration, communication with jurors, verdicts, jury addresses by counsel, presence of the accused, the charge to the jury, sentencing, and note-taking by jurors.

Dufraimont, Lisa. "The Case Against Offence-Based Challenges for Cause in Cases of Violence Against Women and Children." (2000) 44 *Criminal Law Quarterly* 161.

Dufraimont, Lisa. 2006. "*Krieger*: The Supreme Court's Guarded Endorsement of Jury Nullification." (2006) 41 *Criminal Reports* (6th) 209.

Ferguson, Gerry A., and John C. Bouck. *Canadian Criminal Jury Instructions CRIMJI*. Vancouver: Continuing Legal Education Society of British Columbia, 2004, looseleaf.

Gorman, Wayne. "*D.P.P. v. Charles J. Haughey*: A Canadian Perspective on Stays of Proceedings and Pre-Trial Publicity." (2000) 44 *Criminal Law Quarterly* 149.

Granger, Christopher. *The Criminal Jury Trial in Canada*, 2nd ed. Ontario: Carswell, 1996.

Hauschildt, Jordan. "Deadlocked: The Case for Mandatory Pattern Instructions in Criminal Jury Trials." (2005) 50 *Criminal Law Quarterly* 453.

Henry, Miriam, and Frances Henry. "A Challenge to Discriminatory Justice: The *Parks* Decision in Perspective." (1996) 38 *Criminal Law Quarterly* 333.

Holmgren, Janne A. "DNA Evidence and Jury Comprehension." (2005a) 38(3) *Canadian Society of Forensic Science Journal* 123.

Holmgren, Janne A. "It's a Match! Unravelling the Canadian Jury's Interpretation of DNA Evidence." (2005b) 28 *Criminal Reports* (6th) 246.

Hutson Matthew. "Unnatural Selection." *Psychology Today* (01 March 2007) online: www.psychologytoday.com/articles/200703/unnatural-selection; accessed January 4, 2010.

Israel, Mark. "The Underrepresentation of Indigenous Peoples on Canadian Jury Trials" (2003) 25(1) *Law and Policy* 37.

Law Reform Commission of Canada. *Studies on the Jury.* Ottawa, 1979.

Law Reform Commission of Canada. *The Jury in Criminal Trials.* Working Paper #27. Ottawa, 1980.
> The Law Reform Commission undertook a number of opinion polls to assist it in evaluating the jury system in criminal trials. It concluded that a jury of 12 should be retained in criminal proceedings, as should the requirement that the verdict be unanimous. It also recommended changes to the selection process to ensure greater representation of the community and better protection to the jurors from financial hardship.

Law Reform Commission of Canada, *The Jury.* Report #16. Ottawa, 1982.

Law Reform Commission of Canada. *Public Media Access to the Criminal Process.* Working Paper #56. Ottawa, 1987.

Lawrie, Alastair. "Behind Closed Doors." *The Globe and Mail,* June 15, 1990, A4.

Manson, Allan. "The Claim of the *Rose* Case: Jury Addresses and Humble Echoes of Reply." (1999) 20 *Criminal Reports* (5th) 300.

Nathanson, H.S. "Strengthening the Criminal Jury: Long Overdue." (1996) 38 *Criminal Law Quarterly* 217.

Nowlin, Christopher. "The Real Benefit of Trial by Jury for an Accused Person in Canada: A Constitutional Right to Jury Nullification." (2008) 53 *Criminal Law Quarterly* 290.

Ogloff, J.R.P., and N. Vidmar. "The Impact of Pretrial Publicity on Jurors: A Study to Compare the Relative Effects of Television and Print Media in a Child Sex Abuse Case."(1994) 18(5) *Law and Human Behavior* 507.

Ogloff, J.R.P., and V.G. Rose. "The Comprehension of Judicial Instructions." In N. Brewer & K.D. Williams (Eds.). *Psychology and Law: An Empirical Perspective.* New York: Guilford Press, 2005.

Ontario. *Report of the Race Relations and Policing Task Force* (1992) (Chair: Stephen Lewis).

Plaxton, Michael. "Thinking About Appeals, Authority and Judicial Power After *R. v. Sheppard*" (2002) 47 *Criminal Law Quarterly* 59.

Plaxton, Michael. "The Biased Juror and Appellate Review: A Reply to Professor Coughlan." (2001) 44 *Criminal Reports* (5th) 294.

Pomerant, David. *Multiculturalism, Representation and the Jury Selection Process in Canadian Criminal Cases.* WD1994-7e. Ottawa: Department of Justice, 1994.
> This Report describes the selection of juries and their function. It discusses representative panels, representative sources for jury selection, and how to achieve representativeness. Chapter 4 deals with about out-of-court selection, and Chapter 5 with in-court selection. Chapter 6 discusses the unanimous verdict requirement,

Quigley, Tim. *Procedure in Canadian Criminal Law.* Toronto: Carswell, 2005 (with updates).

Quigley, Tim. "*Sheppard*: Functional Standards fo Reasons in Criminal Cases" (2002) 50 *Criminal Reports* (5th) 104.

Quinlan, Paul. "Secrecy of Jury Deliberations—Is the Cost Too High?" (1993) 22 *Criminal Reports* (4th) 127.

Rose, V. Gordon, and James R.P. Ogloff. "Challenges for Cause in Canadian Criminal Jury Trials: Legal and Psychological Perspectives" (2002) 46 *Criminal Law Quarterly* 210.

Salhany, Roger E. *Canadian Criminal Procedure,* 6th ed. Toronto: Canada Law Book Inc., (available online on Criminal Spectrum) .
      Chapter 6, Trial on Indictment, includes a discussion on juries.

Sankoff, Peter, and Hendel, Ursula. "Creating a Right of Reply: *Rose* is Not Without a Few Thorns." (1999) 20 *Criminal Reports* (5th) 305.

Schmitz, Cristin. "Court Changes Procedure for Directing Acquittals." (15 July 1994) *Lawyers Weekly* 9.

Smith, Lynn. "Charter Equality Rights: Some General Issues and Specific Application in British Columbia to Elections, Juries and Illegitimacy." (1984) 18(2) *U.B.C. Law Review* 351.

Spencer, Beverley. "Jury Selection: No Magic, No Science, Just 'Listen to Your Gut' Lawyers are Told." (2 December 1994) *Lawyers Weekly,* 5.

Stuart, Don, Ron Delisle, and Tim Quigley. *Learning Canadian Criminal Procedure*, 9th dd. Toronto: Carswell, 2008.
      Chapter 12 covers jury trials.

Tanovich, David M. "Crown Squarely to Blame." (17 June 2009) *The Windsor Star* A6.

Tanovich, David M., David M. Paciocco, and Steven Skurka, *Jury Selection in Criminal Trials: Skills, Science, and the Law.* Concord, ON: Irwin Law, 1997.
      Chapter 1 provides an overview of jury selection. Chapter 2 is titled "The Jury Panel," Chapter 3, "Pre-Screening the Jury Panel," Chapter 4, "The Meaning and Purpose of a Challenge for Cause," Chapter 5, "Succeeding in Obtaining a Challenge for Cause," Chapter 6, "The Mechanics of the Challenge for Cause Procedure," Chapter 7, "The Triers of the Challenge for Cause," Chapter 8, "The Trial of the Truth of the Challenge for Cause," Chapter 9, "The Mechanics of Peremptory Challenges," Chapter 10, "Practical Jury Selection Tips," and Chapter 11, "Peremptory Challenges and the *Charter*. Appendix III provides excerpts from Provincial Jury Acts.

Todd, Douglas. "Truth and Justice." (31 December 1993) *Vancouver Sun* D10.

Vancouver Sun, Associated Press. "Weeding-out Process Starts for Simpson Jury." (1 October 1994) A9.

CASES

Application under s. 83.28 of the Criminal Code (Re), [2004] 2 S.C.R. 248.

R. v. Barrow, [1987] 2 S.C.R. 694. Pre-screening of jurors must be done in the presence of the accused.

R. v. Born with a Tooth (1993), 81 C.C.C. (3d) 393 (Alta. Q.B.). Jury panels must be both impartial and representative. Artificial means of composing panels that reduce representativeness, even if well intentioned, are impermissible.

R. v. Braich, [2002] 1 S.C.R. 903.

R. v. Budai, [1999] B.C.J. No. 2328 (B.C.C.A.).

R. v. Butler (1984), 63 C.C.C. (3d) 243 (B.C.C.A.). The Court ordered a new trial following allegations that the sheriff's office routinely excluded Aboriginals from juries.

R. v. Campbell (1999), 139 C.C.C. (3d) 258 (Ont C.A.).

Canadian Broadcasting Corp. v. New Brunswick (Attorney General), [1996] 3 S.C.R. 480. Although section 486(1) of the *Criminal Code*, allowing for the exclusion of the public, violates section 2(b) of the *Charter*, it is demonstrably justified under section 1.

Canadian Newspapers Co. v. Canada (Attorney General), [1988] 2 S.C.R. 122.

R. v. Church of Scientology et al. (No. 1) (1992), 74 C.C.C. (3d) 327 (Ont. Ct. Gen. Div); *affd.* [1997] O.J. No. 1548. The Canadian citizenship requirement under the Ontario *Juries Act* violates section 15 of the *Charter*, but is a reasonable limit under section 1. The exclusion of certain occupations and opposite-sex spouses of these persons does not infringe right to a fair trial. The use of computers to randomly select names is substantially in compliance with the legislation, which requires the panel to be drawn randomly from a container. The Court of Appeal expressed the views that the accused had no standing to argue a violation of the equality rights of non-citizens, and that the exclusion of non-citizens did not breach the accused's section 11 rights.

R. v. D. (A.) (2003), 180 C.C.C. (3d) 319 (Ont. S.C.J.).

Dagenais v. Canadian Broadcasting Corp, [1994] 3 S.C.R. 835. A ban on the broadcast of a television program will only be ordered where absolutely necessary to prevent a substantial risk to the fairness of the subsequent criminal trial on which the television story is based. The burden is on the applicant/accused to establish that necessity and that the right to a fair trial in the circumstances outweighs freedom of expression.

R. v. Dinardo, [2008] 1 S.C.R. 788.

R. v. English (1993), 84 C.C.C. (3d) 511 (Nfld. C.A.). Challenges to jurors were permitted on the basis of pre-trial publicity of sexual assault cases from an orphanage. Questions about religious affiliation were not allowed to be asked of jurors. English's application for leave to appeal to the Supreme Court of Canada was dismissed without reasons; [1993] S.C.C.A. No. 465.

R. v. Find, [2001] 1 S.C.R. 863.

R. v. Guess, [2000] B.C.J. No. 2023 (BCCA); application for leave to appeal to the Supreme Court of Canada dismissed; [2000] S.C.C.A. No. 628.

R. v. H.S.B., [2008] 3 S.C.R. 32.

R. v. Hubbert (1975), 29 C.C.C. (2d) 279 (Ont. C.A.); *affd.,* 33 C.C.C. (2d) 207 (S.C.C.). The Ontario Court of Appeal refused to allow defence counsel to ask each prospective juror if knowledge that the accused had previously been incarcerated in a hospital for the "criminally insane" would prejudice them against the accused.

R. v. Keegstra (1991), 63 C.C.C. (3d) 110 (Alta. C.A.); leave to appeal to the S.C.C. refused. Pre-trial publicity may support a challenge of prospective jurors for cause. The test is whether there is any doubt that the impartiality of any of the prospective jurors may have been affected by the publicity in a way that cannot be otherwise remedied (such as by judicial admonition).

R. v. Kenny (1992), 68 C.C.C. (3d) 36 (Nfld. S.C., Trial Division). Even extensive pre-trial publicity will not necessarily result in a stay of proceedings because of the impossibility of obtaining a fair trial; the prejudice can generally be adequately addressed by such remedies as challenges of prospective jurors for cause and judicial admonitions.

R. v. Kirkham (1998), 126 C.C.C. (3d) 397 (Sask. Q.B.). The prosecutor is acquitted of attempting to obstruct justice in the selection of jurors for the Latimer prosecution.

CHAPTER 5: *Juries and Procedure at Trial*

R. v. Koh (1998), 131 C.C.C. (3d) 257 (Ont. C.A.).

R. v. Krieger, [2006] 2 S.C.R. 501.

R. v. Latimer, [1997] 1 S.C.R. 217.

R. v. Latimer, [2001] 1 S.C.R. 3. Contains a discussion of jury nullification.

R. v. Laws, [1998] O.J. No. 3623 (Ont. C.A.). Citizenship requirements for jurors do not violate section 15 the *Charter*.

R. v. Lee, [1989] 2 S.C.R. 1384. Section 598, which allows the court to proceed by judge alone where an accused fails to attend trial, violates an accused's right under section 11(f), but is justified under section 1.

R. v. Makow (1974), 20 C.C.C. (2d) 513 (B.C.C.A.). Before a party is permitted to challenge a prospective juror for cause on the basis of lack of indifference, there must be external evidence to support the challenge. Where the concern results from pre-trial publicity, it may be sufficient for the court to address questions to the jury panel generally.

R. v. Mentuck, [2001] 3 S.C.R. 442.

R. v. Morgentaler, [1988] 1 S.C.R. 30.

R. v. Morgan (1997), 125 C.C.C. (3d) 478 (Ont. Ct. Gen. Div.).

R. v. Naglik (1993), 83 C.C.C. (3d) 526 (S.C.C.). While juries must be unanimous to reach an effective verdict, they must not be given the impression that they have no right to disagree.

R. v. Pan, [1999] O.J. No. 1214 (Ont. C.A.). Both the common law rule precluding a court from considering evidence of jury deliberations for the purpose of impugning the jury's verdict, and section 649 of the *Code*, making it illegal for jurors to discuss deliberations except in the context of an investigation or trial for obstruction of justice, do not offend the *Charter*.

R. v. Pan; R. v. Sawyer, [2001] 2 S.C.R. 344.

R. v. Parks (1993), 84 C.C.C. (3d) 353 (Ont. C.A.) [1993] O.J. No. 2157; leave to appeal refused [1993] S.C.C.A. No. 481. Limited questioning of potential jurors should be permitted to support challenges for cause based on partiality, in situations where there is a realistic prospect of bias, such as on racial grounds, and in which the traditional safeguards would not suffice.

R. v. Pearson (1994), 89 C.C.C. (3d) 535 (Que. C.A.). Although a jury must be unanimous as to guilt to convict, they need not be unanimous as to *how* the accused committed the offence.

Phillips v. Nova Scotia (Commission of Inquiry into the Westray Mine Tragedy), [1995] 2 S.C.R. 97.

R. v. R.E.M., [2008] 3 S.C.R. 3.

R. v. Rose, [1998] 3 S.C.R. 262. Sections 651(3) and (4) of the *Criminal Code*, that require the accused to address the jury first if he or she calls evidence at trial, do not violate the accused's rights under section 7 or 11(d) of the *Charter*.

R. v. Rowbotham; R. v. Roblin, [1994] 2 S.C.R. 463. In cases where the trial judge would formerly have directed the jury to return a verdict of "not guilty," the trial judge should withdraw the case from the jury and enter the acquittal himself or herself.

R. v. Sheppard, [2002] 1 S.C.R. 869. The Court affirmed the requirement for trial judges to provide reasons for their decisions.

Sherratt v. The Queen, [1991] 1 S.C.R. 509. Trial judge was correct in not allowing each juror to be questioned on their partiality based on pre-trial publicity. There must be an "air of reality," or a "realistic potential for the existence of partiality, on a ground sufficiently articulated in the application."

R. v. Spence, [2005] 3 S.C.R. 458.

Thatcher v. The Queen, [1987] 1 S.C.R. 652. Although a jury must be unanimous as to guilt to convict, they need not be unanimous as to *how* the accused committed the offence. Thus, a jury could still convict while not unanimous as to whether the accused was guilty as a principal or a party.

R. v. Turpin (1989), 48 C.C.C. (3d) 8 (S.C.C.). Section 11(f) of the *Charter* does not entitle an accused to be tried without a jury when one is otherwise required under the *Criminal Code.*

Vancouver Sun (Re), [2004] 2 S.C.R. 332.

R. v. Walker, [2008] 2 S.C.R. 245.

R. v. Williams, [1998] 1 S.C.R. 1128. Failure to allow challenge for cause where there is a realistic potential of partiality is a breach of section 11(d) of the *Charter.* Thus, a reasonably liberal approach should be adopted in deciding whether to allow challenge for cause based on prejudice. Where prejudice is shown on a national or provincial basis, it will often be reasonable to assume prejudice at a local level. Where prejudice has been established, it may be taken judicial notice of. The Court contrasts the Canadian and U.S. approaches.

R. v. Wilson (1996), 107 C.C.C. (3d) 86 (Ont. C.A.).

R. v. Zundel (1987), 31 C.C.C. (3d) 97 (Ont.C.A.); leave to appeal to the S.C.C. refused. The trial judge erred in not allowing challenges for cause on the basis of extensive pre-trial publicity and notoriety. The test is whether either "could potentially have the effect of destroying the prospective juror's indifference between the Crown and the accused."

CHAPTER 6: *Sentencing and Appeals*

CHAPTER OBJECTIVES

In studying this chapter, you should develop an understanding of the following topics and concepts:

- alternative measures
- the nature of the sentencing hearing
- input into the sentencing process, such as pre-sentence reports and victim impact statements
- the forms of disposition available to the court, and the conditions for and restrictions on their use
- the issues involved in permitting criminals to profit from stories of their crimes
- what rights of appeal exist, and how they differ between the Crown and defence, indictable and summary conviction offences, conviction and sentence, and issues of fact and issues of law
- when fresh evidence can be introduced on appeal
- the powers of the Minister of Justice to grant or refuse mercy or to refer matters directly to the courts for consideration

ALTERNATIVE MEASURES

Provinces may establish alternative measures under section 717 of the *Code*, to deal with persons alleged to have committed offences who accept responsibility for their acts or omissions, and who consent to participation in such alternatives. Alternative measures can be used only if their use "is not inconsistent with the protection of society" (section 717(1)). They should not be used unless there is sufficient evidence to prosecute the alleged offender, and persons contemplating their use should consider the needs of alleged offender and the interests of society and the victim (section 717(1)(b) and (f)).

Although the offender does not receive a criminal record, a record will be kept and alternative measures will be referred to in a probation officer's report for subsequent offences (section 721(3)(c)). While admissions for the purposes of alternative measures are not admissible against the person in any criminal or civil proceedings (section 717(3)), some provinces include them in criminal records checks done for the purposes of specified employment (British Columbia, Ministry of Public Safety and Solicitor General).

SENTENCING PURPOSE, OBJECTIVES, AND PRINCIPLES

The purpose and objectives of sentencing, once exclusively within the purview of the common law, were specified in the *Criminal Code* in 1996. Section 718 states that:

> The fundamental purpose of sentencing is to contribute, along with crime prevention initiatives, to respect for the law and the maintenance of a just, peaceful and safe society by imposing sanctions that have one or more of the following objectives:
>
> a) to denounce unlawful conduct;
> b) to deter the offender and other persons from committing offences;
> c) to separate offenders from society, where necessary;
> d) to assist in the rehabilitation of offenders;

e) to provide reparation for harm done to victims or to the community; and

f) to promote a sense of responsibility in offenders, and acknowledgment of the harm done to victims and to the community.

Section 718.01 requires the sentencing judge to "give primary consideration to the objectives of denunciation and deterrence" for offences involving the abuse of persons under 18 years of age. A "fundamental principle" of sentencing is that the sentence "must be proportionate to the gravity of the offence and the degree of responsibility of the offender" (section 718.1). Further principles of sentencing in section 718.2 specify circumstances deemed to be aggravating, such as whether the crime was motivated by bias or hate based on race, religion, sex, age, sexual orientation, or other similar factors (see Carter 2001), whether the offender abused his or her child or spouse or abused a position of trust or authority in committing the crime, or that the offence involved a criminal organization or was a terrorism offence. Section 718.21 lists additional factors for the court to consider when imposing a penalty on an organization.

Section 718.2(d) states that "an offender should not be deprived of liberty, if less restrictive sanctions may be appropriate in the circumstances," and section 718.2(e) states that the court shall take into consideration all reasonable sanctions other than imprisonment, "with particular attention to the circumstances of aboriginal offenders." In April 1999, the Supreme Court of Canada identified the aim of subsection (e) as the reduction of "the tragic overrepresentation of aboriginal peoples in prison" (*Gladue* para. 87), and provided principles to guide sentencing courts in dealing with aboriginal offenders (see Turpel-Lafond 1999). According to Quigley (2009), the section has not halted an increase in the over-representation of Aboriginals incarcerated in Canada (also see Pfefferle, 2008). Roach (2008) suggests that recent amendments to the *Code* will actually increase Canada's rate of imprisonment.

In 2004, the federal government amended the *Code* to deal more harshly with securities offences and other types of fraud. Section 380.1(1) now lists circumstances that the court shall consider to be aggravating:

(a) the value of the fraud committed exceeded one million dollars;

(b) the offence adversely affected, or had the potential to adversely affect, the stability of the Canadian economy or financial system or any financial market in Canada or investor confidence in such a financial market;

(c) the offence involved a large number of victims; and

(d) in committing the offence, the offender took advantage of the high regard in which the offender was held in the community.

Section 380.1(2) states that the court "shall not consider as mitigating circumstances the offender's employment, employment skills or status or reputation in the community if those circumstances were relevant to, contributed to, or were used in the commission of the offence."

THE SENTENCING HEARING

Section 723(1) of the *Criminal Code* requires a judge to provide both the prosecutor and the person being sentenced "an opportunity to make submissions with respect to any facts relevant to the sentence to be imposed." The judge "shall hear any relevant information" (section 723(2)), may "require the production of evidence that would assist…in determining the appropriate sentence" (section 723(3)), and may compel anyone who is a compellable witness to assist the court (section 723(4)). Although hearsay evidence is admissible, the judge (if she or he thinks it is in the "interests of justice") may compel a person to testify if the person has personal knowledge, is reasonably available, and is compellable (section 723(5)).

Section 724 stipulates which facts the judge can rely on in determining a sentence, and how to resolve disputes over facts. Prior to these amendments, the Supreme Court of Canada had decided that if the accused disputed any of the facts referred to by the prosecutor, the prosecutor was obliged to prove them beyond a reasonable doubt (*Gardiner*). Now sections 724(3)(d) and (e) state that the court must be satisfied on a balance of probabilities when facts are in dispute, although it must be satisfied beyond a reasonable doubt of "the existence of any aggravating fact or any previous conviction by the offender."

If the facts are in dispute after a guilty plea (and the Crown accepts the guilty plea), the judge will hold a hearing to determine the facts and then a hearing to determine the sentence. For example, when Frank Biller plead guilty to five counts of fraud and theft involving the sale of syndicated mortgages, in which investors lost millions of dollars, the judge held a 25-day sentencing hearing, to determine the degree of Biller's culpability (*Biller* 2005a). Following findings on this issue, the judge heard three days of submissions on the appropriate sentence, and then sentenced Biller to what amounted to a three year term of imprisonment (*Biller* 2005b).

Section 727 provides the procedures for giving notice to the offender if the Crown is seeking a more serious penalty because of previous convictions. Section 725 allows the sentencing judge to accept guilty pleas for other offences, and to take them into account when sentencing the offender, unless the judge thinks that a separate prosecution would be in the public interest.

Section 726 requires the judge to "ask whether the offender, if present, has anything to say." Failure to provide an offender with an opportunity to speak at the sentencing hearing may violate the offender's rights under section 7 of the *Charter* (*Dennison*). The sentencing judge is required to consider any relevant evidence, and submissions made by or on behalf of the prosecutor and the offender (section 726.1). Crown counsel and defence counsel might also make a **joint submission** as to the type or length of sentence that the judge ought to impose. The judge is not bound to follow joint submissions, and may consider a number of other factors, as discussed below. However, in *Ly*, the Ontario Court of Appeal found that the trial judge had erred in imposing a custodial sentence when Crown and defence had provided a joint submission that a custodial sentence was not appropriate in the case.

Subject to the above provisions, the judge must sentence an accused for the offence in respect of which the accused was found guilty. If an accused is charged with dangerous driving causing death and the jury convicts the accused of the lesser included offence of dangerous driving, the judge must impose a sentence for dangerous driving (*Brown*). The sentencing judge is required to give reasons for the sentence imposed (section 726.2).

Probation Reports

Section 721 allows the judge to order a **probation report,** or **pre-sentence report (PSR),** to be prepared before sentencing. The request for such a report may be initiated by the court, the offender (or defence counsel), or the prosecutor. Section 721(3) lists what information the report must contain, when possible, unless the court orders otherwise:

(a) the offender's age, maturity, character, behaviour, attitude and willingness to make amends;

(b) . . . the history of previous dispositions under the *Young Offenders Act* . . .and of previous findings of guilt under [the *Criminal Code*] and any other Act of Parliament;

(c) the history of any alternative measures used to deal with the offender, and the offender's response to those measures; and

(d) any matter required, by any regulation made under subsection (2), to be included in the report.

Under section 721(4), the report "must also contain information on any other matter required by the court, after hearing argument from the prosecutor and the offender, to be included in the report, subject to any contrary regulation made under subsection (2)." When such a report is filed, the clerk of the court must send a copy to the offender (or defence counsel), and to the prosecutor (section 721(5)). Again, if the offender disputes any of the facts on which the report relies, the facts must be proved or will be disregarded by the sentencing judge. Some lower court decisions have stated that the report is not supposed to provide the court with information about the offence itself (see *Urbanovick* and *Rudyk*). For a discussion of the contents of PSRs and their trend towards risk assessment, see Cole and Angus (2003).

Victim Impact Statements

In 1988, Parliament added a section to the *Criminal Code* allowing **victim impact statements (VIS)** to be introduced at the sentencing hearing. In 1996, the section was amended to require the court to consider such statements from the victims of an offence (sections 722(1)), and after further amendments in 1999, the court now must allow the victim to read a filed impact statement if the victim so requests (section 722(2.1)). Section 722(4) defines a victim as "a person to whom harm is done or who suffers physical or emotional loss as a result of the commission of the offence." The statement is to be in writing and filed with the court. Copies will be sent by the clerk of the court to the offender or the offender's counsel, and to the prosecutor (section 722.1). Section 722.2 requires the court to "inquire of the prosecutor or a victim of the offence, or any person representing the victim of the offence, whether the victim or victims have been advised of the opportunity to prepare" a victim impact statement. Section 722(3) allows the court to consider evidence from a victim, even if it is not filed in accordance with section 722(2). Although the victims may testify if called as witnesses by the Crown, or if defence counsel wants to cross-examine them on their statements, the sentencing hearing is conducted by the Crown, who decides what information will be presented to the court (subject to the requirements of section 722). In some locations, Victims' Services workers assist the victims in filling out the impact statement forms.

The role of the victim in criminal proceedings is a much debated topic. Victim impact statements provide the court with information on how a crime affects the victim. They may also have a cathartic effect on the victim, provide the victim with a means of reassessing a relationship with the accused, assist in the rehabilitation of the accused, and provide parole boards and probation officers with relevant information in their work (Roberts 1992, 1–2). There is, however, a concern that victim impact statements might infringe upon the accused's right to an impartial hearing (Roberts 1992, 8). For a more detailed discussion on the changing role of the victim, see Roach (1999) and Roberts (2003). In addition to victim impact statements, victims may also seek redress directly from the offender, through civil actions (see discussion in the Introduction).

Previous Criminal Record

The sentencing judge will consider the accused's **criminal record**, which (if one exists) is usually submitted to the court by the prosecutor. The court will consider only those offences for which the accused has been convicted, or for which absolute or conditional discharges have been imposed. Convictions alleged by the Crown but not admitted by the accused must be proved by the Crown beyond a reasonable doubt (section 724(3)(e)).

Time in Custody

Section 719(3) allows the sentencing judge to take into account any time already spent in custody in relation to the offence (awaiting trial or sentencing) when imposing a sentence on the accused. The practice is to give more credit for pre-trial or pre-sentence custody than the actual time spent in such custody (*Tallman*). For example, a month in custody awaiting trial is often considered to be equivalent to two months of sentenced incarceration. This is because of the quality of pre-trial custody (Manson 2004, 303), and because offenders will often serve much less than their specified custodial sentence, due to credit for good behaviour (307).

SENTENCES AND OTHER ORDERS

The sentencing judge has several options in sentencing. There are, however, some statutory limits on the judge's discretion. Some argue that judges have far too much discretion when it comes to sentencing, and that legislation ought to restrict that discretion through, for example, mandatory minimum sentences. Others argue in favour of even more judicial discretion and less harsh punishment.

Absolute and Conditional Discharges

The most lenient penalty for an accused who pleads guilty or is found guilty, in limited circumstances, is an **absolute discharge**; the second most lenient penalty is a **conditional discharge**. Both of these discharges are dealt with under section 730(1) of the *Criminal Code*. In neither of these cases is the accused "convicted." The difference between an absolute and conditional discharge is that the absolute discharge is unconditional, while the conditional discharge has a **probation order** attached to it (described in section 731). Thus, a person may be given a conditional discharge and put on probation for six months, with a condition requiring the performance of community work service, or prohibiting him or her from being found in a particular section of town. A person cannot receive an absolute or conditional discharge for an offence if there is a minimum penalty specified in the legislation, or if the accused is charged with an offence punishable by imprisonment for 14 years or life. Therefore, it is always necessary to look at the charging section to determine whether the judge has the option of discharging the accused.

The fact that a discharge is available does not mean that the court will impose one. The court must consider whether a discharge is in the "best interests of the accused and not contrary to public interest" (section 743(1)). The "best interests of the accused," according to the British Columbia Court of Appeal, presupposes:

> that the accused is a person of good character, without previous conviction, that it is not necessary to enter a conviction against him in order to deter him from future offences or to rehabilitate him, and that the entry of a conviction against him may have significant adverse repercussions (*Fallofield* 454-55).

The court was also of the view that a discharge should not be granted routinely for any particular offence (455).

The fact that an accused has been granted a discharge in the past will clearly count against that accused when seeking another discharge (*Tan*), however, a person who has previously entered a diversion program will not automatically be disqualified from being discharged (*Drew*). The fact that an accused's immigration status is in jeopardy is a factor the court will consider, but that will not by itself cause the court to grant a discharge (*Chiu*), although there is a somewhat contrary opinion from the Alberta Court of Appeal (*Kerr*). Whether a discharge would be in the "public interest" requires a consideration of deterrence to others, but this factor should "not preclude the judicious use of the discharge provisions" (*Fallofield*).

Section 730(3) states that where a discharge is granted, "the offender shall be deemed not to have been convicted of an offence," except for the purposes of appeal and a few other circumstances specified in the subsection. The discharge is still recorded however, so although the *Code* contemplates avoiding a conviction, a person is only technically without a criminal record.

Section 6.1 of the *Criminal Records Act*, R.S.C. 1985, Chap. C-47 provides that a discharge is automatically removed from a person's record and from the automated criminal records retrieval system of the RCMP, after one year in the case of an absolute discharge, and after three years in the case of a conditional discharge. The *Act* allows the Minister of Public Safety and Emergency Preparedness to disclose a purged record, and additionally, section 6.2 allows:

> the name, date of birth and last known address of a person who has received a pardon or a discharge . . . [to be] disclosed to a police officer if a fingerprint, identified as that of the person, is found
> a) at the scene of a crime during an investigation of the crime; or
> b) during an attempt to identify a deceased person or a person suffering from amnesia.

Absolute and conditional discharges can also be brought up at sentencing hearings for subsequent offences. The fact that a person has already received an absolute or conditional discharge means the person is much less likely to receive another one in the future.

If the conditions attached to a discharge are breached, section 733.1 makes such a breach a hybrid offence, with a maximum penalty of two years imprisonment if the Crown proceeds by indictment, and a maximum of 18 months with a maximum fine of $2000 on summary conviction. Section 730(4) allows the court to revoke the discharge, enter a conviction, and sentence the accused for the original offence as well as for the offence of **breach of probation.** According to the Alberta Court of Appeal, this section does not violate an accused's right under section 11(h) of the *Charter* to not be tried or punished for an offence for which the accused has already been punished (*Elendiuk*). A conditional discharge reserves the court's right to convict and sentence an offender who violates the conditions of the discharge.

Suspended Sentence and Probation

Section 731 allows the court to convict the accused, suspend the passing of sentence, and place the person on probation for a term not exceeding three years (section 732.2(2)(b)). Probation cannot be used if there is a minimum sentence required by statute. Again, the charging section sets out whether there is a minimum sentence. Section 731(1)(b) allows the judge to impose a probation order, in addition to fining or imprisoning the offender for two years or less.

A **suspended sentence** includes a probation order, so this sentence is often referred to as simply "probation" (see Form 46). All probation orders must contain the conditions that the accused "keep the peace and be of good behaviour," appear in court as required by the court, and notify the court or probation officer of any change of name, address or employment (section 732.1(2)). In addition, the court may impose one or more of the "optional conditions" listed in section 732.1(3), including reporting to a probation officer, remaining within the jurisdiction of the court, abstaining from alcohol or drugs, abstaining from owning weapons, performing up to 240 hours of community service, participating in various treatment programs for drug or alcohol abuse, and complying with such reasonable conditions "as the court considers desirable...for protecting society and for facilitating the offender's successful reintegration into the community." Under the last provision, judges sometimes impose geographical restrictions on offenders. For example, a person convicted of shoplifting may be ordered not to be found within one block of the store from which she or he shoplifted. Under certain circumstances, the courts will sanction banishment from a community (*Taylor*). A probation order cannot authorize a search and seizure of bodily substances in order to enforce an order to abstain from alcohol or drugs (*Shoker* para. 3).

Organizational probation was added to the *Criminal Code* in 2004. If the accused is an organization, the additional conditions that may be added to a probation order include establishing policies to reduce the likelihood of a repeat offence, identifying senior officers who are responsible for those policies, publicizing the offence, sentence and measures taken to reduce the likelihood of conviction, and "any other reasonable conditions that the court shall consider desirable to prevent the organization from committing subsequent offences or to remedy the harm caused by the offence" (section 732.1(3.1)). The court shall first consider "whether it would be more appropriate for another regulatory body to supervise the development or implementation of policies, standards and procedures" referred to in the optional conditions.

Under section 732.2(3), the court may alter the optional conditions of a probation order on application by the offender, the probation officer, or the prosecutor. The sentence of the offender is suspended during the period of probation, and a breach of the terms of the probation order is an offence under section 733.1. If an offender on probation breaches probation or commits another offence, section 732.2(5) also allows the court to revoke an earlier suspended sentence and to impose any sentence available for the offence for which the person was placed on probation. The optional conditions of the probation order may also be altered at this time, and the period of probation may be extended up to one year.

Fines

Section 734 of the *Criminal Code* allows for **fines,** in addition to or instead of any other punishment authorized, for offences where there is no minimum penalty. If there is a minimum penalty, the court may impose it and a fine. Fines that can be imposed following a conviction for an indictable offence usually have no monetary limit. Section 787 states that the maximum penalty for a summary conviction offence is $5000, six months, or both, unless otherwise specified in the legislation. Section 735 provides that organizations convicted of summary conviction offences can be fined an amount not in excess of $100,000, unless otherwise provided by legislation.

Some legislation specifies other maximum and minimum penalties. For example, section 238 of the *Income Tax Act* provides for a minimum fine of $1000 and a maximum of $25,000 on conviction of a summary conviction offence, or to such a fine along with imprisonment for a term not more than 12 months. Section 239 of the *Act* allows, in the case of other specified offences on summary conviction, "a fine not less than 50%, and not more than 200%, of the amount of the tax that was sought to be evaded; or both the fine…and imprisonment for a term not exceeding two years."

When imposing a fine for an offence that does not specify a minimum fine, the court must consider the offender's ability to pay or discharge it under the fine option program before imposing a fine (section 734(2) of the *Criminal Code*). Sections 734(4) and (5) detail the calculation of the time to be served in **default,** if the fine is not paid. The court can allow the accused **time to pay** a fine. On application by the offender, under section 734.3, the court may extend the time allowed for payment of the fine. These provisions now result in fewer people spending time in jail because of an inability to pay a fine. Section 734.6 allows the Attorney-General to recover a fine through entering it as a judgment and enforcing it through civil proceedings.

Conditional Sentences of Imprisonment

In 1996, the federal government introduced **conditional sentences** of imprisonment as another option for sentencing. When a person is convicted of an offence for which there is no minimum term of imprisonment, the court may order that the offender serve the sentence in the community, if the sentence imposed is for imprisonment for less than two years, and if the court is satisfied that this "would not endanger the safety of the community, and would be consistent with the fundamental purpose and principles of sentencing set out in sections 718–718.2" (section 742.1). In 2007, the federal government removed this option for serious personal injury offences (defined in section 752), and terrorism and criminal organization offences where the maximum penalty is ten years or more. In addition, the court must first consider whether sections 109 and 110 (mandatory and discretionary prohibition orders on weapons and explosives) are applicable (section 742.2(1)). A conditional sentence under section 742.3(2)(b) does not affect orders under sections 109 or 110 (section 742.2(2)).

Section 742.3(1) lists the compulsory conditions of a conditional sentence (slightly more stringent than a probation order). In addition to the compulsory conditions of a conditional sentence order, optional conditions suggested in section 742.3(2) are similar to those found in a probation order. A breach of a conditional sentence order can result in the offender being brought back to court under section 742.6 for a hearing. The court's powers under section 742.6(9) include the option of directing that the offender serve the remainder of the sentence in custody, "if satisfied, on a balance of probabilities, that the offender has without reasonable excuse, the proof of which lies on the offender, breached a condition of the conditional sentence." If a person is imprisoned for another offence during a conditional sentence, the conditional sentence is suspended during the period of incarceration, and resumes upon the offender's release (section 742.7).

The Supreme Court of Canada provided a detailed analysis of the conditional sentencing provisions in *Proulx*. Conditional sentences "were enacted both to reduce reliance on incarceration as a sanction and to increase the use of principles of restorative justice in sentencing" (para. 127). Unlike probation, which is primarily concerned with rehabilitation, conditional sentences are concerned with both punishment and rehabilitation. The Court wrote, "Therefore, conditional sentences should generally include punitive conditions that are restrictive of the offender's liberty. Conditions such as house arrest should be the norm, not the exception" (para. 127). The Court provides further guidance in paragraph 127 of its decision.

A major concern with conditional sentences is the possibility that they are widening the net of social control; that is, although designed to have offenders who would otherwise be in jail, serve time in the community, they might be increasing the number of offenders subjected to additional control by imposing conditional sentences on those who would otherwise not be subjected to incarceration (see discussion by North 2001; Roach 2000; Roberts and Gabor 2003). Conditional sentences are also discussed and analyzed by Gemmell 1997; Manson 1997a; Manson 1997b; Manson 1998a; Manson 1998b; Manson 2001a; Mazey 2002; Roberts 2002; Roberts and Cole 1999; Roberts and Healy 2001; Roberts and von Hirsch 1998).

Incarceration

The maximum terms of **imprisonment** allowed are usually prescribed in the charging sections of the statute that creates the offence. If there is no penalty specified in the charging section, section 787 provides a maximum penalty for summary conviction offences of six months **incarceration,** a $5000 fine, or both. The maximum penalty for an indictable offence, where no maximum is specified, is a term of imprisonment not exceeding five years (section 743).

Minimum penalties are set out in some legislation for first offences, and for second and subsequent offences. For example, section 85 of the *Code* stipulates a minimum penalty of one year for using a firearm in the commission of specified offences. For a second such offence, the minimum penalty is three years. Section 745 prescribes a sentence of life in prison when a person is convicted of murder or high treason. A person who commits manslaughter with a firearm is subject to a minimum penalty of four years (section 236). In relation to various book-making and betting offences, sections 202 and 203 require a minimum sentence of 14 days for a second offence, and three months for third and subsequent offences. Crutcher (2001) reported that in 1999, 29 of the over 400 offences in the *Code* carried a minimum penalty, and 18 of these 29 had been introduced in 1995 under the *Firearms Act*.

Despite strong arguments against mandatory minimum sentences (see, for example, Malik 2007), they have been found to be constitutionally valid (*Ferguson* 2008). If a mandatory minimum sentence does not violate section 12 of the *Charter*, an accused cannot be exempt from the mandatory minimum because of the circumstances of the case (*Ferguson* 2008; see discussion in Calarco 2008 and Coughlan 2008).

Intermittent Sentences

Section 732 allows the court, when imposing a sentence of imprisonment of 90 days or less, to order the offender to serve time intermittently, and to comply with a probation order during the time the offender is not confined. Intermittent sentences are often used to allow the offender to continue employment while serving time on weekends.

Box 6.1 The Meaning of Incarceration

Judge Katie McGowan of St. Catharine's, Ontario had the following exchange with an accused during a sentencing hearing:

The Court:	Mr. M—is there anything you wish to say on your own behalf?
The Accused:	What's this?
The Court:	Stand up, please.
The Accused:	Yeah, what's this incarcer—car—incar—what's this mean, she's talking about incarceration. What's that mean?
The Court:	The Crown is requesting that you be sent to jail. That is what incarceration means.
The Accused:	Yea. Yeah.
The Court:	For a period of four to six months.
The Accused:	I thought they were going to operate on me.
The Court:	No, not in this jurisdiction, sir (Advocate (1997) 55(5) at 801).

Dangerous Offender Declaration

On application by the Crown, section 752.1 allows the court, prior to sentencing a person convicted of a "serious personal injury offence" (defined in section 752), or a list of offences in section 753.1(2)(a) (various sexual offences), to remand the person for up to 60 days for an assessment, if the court is of the opinion "that there are reasonable grounds to believe that the offender might be found to be a **dangerous offender** under section 753 or a **long-term offender** under section 753.1." If the court finds that the person is a dangerous offender, the offender will be sentenced to an **indeterminate sentence** under section 753(4). If the offender is found to be a long-term offender, the court shall impose a minimum term of imprisonment for two years, and order the offender, thereafter, to be supervised in the community for a period not exceeding 10 years (section 753.1(3)). Sections 754 through 758 set out the procedures to be followed in such hearings, and section 759 allows for the appeal from such an order, or from the refusal to make an order. Victims are allowed to apply to attend such hearings. The Supreme Court of Canada has decided that an indeterminate sentence does not violate a person's rights under the *Charter* (*Lyons*).

Section 761 requires the National Parole Board to review "the condition, history and circumstances" of a person subject to an indeterminate sentence seven years after the declaration, and every two years thereafter.

Restitution

Section 738(1) allows the sentencing judge, on application by the Attorney-General or on his or her own motion, to order that the offender pay **restitution** to another person, in the amount necessary to replace property damaged (lost or destroyed in the commission of the offence), to cover pecuniary damage (including loss of income) for offences causing bodily harm, and to compensate reasonable expenses incurred by a spouse or child of an offender who moves out of a home because of an offence involving threats of bodily harm or bodily harm. All of these provisions are limited to situations "where the amounts are readily ascertainable." These provisions were added in response to a Supreme Court of Canada decision that such orders were within the jurisdiction of the federal government, so long as the issue of assessment was simplistic and the amount could be arrived at expeditiously (*Zelensky*). If restitution is ordered and not paid, the restitution order can be registered in court as a civil judgment, and the person can enforce the debt through civil proceedings (section 741). Section 739 allows the court to compensate innocent purchasers of stolen property. A restitution order does not preclude a victim from seeking civil remedies through the courts (section 741.2).

Victim Fine Surcharge

Section 737 requires a person convicted or discharged under section 730 to pay a **victim fine surcharge**, in addition to any other penalty, for the offences specified in the section (all *Criminal Code* offences and offences under the *Controlled Drugs and Substances Act*). The amount of the surcharge is 15 percent of the fine imposed, or if no fine is imposed, $50 for a summary conviction offence and $100 for an indictable offence. Section 737(3) allows the court to impose a higher amount if it is appropriate, and if the offender is capable of paying the higher amount. Section 737(5) allows the court to exempt an offender from paying a surcharge if it would cause "undue hardship" to the offender or his or her dependents; subsection (6) requires the court to provide reasons for any exemption.

Prohibitions

There are a number of provisions scattered throughout the *Criminal Code* that allow or require a court to impose **prohibition orders** on an accused. Violations of those orders are offences. For example, section 447.1(1) allows the court to prohibit persons convicted of cruelty to animals from owning or having care or control of an animal or bird, and subsection (2) creates an offence for breach of the prohibition. Section 111 contains provisions regarding firearms prohibitions, and section 259 involves driving prohibitions in connection with alcohol-related driving offences. In *Wiles*, the Supreme Court of Canada upheld the mandatory firearms prohibition order pursuant to section 109(1)(c) following a conviction of unlawfully producing cannabis. The trial judge had erroneously found that it violated section 12 of the *Charter* ("cruel and unusual punishment"), and was not justifiable under section 1.

Sex Offender Registration

The *Sex Offender Information Registration Act*, S.C. 2004, c. 10, designed to "to help police services investigate crimes of a sexual nature by requiring the registration of certain information relating to sex offenders" (section 1), came into effect on December 15, 2004. A data base of sexual offenders is maintained by the RCMP (section 14). For a discussion of some of the issues raised by such registries, see Davies (2004) and Laine (2007). Section 490.011 of the *Criminal Code* defines a "designated offence" for which an order may be imposed, and subsequent sections deal with the procedure, duration and appeals of such orders.

Proceeds of Crime and Forfeiture

The anti-profiteering legislation, also referred to as the **Proceeds of Crime** legislation (see Part XII.2 of the *Code*, sections 462.3 to 462.5), was proclaimed to be in force as of January 1, 1989. These provisions allow the Attorney-General to confiscate the "proceeds of crime," as defined in that Part, for designated offences. Section 462.37 sets out the procedure to obtain a forfeiture order upon conviction. There are other forfeiture provisions scattered throughout the *Code* and in the *Controlled Drugs and Substances Act*, S.C. 1996, c. 19. See German (1998) for a comprehensive discussion of this legislation. Some provinces have passed their own civil forfeiture legislation; see, for example, the British Columbia *Civil Forfeiture Act,* S.B.C. 2005, Chapter 29. Similar legislation in Ontario (*Civil Remedies Act*, S.O. 2001, c. 28) was found to be constitutional by the Supreme Court of Canada, and not to interfere with the *Criminal* forfeiture provisions (*Chatterjee* 2009 para. 42). Binnie, J. stated, "If such operational interference were demonstrated, of course, or if it were shown that the CRA frustrated the federal purpose underlying the forfeiture provisions of the *Criminal Code*, the doctrine of federal paramountcy would render inoperative the CRA to the extent of the conflict or interference" (para. 42).

Forfeiture orders in respect of real estate under the *Controlled Drugs and Substances Act,* governed by the factors set out in subsections 19.1(3) and (4), are to be considered independently of the sentencing provisions of the *Criminal Code*, as this prevents "the unpalatable possibility of trading property for jail time" (*Craig*, 2009 para. 3). The sentencing inquiry under the *Code* "focuses on the individualized circumstances of the offender; the main focus of forfeiture orders, on the other hand, is on the property itself and its role in past and future crime" (para. 40). It is also possible for the court to order partial forfeiture of real estate (*Ouellette*).

PROCEEDS FROM PUBLICATIONS ABOUT CRIMES

Sensationalism surrounding several criminals and the profits they made selling their stories resulted in public pressure to prevent criminals from profiting from publicity about their crimes. The first law in the United States, dubbed the "Son of Sam" law, was introduced in New York in 1977 in response to public outcry, when McGraw-Hill Book Company bought the story of David Berkowitz in an arrangement that included a $250 000 advance, $150 000 for the ghost writer, and $75 000 to Berkowitz ("Son of Sam"), a serial killer (Okuda 1988, 1354). The law, which was designed to funnel profits to the victims rather than prevent the publication of the material, was used as a model in 47 other states (Yager 2004, 435). The law was subsequently struck down by the United States Supreme Court in 1991, under the United States' First Amendment, on the basis that it was too broad, and imposed "a financial burden that it places on no other speech and no other income" (441). Various efforts have been made by state legislators to revamp Son of Sam laws in order not to violate the United States' Constitution (Yager 2004; Malecki 2006).

In Canada, there is no federal legislation in place to deal with proceeds from the publication about crime. On February 22, 1995, Scarborough West Member of Parliament Thomas Wappel introduced a private member's bill to amend both the *Criminal Code* and the *Copyright Act,* such that a person convicted of an indictable offence could not profit from selling her or his story. Under his proposed bill, any money made from a book, article, screenplay, recording, and so forth would be seized by the Crown as the proceeds of crime, and the Crown would hold copyright to the material, which could be enforced internationally (press release February 22, 1995). Wappel introduced his bill following media speculation that Kara Homalka would be allowed to profit by selling her story of the sex slayings of Kristen French of St. Catharines, Ontario, and Leslie Mahaffy of Burlington, Ontario (personal correspondence). Opponents to such legislation believe that the existing avenue of redress (civil action by the victim against the person convicted) is sufficient. Wappel finally managed to get a profits-of-crime bill through the House of Commons in the fall of 1997; however, his bill was killed by the Senate in June 1998 (see Gaucher and Elliott 2001 for a discussion of some of the issues surrounding such legislation).

In 2003, the Ontario government replaced its *Victims' Right to Proceeds of Crime Act* of 1994 with the *Prohibiting Profiting From Recounting Crimes Act*, S.O. 2002, c. 2. The purpose of the latter Act is not to prohibit publication, but "to use proceeds of contracts for recounting crime to compensate persons who suffer pecuniary or non-pecuniary losses as a result of designated crimes and to assist victims of crime" (section 1). The legislation applies only in Ontario, and so it cannot be enforced elsewhere. In 2009, the Saskatchewan government introduced *The Profits of Criminal Notoriety Act*, requiring profits from recounting crime to be paid to the government.

APPEALS

There are no **rights of appeal** except as provided by statute, discussed below under "Appeals—Procedure by Indictment," and "Appeals—Summary Conviction Procedure." Other sections of the *Code* allow for appeals from findings, such as that an accused is a dangerous mentally disordered offender under Part XX.1 (see section 672.72 to 672.78). In addition, Part XXVI (sections 774–85) of the *Code* allows for remedies for jurisdictional errors.

Box 6.2 The Need for Finality in Criminal Proceedings

According to the Supreme Court of Canada, "unless the accused is still "in the judicial system," an accused is unable to reopen his or her case and rely on subsequently decided judicial authorities, even where the provision under which the accused was convicted is subsequently declared constitutionally invalid" (*Sarson* para. 26). Note: this was not a case of wrongful conviction. The Court was convinced that Sarson was legally and morally responsible for the murder he had been convicted of.

In considering the grounds of appeal for summary and indictable proceedings, it is important to distinguish between questions of law, questions of fact, and questions of mixed fact and law. In addition to being important for determining whether an accused or the Crown can appeal, the distinction between questions of law and questions of fact are also important in jury trials. If a question of law arises, the judge will decide it, whereas the jury will decide all questions of fact. In the *Southam* case, the Supreme Court of Canada explained the differences: "questions of law are questions about what the correct legal test is; questions of fact are questions about what actually took place between the parties; and questions of mixed law and fact are questions about whether the facts satisfy the legal tests" (para. 35). Having stated the differences, the Court then went on to say, "the distinction between law on the one hand and mixed law and fact on the other is difficult. On occasion, what appears to be mixed law and fact turns out to be law, or vice versa" (para. 35).

Examples of questions of law include the interpretation of legislation (for example, what the meaning of "sexual" is in sexual assault provisions). In a jury trial, the judge will instruct the jury on the meaning of the words in the legislation, as these are questions of law. Whether the judge has misdirected the jury, or himself or herself, on what the law is, is a question of law. The question of whether evidence is admissible or should be excluded (discussed in Part II of this text) is also a question of law. In *Skalbania*, the Supreme Court of Canada held that the mental state required to establish guilt under section 332(1) of the *Criminal Code* (misappropriation of money) is a question of law. In *Buhay*, the Supreme Court of Canada stated that the "appreciation of whether the admission of evidence would bring the administration of justice into disrepute is a question of mixed fact and law as it involves the application of a legal standard to a set of facts" (para. 45). In another case, the Court wrote, "Whether a conviction can be said to be unreasonable, or not supported by the evidence, imports in every case the application of a legal standard. ...the application of that legal standard is enough to make the question a question of law" (*Biniaris* para. 23).

Examples of questions of fact include questions of the credibility of witnesses, and whether the Crown has proved the guilt of the accused beyond a reasonable doubt. At the end of the trial, after the defence has called any evidence it chooses, the defence may argue that the Crown has not proved its case beyond a reasonable doubt. This is a question of fact for the trier of fact (jury) to decide.

It is also important to know that there are situations in which there is an **appeal as of right**, such that the permission of a court is not required. In other circumstances, there can be no appeal without the court's permission—that is, without its granting **leave to appeal**.

Appeals—Procedure by Indictment

Part XXI of the *Criminal Code* (sections 673 to 696) deals with appeals relating to indictable offences, including indictable offences for which the accused elects to be tried by a provincial court judge, and all hybrid offences where the Crown proceeds by way of indictment. If an accused is charged with an indictable offence but is convicted of a lesser included offence that is a summary conviction offence, the appeal is governed by the nature of the proceedings; that is, since they were prosecuted by indictment, the appeal is under Part XXI. A person may also appeal a summary conviction offence under this Part "if the summary conviction offence was tried with an indictable offence and there is an appeal in respect of the indictable offence" (section 675(1.1)).

Section 675 sets out the rights of appeal for a person convicted in proceedings by indictment. The accused can appeal as of right to the Court of Appeal from a conviction on a question of law, with leave of the Court of Appeal or trial judge on a question of fact or a question of mixed fact and law, or with leave of the Court of Appeal on any other ground that appears to the court "to be a sufficient ground of appeal." The accused can also appeal from sentence with leave of the Court of Appeal, unless the sentence is mandatory.

The Crown's rights of appeal are more restricted (see section 676). It may appeal from a judgment or acquittal as of right on a question of law. Otherwise, the Crown cannot appeal an acquittal. It may appeal from sentence on the same basis as the accused (section 676(1)(d)).

Section 679 allows for **release pending appeal**. Notice that such bail is not automatic, and the decision is made by a judge of the Court of Appeal. Subsection 679(3) places the burden on the convicted person seeking to be released pending an appeal against conviction to show that:

   a) the appeal or application for leave is not frivolous;
   b) he will surrender himself into custody in accordance with the terms of the order; and
   c) his detention is not necessary in the public interest.

Both the Ontario Court of Appeal (*Farinacci*) and the British Columbia Court of Appeal (*Branco*; leave to appeal to the Supreme Court of Canada refused) have examined the requirement that the accused show "his detention is not in the public interest," in light of the Supreme Court of Canada decision in *Morales* (see discussion in Chapter 2). Both courts found that section 11(e) of the *Charter* (the right "not to be denied reasonable bail without just cause") did not apply to a bail hearing pending appeal, as the presumption of innocence no longer applies after an accused has been found guilty of committing an offence.

Section 679(4) specifies what the accused must establish to gain interim release if the appeal is against sentence only. Section 679(5) lists the conditions or terms under which a judge of the Court of Appeal can release an offender pending appeal.

Section 686 delineates the powers of the Court of Appeal. The Court may allow an appeal from conviction if it is of the opinion that:

   (i) the verdict should be set aside on the ground that it is unreasonable or cannot be supported by the evidence,
   (ii) the judgment of the trial court should be set aside on the ground of a wrong decision on a question of law, or
   (ii) on any ground there was a **miscarriage of justice.**

The Court of Appeal is not supposed to substitute its opinion for that of the trial judge, or of the jury. If it is a case that should have been left to the jury, it is up to the jury to weigh the evidence and reach a verdict. However, whether a verdict is unreasonable or cannot be supported by the evidence is a question of law (*Yebes*). In *(W.)R.*, Madame Justice McLachlin of the Supreme Court of Canada reviewed the *Yebes* decision, and stated the test to be applied under section 686 as follows:

> It is thus clear that a court of appeal, in determining whether the trier of fact could reasonably have reached the conclusion that the accused is guilty beyond a reasonable doubt, must re-examine, and to some extent at least, reweigh and consider the effect of the evidence. The only question remaining is whether this rule applies to verdicts based on findings of credibility. In my opinion, it does. The test remains the same: could a jury or judge properly instructed and acting reasonably have convicted? That said, in applying the test the Court of Appeal should show great deference to the findings of credibility made at trial. This court has repeatedly affirmed the importance of taking into account the special position of the trier of fact on matters of credibility…[authorities omitted]. The trial judge has the advantage, denied to the appellate court, of seeing and hearing the evidence of the witnesses. However, as a matter of law it remains open to an appellate court to overturn a verdict based on findings of credibility where, after considering all the evidence and having due regard to the advantages afforded the trial judge, it concludes that the verdict is unreasonable (para. 20).

In *Burke*, the Supreme Court of Canada quashed convictions for indecent assault against a former Christian Brother from Mount Cashel Orphanage in Newfoundland, on the basis that the convictions were unreasonable. One of the witnesses had appeared on TV and told of horrible abuses. Later when he appeared before a Commission of Inquiry, he admitted that his statements on TV were untrue.

Section 686(1)(b) allows the Court of Appeal to dismiss an appeal from conviction where (among other things) the Court of Appeal believes that there was "no substantial wrong or miscarriage of justice," or "notwithstanding any procedural irregularity at trial…the appellant suffered no prejudice thereby."

If the Court of Appeal allows an appeal from conviction, it shall quash the conviction and either direct an acquittal or order a new trial under section 686(2). If the Crown successfully appeals an acquittal, the Court can order a new trial or substitute a conviction, unless the acquittal was made by a jury (see section 686(4)(b)(ii)). This exception was added in 1976, after Morgentaler was acquitted by a jury of performing illegal abortions only to have the Quebec Court of Appeal enter a conviction. This case raised a major issue regarding the role of the jury (see Chapter 5—one of the functions of the jury is to protect citizens against oppressive laws). Nelson Skalbania, acquitted of misappropriation of money by a trial judge sitting alone, challenged the constitutionality of section 686(4)(b)(ii) when the Alberta Court of Appeal substituted a guilty verdict. The Supreme Court of Canada held that the section was constitutional (*Skalbania*).

Section 687 deals with the powers of the Court of Appeal when hearing an appeal from sentence. The Court of Appeal cannot merely substitute its own view on a sentence appeal and will interfere only where there has been an error in principle or where the sentence is unusually high or low ("outside the range"). In a controversial decision, the Supreme Court of Canada split five to four in *McDonnell*, holding that the Alberta Court of Appeal was wrong to substitute a four year (and one year consecutive) sentence of imprisonment for the one year (and six months concurrent) imposed by the trial judge. McDonnell was found guilty of sexually assaulting his children's 14-year-old babysitter.

Section 691 governs appeals to the Supreme Court of Canada. An accused may appeal as of right on a question of law from a conviction confirmed by the Court of Appeal where there is a dissenting judgment. The accused may appeal from conviction on any questions of law with leave of the Supreme Court of Canada. A person whose acquittal at trial is set aside by a Court of Appeal may appeal by right to the Supreme Court of Canada on a question of law, if there was a dissenting judgment at the Court of Appeal. If the Court of Appeal enters a guilty verdict, the accused has a right of appeal without the need for a dissenting judgment. The accused may appeal any other question of law with leave of the Supreme Court of Canada.

The Crown can appeal as of right to the Supreme Court of Canada on questions of law where there is a dissent at the Court of Appeal, or on other questions of law with leave of the Supreme Court of Canada (section 693).

Box 6.3 How Many Trials Can an Accused Have for the Same Offence?

There are no limits in the *Criminal Code* on the number of trials that an accused can have as a result of appellate courts ordering new trials. However, in a 1997 case, the Supreme Court of Canada agreed with the Manitoba Court of Appeal that a stay of proceedings was warranted because of the "numerous trials and appeals" that former Winnipeg Blue Bomber Brian Jack had been through. He had been convicted three times of killing his wife, and all three convictions were ultimately overturned on appeal. His lawyer, Richard Wolson, was reported to have said that "There have been no fourth trials that I'm aware of in this country" ("Ex-CFLer Goes Free After Three Convictions," *The Globe and Mail,* June 21, 1997, A6). In 2001, the Ontario Court of Appeal decided that ordering a fifth trial for Neville Hunter, who had spent seven months in pre-trial custody and three and a half years in a penitentiary, would be an abuse of process under section 7 of the *Charter*. The Supreme Court of Canada cited this case with approval in *Duguay* (para. 132).

Appeals—Summary Conviction Procedure

Part XXVII (sections 812 through 839) deals with appeals in relation to summary conviction offences, which include hybrid offences where the Crown proceeds by way of summary conviction. Prior to 1977, all summary conviction appeals were by way of **trial *de novo***–that is, there was a new trial, and witnesses were called as if the first trial had not occurred. Now, under 822(4), a trial *de novo* occurs only in exceptional circumstances.

There are two general avenues of appeal for summary conviction offences; one is under section 813, to an "Appeal Court" as defined in section 812 (e.g., the British Columbia Supreme Court, the Superior Court of Justice in Ontario, the Saskatchewan or Alberta Court of Queen's Bench). Under section 813, the defendant or the Crown may appeal from conviction, acquittal, or sentence. The appeal can be on any ground, unless there is legislation prohibiting the appeal.

The second avenue of appeal, under section 830, provides for a more limited appeal of summary conviction proceedings, with the appeal based on the trial transcripts and an **agreed statement of facts** (hence the reference to "Summary Appeal on Transcript or Agreed Statement of Facts"). The section provides for what used to be called an **"appeal by way of stated case."** This appeal is limited to questions of law or questions of jurisdiction. Section 830 describes the procedure, and section 836 prohibits an appeal under section 813 if an appeal has been taken under section 830.

Section 839 provides for further appeals to the Court of Appeal, with leave, on questions of law. Appeals to the Supreme Court of Canada on summary conviction matters are allowed only under limited circumstances, under section 40 of the *Supreme Court Act*, R.S.C. 1985, Chap. S-26.

FRESH EVIDENCE ON APPEAL

Section 683(1)(b) allows the Court of Appeal to admit what is referred to as **fresh evidence**, if it is in the "interests of justice" to do so. If the new evidence is admitted, and if it is conclusive, the Court of Appeal may deal with the matter. If the evidence is not conclusive, a new trial should be ordered.

In *Palmer and Palmer*, the Supreme Court of Canada outlined four principles for admitting fresh evidence at the Court of Appeal:

> (1) The evidence should generally not be admitted if, by due diligence, it could have been adduced at trial provided that this general principle will not be applied as strictly in a criminal case as in civil cases.
> (2) The evidence must be relevant in the sense that it bears upon a decisive or potentially decisive issue in the trial.
> (3) The evidence must be credible in the sense that it is reasonably capable of belief, and
> (4) It must be such that if believed it could reasonably, when taken with the other evidence adduced at trial, be expected to have affected the result (775).

Palmer and Palmer was followed by the Supreme Court of Canada in *Warsing*, which added that the failure to meet the "due diligence" branch of the test is not absolutely determinative of the issue, but should instead be considered in light of all the circumstances. "If the evidence is compelling and the interests of justice require that it be admitted, then the failure to meet the test should yield to permit its admission" (para. 51).

Fresh evidence is generally presented by the accused or defence counsel on an appeal from a conviction. The rules against double jeopardy prohibit the Crown from introducing such evidence when the accused is acquitted. However, the Law Commission of the United Kingdom has recommended that High Court be allowed to order a new trial for more serious offences if new evidence comes to light that makes the prosecutor's case "substantially stronger," if the evidence could not have been found in time with "due diligence," and if it is in the "interests of justice" to have a new trial (*Lawyers Weekly*, October 29, 1999, 2).

THE "FAINT-HOPE" CLAUSE

Sections 745.6 provides a "faint hope" of early release to those offenders who have been sentenced to life imprisonment for a single murder, or for high treason with a minimum period before parole eligibility of at least 15 years, and who have served at least 15 years in prison. Section 745.6(2) specifically exempts offenders who have been convicted of more than one murder. This change was apparently in response to the prospect of such scenes as serial child killer Clifford Olsen appearing in court in front of the families of many of his victims, applying for a reduction in parole eligibility that was almost certain not to be granted (see Hall 1997, A1). In the same way, section 745.61 was added to ensure that any application for review of minimum parole eligibility only occurs after the Chief Justice or his or her designate has determined that there is a reasonable prospect of the application succeeding. Sections 745.62 to 745.64 govern the procedure for these applications.

APPLICATIONS FOR MINISTERIAL REVIEW—MISCARRIAGES OF JUSTICE

The phenomenon of wrongful conviction is an under-researched area in the Canadian criminal justice system. What is known is that it is very difficult and time-consuming process for a person who is wrongfully conviction to seek vindication:

> After 37 years, Steven Truscott finally has the opportunity to prove that he did not rape and kill twelve-year-old Lynne Harper in 1959, when he was fourteen years old. It took David Milgaard and his tenacious supporters 28 years to prove he did not kill Gail Miller in 1969, when he was seventeen years old. This was after he had been in prison for 22 years. Donald Marshall spent eleven years in jail for the 1971 murder of Sandy Seale, a murder he was wrongly convicted of at the age of eighteen. Guy Paul Morin, after eighteen months in jail for the 1984 murder of nine-year-old Christine Jessop, was finally able to clear his name in 1997, through the most advanced DNA technology (Braiden and Brockman 1999, 4; also see Anderson and Anderson 2009; and the Kaufman Report 2004 in Appendix A).

Despite calls for reform, such as from the Commission of Inquiry into the Donald Marshall Jr. prosecution, which recommended the establishment of an independent review body with full investigative powers to investigate claims of wrongful conviction, no such body has been established. In 1998, the federal Department of Justice issued a discussion paper on reforming this process, and in 2002 it amended the *Criminal Code* to add Part XX1.1 (Applications for Ministerial Review—Miscarriages of Justice). Applications made to the Minister of Justice under section 696.1 are reviewed by the Minister or the Minister's delegate under section 696.2. Section 696.3 allows the Minister to refer the matter to the Court of Appeal for an opinion, direct a new trial, or refer the matter to the Court of Appeal as if it were an appeal. A decision by the Minister is not subject to appeal (section 696.3(4)). Section 696.4 states that the Minister shall take into account all matters the Minister considers relevant, including any new matters not considered by previous courts, "the relevance and reliability" of the information provided by the applicant, and "the fact that an application under this Part is not intended to serve as a further appeal and any remedy available on such an application is an extraordinary remedy."

Box 6.4 Minister of Justice's Response to Applications for Ministerial Review

According to the 2009 Annual Report, since 2003 the Minister of Justice has referred five cases to the courts for a new trial, and seven cases to the courts of appeal. In eight of the cases, the charges were stayed or acquittals were entered, and three cases were still before the courts (Minister of Justice, 2009). The 2008 Annual Report provided more details of the eleven cases that were referred to the courts: "six dealt with disclosure issues (Truscott, Wood, Driskell, Walsh, Tremblay, Phillion); three involved faulty or poor science, (Truscott, Driskell, Mullins-Johnson); and the remaining cases (Cain, Bjorge, Kaminski and the Alberta case) dealt with other evidentiary issues including recantations, new evidence concerning possible witness tampering, and new evidence concerning witness credibility"(Minister of Justice, 2008).

Plaxton (2002) raises the question: "Are wrongful convictions wrong?" He describes two approaches to this question. One is that it is better that 10 or 20 guilty people escape punishment than for one innocent person to be convicted. Wrongful convictions might happen, but they should not happen, and we should install safeguards to ensure that they do not. The other perspective is that wrongful convictions are the price of a criminal justice system. Just as soldiers get killed in war, some innocent persons are convicted in our criminal justice system. We expect some wrongful convictions. Plaxton advocates the first approach. He argues that the latter approach may actually encourage wrongful convictions. Politically unpopular people may be arrested and charged "in the hope that an admittedly malfunctioning criminal justice system will convict them" (426-27).

In 2005, the Federal-Provincial-Territorial Heads of Prosecutions Committee Working Group released the *Report on the Prevention of Miscarriages of Justice* (2004), suggesting that wrongful convictions can be reduced by addressing some of their causes: tunnel vision, mistaken eyewitness identification and testimony, false confessions, in-custody informers, and faulty forensic procedures. The topic of wrongful convictions was canvassed in a Special Issue of the *Canadian Journal of Criminology and Criminal Justice* (see Campbell and Denov 2004; Grounds 2004; Hickman 2004; Huff 2004; Kennedy 2004 Scullion 2004; and Weisman 2004) and a recent issue of the *Criminal Law Quarterly* (see Sherrin 2007; Baxter 2007; Santoro 2007; Roach 2007 and Furgiuele 2007).

Box 6.5 Section 696.1 The Truscott Referrals

In 1959, at the age of 14, Steven Truscott was sentenced to be "hanged by the neck until you are dead" for the murder of 12-year old Lynne Harper in Goderich, Ontario, but the federal cabinet later commuted the sentence to life in prison (see Sher's book in Appendix C). In 1966, the Supreme Court of Canada, in a 6-1 ruling, refused to order a new trial in the case. Truscott was paroled in 1969, married and raised three children. In 2000, he decided to take his case public through a documentary on the Fifth Estate, and through AIDWYC (Association in Defence of the Wrongfully Convicted) which filed an application for retrial under section 690 (now section 696.1). The Minister of Justice asked the Honourable Fred Kaufman to review the case and his report was released to the public in 2005 (see Appendix A). The report led the federal Minister of Justice to conclude that there was likely a "miscarriage of justice" in the conviction. The Minister referred the case to the Ontario Court of Appeal which, in August 2007, unanimously overturned Truscott's conviction as a miscarriage of justice, and entered an acquittal (Minister of Justice 2008).

SUMMARY

Sentencing follows conviction. The accused must be given a chance to address the court before sentencing. The court may also receive the input of probation officers in a pre-sentence report and is required to hear victims, through their impact statements. The accused's criminal record will also be considered, as will any time the accused has spent in custody awaiting trial.

The court may discharge an accused, either absolutely or on conditions, for all but the most serious offences, and those with a specified minimum penalty. A discharge is available where it would be in the best interests of the accused and not contrary to the public interest. Although a discharge is not a conviction in the strict sense, discharges may be alleged as part of a criminal record in future criminal proceedings. Further, should a person breach the conditions of a discharge, the discharge may be revoked and the person sentenced to any penalty that was available at the time of the original discharge.

The court may suspend the passing of sentence and place a person on probation, following conviction, unless there is a minimum penalty provided by law. Breaching the conditions of the probation order is not only a separate criminal offence, but one consequence of such a breach is that the person may be brought back to court and sentenced for the original offence.

Conditional sentences of imprisonment allow offenders to serve their time outside prison, subject to conditions imposed by the *Criminal Code* and by the sentencing judge. Restitution may be ordered paid by the offender, to the victim of his or her crime. A victim fine surcharge is automatically applied to a fine, which goes to fund provincial programs to assist the victims of crime. In some cases, prohibition orders may be made, prohibiting the offender from possessing weapons, driving motor vehicles, and so on.

Offenders who are convicted of serious personal injury offences may be declared to be dangerous offenders and sentenced to an indeterminant period of incarceration. Their status will be reviewed periodically, pursuant to legislation.

The proceeds of crime may be confiscated. At present, there is no federal law in Canada preventing criminals from profiting from their crimes after the fact, through the sale of their stories, although a number of provinces have implemented such legislation.

Rights of appeal vary, depending on whether the offence involved was indictable or summary conviction, whether the appeal is from conviction or sentence, and whether the appellant is the Crown or the defence. Rights of appeal are often limited in respect of whether the appeal involves questions of law alone, questions of fact, or questions of mixed fact and law.

The appellate court may consider fresh evidence on appeal where it is in the interests of justice to do so. Generally, such evidence will only be permitted where it was unavailable at trial, where it is relevant, credible, and where it could, if believed, have affected the result of the trial.

Those who believe they are wrongfully convicted may apply to the Minister of Justice for a ministerial review.

QUESTIONS TO CONSIDER

(1) What is the fundamental purpose of sentencing, and what objectives are sanctions suppose to have?

(2) What is the fundamental principle of sentencing and what other principles is it suppose to achieve?

(3) Section 718.21 of the *Code* sets out additional factors for the court to consider when imposing a penalty on an organization. How might these factors deter or not deter corporate crime?

(4) In 2004, section 380.1 was added to deal more harshly with securities offences and other types of fraud. How might this section deter or not deter such crimes?

(5) The *Criminal Code* states that victim impact statements shall be in writing, and victims shall be allowed to read their statements to the court. The judge may also consider any other

evidence concerning the victim. Should victims, at a sentencing hearing, be allowed to give oral evidence that is not in their written victim impact statement? Why or Why not?

(6)   The *Criminal Code* defines a victim, for the purposes of victim impact statements, as "a person to whom harm was done or who suffered physical or emotional loss was a result of the commission of the offence." Should those who are indirectly affected by the crime be allowed to file a victim impact statement? Why or why not?

(7)   Can a hungry 20 year old, with no previous record, who is convicted of breaking into a house with the intention of stealing some food (section 348(d)), be given an absolute discharge?

(8)   How might the optional conditions of a probation order in section 732.1(3.1) prevent future corporate crime?

(9)   Should convicted persons be allowed to sell their stories for profit? What are the alternatives, and what would you recommend as a resolution of the debate?

(10)   On what grounds can a person appeal a conviction for breaking and entering with intent under section 348(1)(d)?

(11)   On what basis will fresh evidence be admitted upon appeal?

(12)   Can the Court of Appeal enter a conviction on appeal, where an accused was charged with murder and acquitted by a jury? What if the person is acquitted by a judge?

BIBLIOGRAPHY

Anderson, B. and D. Anderson. *Manufacturing Guilt: Wrongful Convictions in Canada*. Halifax, Fernwood Publishing, 2009, 2nd.

Baxter, Angela. "Identification Evidence in Canada: Problems and Potential Solutions." (2007) 52 *Criminal Law Quarterly* 175.

Braiden, Patricia, and Joan Brockman. "Remedying Wrongful Convictions Through Applications to the Minister of Justice Under Section 690 of the *Criminal Code*." (1999) 17 *Windsor Yearbook of Access to Justice* 3.

British Columbia. Ministry of Public Safety and Solicitor General; www.pssg.gov.bc.ca/criminal-records-review/; accessed September 26, 2009.

Campbell, Kathryn M. and Myriam S. Denov. "The Burden of Innocence: Coping with a Wrongful Imprisonment." (2004) 46(2) *Canadian Journal of Criminology and Criminal Justice* 139.

Canadian Sentencing Commission. *Sentencing Reform: A Canadian Approach*. Ottawa: Minister of Supply and Services, 1987.

Carter, Mark. "Addressing Discrimination Through the Sentencing Process: Criminal Code s.718.2(a)(i) in Historical and Theoretical Context." (2001) 44 *Criminal Law Quarterly* 399.

Calarco, Paul. "*R. v. Ferguson*: An Opportunity for the Defence." (2008) 54 *Criminal Reports* (6th) 223.

Clewley, Gary R., and Paul G. McDermott. *Sentencing: The Practitioner's Guide*. Ontario: Canada Law Book Inc., available on Criminal Spectrum.

Cole, David P. and Glenn Angus. "Using Pre-Sentence Reports to Evaluate and Respond to Risk." (2003) 47 *Criminal Law Quarterly* 302.

Coughlan, Steve. "The End of Constitutional Exemptions." (2008) 54 *Criminal Reports* (6th) 220.

Crutcher, Nicole. "Mandatory Minimum Penalties of Imprisonment: An Historical Analysis." (2001) 44 *Criminal Law Quarterly* 279.

Davies, Heather. "Sex Offender Registries: Effective Crime Prevention Tools or Misguided Responses?" (2004) 17 *Criminal Reports* (6th) 156.

Department of Justice. *Addressing Miscarriages of Justice: Reform Possibilities for Section 690 of the Criminal Code: A Consultation Paper*. Ottawa: Minister of Public Works and Government Services Canada, 1998. Available on the Department of Justice Web site <canada.justice.gc.ca>.

Department of Justice. "Minister of Justice Refers Murder Case to Alberta Court of Appeal." (15 February 2005a); www.justice.gc.ca/en/news/nr/2005/doc_31390.html; accessed September 19, 2005.

Department of Justice. " Minister of Justice Finds Miscarriage of Justice: Refers Murder Case to Quebec Court of Appeal." (12 July 2005b); www.justice.gc.ca/en/news/nr/2005/doc_31574.html; accessed September 19, 2005.

Federal-Provincial-Territorial Heads of Prosecutions Committee Working Group. *Report on the Prevention of Miscarriages of Justice* (September 2004); www.justice.gc.ca/en/dept/pub/hop/toc.html; accessed September 19, 2005.

Furgiuele, Andrew. "The Self-Limiting Appellate Courts and Section 686." (2007) 52 *Criminal Law Quarterly* 237.

Ferris, T.W. *Sentencing—Practical Approaches*. Toronto: LexisNexis Canada. 2005.

Gaucher, Robert and Liz Elliott. "'Sister of Sam': The Rise and Fall of Bill C-205 /220." (2001) 19 *Windsor Yearbook of Access to Justice* 72.

Gemmell, Jack. "The New Conditional Sentencing Regime." (1997) 39 *Criminal Law Quarterly* 334.

German, Peter M. *Proceeds of Crime and Money Laundering* . Toronto: Carswell, 1998 (with updates).

Grounds, Adrian. "Psychological Consequences of Wrongful Conviction and Imprisonment." (2004) 46(2) *Canadian Journal of Criminology and Criminal Justice* 165.

Hall, Neal. "Victims' Families Brace themselves to face Olson: The Disembodied Voice of the Serial Killer will be heard in the Vancouver Law Courts Tuesday." (10 March 1997) *Vancouver Sun* A1.

Healy, Patrick. "Questions and Answers on Conditional Sentencing in the Supreme Court of Canada." (1999) 42 *Criminal Law Quarterly* 3.

Hickman, Alexander. "Wrongful Convictions and Commissions of Inquiry." (2004) 46(2) *Canadian Journal of Criminology and Criminal Justice* 183.

Hill, S. Casey, David M. Tanovich and Louis P. Strezos. *McWilliams' Canadian Criminal Evidence*. Aurora, ON: Canada Law Book Limited (available online on Criminal Spectrum).
    Chapters 34 and 35 talk about evidence at sentencing and appellate proceedings, respectively.

Huff, C. Ronald. "Wrongful Convictions: The American Experience." (2004) 46(2) *Canadian Journal of Criminology and Criminal Justice* 107.

Kennedy, Jerome. "Righting the Wrongs: The Role of Defence Counsel in Wrongful Convictions." (2004) 46(2) *Canadian Journal of Criminology and Criminal Justice* 197.

Laine, Yeshe. "The Interplay between *Christopher's Law* and the *Sex Offender Information Registration Act*." (2007) 52 *Criminal Law Quarterly* 470.

Malecki, Melissa J. "Son of Sam: Has North Carolina Remedied the Past Problems of Criminal Anti-Profit Legislation?" (2006) 89 *Marquette Law Review* 673.

Manson, Allan. "Finding a Place for Conditional Sentences." (1997a) 3 *Criminal Reports* (5th) 283.

Manson, Allan. "The Appeal of Conditional Sentences of Imprisonment." (1997b) 5 *Criminal Reports* (5th) 279.

Manson, Allan. "*McDonnell* and the Methodology of Sentencing." (1997c) 6 *Criminal Reports* (5th) 277.

Manson, Allan. "A Brief Reply to Professors Roberts and von Hirsch." (1998a) 10 *Criminal Reports* (5th) 232.

Manson, Allan. "Conditional Sentences: Courts of Appeal Debate the Principles." (1998b) 15 *Criminal Reports* (5th) 176.

Manson, Allan. "The Conditional Sentence: A Canadian Approach to Sentencing Reform, Or Doing the Time Warp Again." (2001a) 44 *Criminal Law Quarterly* 375.

Manson, Allan. *The Law of Sentencing.* Toronto: Irwin Law, 2001b.

Manson, Allan. "Pre-Sentence Custody and the Determination of a Sentence (Or How to Make a Mole Hill out of a Mountain)" (2004) 49 *Criminal Law Quarterly* 292.

Mazey, Edward. "Conditional Sentence Under House Arrest." (2002) 46 *Criminal Law Quarterly* 246.

Minister of Justice. *Applications for Ministerial Review–Miscarriages of Justice*, Annual Report, 2008 (Ottawa, 2008).

Minister of Justice. *Applications for Ministerial Review–Miscarriages of Justice*, Annual Report, 2009 (Ottawa, 2009).

North, Dawn. "The 'Catch 22' of Conditional Sentencing." (2001) 44 *Criminal Law Quarterly* 342.

Okuda, Sue S. "Criminal Antiprofit Laws: Some Thoughts in Favor of Their Constitutionality." (1988) 76 California Law Review 1353.

Pfefferle, Brian R. "Gladue Sentencing: Uneasy Answers to the Hard Problem of Aboriginal Over-Incarceration." (2008) 32 *Manitoba Law Journal* 113.

Plaxton, Michael. "Are Wrongful Convictions Wrong? The Reasonable Doubt Standard and the Role of Innocence in Criminal Procedure." (2002) 46 *Criminal Law Quarterly* 407.

Quigley, Tim. *Procedure in Canadian Criminal Law.* Toronto: Carswell, 2006.

Quigley, Tim. "Are We doing anything about the Disproportionate Jailing of Aboriginal People?" (1999) 42 *Criminal Law Quarterly* 129.

Quigley, Tim. 2009. "Pessimistic Reflections on Aboriginal Sentencing in Canada." (2009) 64 *Criminal Reports* (6th) 135.

Roach, Kent. *Due Process and Victims' Rights: The New Law and Politics of Criminal Justice.* Toronto: University of Toronto Press, 1999.

Roach, K. "Conditional Sentencing and Widening the Net." (2000) 43 *Criminal Law Quarterly* 273.

Roach, Kent. 2008. "Editorial: Rates of Imprisonment and Criminal Justice Policy." (2008) 53 *Criminal Law Quarterly* 273.

Roach, Kent. "Unreliable Evidence and Wrongful Convictions: The Case for Excluding Tainted Identification Evidence and Jailhouse and Coerced Confessions." (2007) 52 *Criminal Law Quarterly* 210.

Roberts, Julian V. "The Hunt for the Paper Tiger: Conditional Sentencing after Brady." (1999) 42 *Criminal Law Quarterly* 38.

Roberts, Julian V. "The Evolution of Conditional Sentencing: An Empirical Analysis." (2002) 3 *Criminal Reports* (6th) 267.

Roberts, Julian V. "Victim Impact Statements and the Sentencing Process: Recent Developments and Research Findings." (2003) 47 *Criminal Law Quarterly* 365.

Roberts, Julian V., and David P. Cole (eds.). *Making Sense of Sentencing.* Toronto: University of Toronto Press, 1999.

Roberts, Julian V. and T. Gabor. "The Impact of Conditional Sentencing: Decarceration *and* Widening of the Net." (2003) 8 *Canadian Criminal Law Review* 33.

Roberts, Julian and Patrick Healy (2001). "The Future of Conditional Sentencing." (2001) 44 *Criminal Law Quarterly* 309.

Roberts, Julian V., and Carole LaPrairie. "Sentencing Circles: Some Unanswered Questions." (1997) 39 *Criminal Law Quarterly* 69.

Roberts, Julian V., and Andrew von Hirsch. "Conditional Sentences of Imprisonment and the Fundamental Principle of Proportionality in Sentencing." (1998) 10 *Criminal Reports* (5th) 222.

Roberts, Tim. *Assessment of the Victim Impact Statement Program in British Columbia.* WD1992-5e. Ottawa: Department of Justice, February 1992.

Ruby, Clayton C. *Sentencing*, 6th ed. Toronto and Vancouver: Butterworths Canada Ltd., 2004.

Salhany, Roger E. *Canadian Criminal Procedure,* 6th ed. Toronto: Canada Law Book Inc., (available online on Criminal Spectrum).
    Chapter 8 examines sentencing, sentencing principles, and the various sentencing options. Chapter 9 covers appeal procedures for both indictable and summary conviction offences. Chapter 10 talks about extraordinary remedies.

Santoro, Daniel C. "A Legal Argument for the Mandatory Videotaping of Photo Line-up Identification Interviews." (2007) 52 *Criminal Law Quarterly* 190.

Scullion Kerry. "Wrongful Convictions and the Criminal Conviction Review Process Pursuant to s.696.1 of the Canadian *Criminal Code*." (2004) 46(2) *Canadian Journal of Criminology and Criminal Justice* 189.

Sherrin, Christopher. 2007. "Comments on the Report on the Prevention of Miscarriages of Justice." (2007) 52 *Criminal Law Quarterly* 140.

Stuart, Don, Ron Delisle, and Tim Quigley. *Learning Canadian Criminal Procedure*, 9th Ed. Toronto: Carswell, 2008.
    Chapter 14 contains information on appeals, and Chapter 15 on Ministerial Review and Claims of Injustice.

Turpel-Lafond, M.E. "Sentencing Within a Restorative Justice Paradigm: Procedural Implications of *R.v. Gladue*." (1999) 43 *Criminal Law Quarterly* 34.

Weisman, Richard. "Showing Remorse: Reflections on the Gap between Expression and Attribution in Cases of Wrongful Conviction." (2004) 46(2) *Canadian Journal of Criminology and Criminal Justice* 121.

Yager, Jessica. (2004). "Investigating New York's Son of Sam Law: Problems with the Recent Extension of Tort Liability fo People Convicted of Crimes." (2004) 48 *New York Law School Law Review* 433.

CHAPTER 6: *Sentencing and Appeals*

CASES

R. v. Biller, [2005a] BCSC 1066. Findings on degree of culpability, after a guilty plea to fraud and theft.

R. v. Biller, [2005b] BCSC 1278. Sentencing.

R. v. Biniaris, [2000] 1 S.C.R. 381.

R. v. Branco (1994), 87 C.C.C. (3d) 71 (B.C.C.A.). application for leave to the Supreme Court of Canada dismissed without reasons; 90 C.C.C. (3d) vi. Section 11(e) of the *Charter* does not apply to bail pending appeal, since the presumption of innocence is extinguished on conviction.

R. v. Brown, [1991] 2 S.C.R. 518. A judge may only sentence an offender in relation to the offence on which a conviction is entered. If a jury convicts of a lesser included offence, the court cannot consider in sentence as aggravating circumstances those facts implicitly rejected by the verdict.

R. v. Buhay, [2003] 1 S.C.R. 631.

R. v. Burke, [1996] 1 S.C.R. 474. An appellate court should not overturn a conviction under section 686(1)(a)(i) unless it has considered, on all of the evidence before the trier of fact, that no jury, properly instructed and acting in a judicial manner, could have reasonably convicted on that evidence.

Chatterjee v. Ontario (Attorney General), [2009] 1 S.C.R. 624.

R. v. Chiu (1984), 31 Man. R. (2d) 15 (Man. C.A.). The possible consequences under the *Immigration Act* to the accused resulting from a criminal conviction can properly be considered as one factor in deciding whether to grant a discharge.

R. v. Craig, 2009 SCC 23. The forfeiture provisions of the *Controlled Drugs and Substances Act* are to be considered independently of the sentencing provisions of the *Criminal Code*.

R. v. Dennison (1990), 60 C.C.C. (3d) 342 (N.B.C.A.).

R. v. Drew (1978), 45 C.C.C. (2d) 212 (B.C.C.A.). Prior assignment to and completion of a diversion program is irrelevant to sentence or discharge for a new offence.

R. v. Duguay, [2003] 3 S.C.R. 307.

R. v. Elendiuk (1986), 27 C.C.C. (3d) 94 (Alta. C.A.). A conditional discharge is not final punishment; the right to sentence the accused for the offence should the conditions of the discharge be breached, including by the commission of a further offence, is reserved.

R. v. Fallofield (1973), 13 C.C.C. (2d) 450 (B.C.C.A.). In granting a discharge, it must be in the accused's best interests, which requires the accused generally to be of good character and without previous conviction, and to show that a conviction will possibly have significant negative consequences. Also, it must not be contrary to the public interest, which involves a consideration of general deterrence, but not to the extent that it automatically precludes a discharge.

R. v. Farinacci (1994), 86 C.C.C. (3d) 32 (Ont. C.A.). Section 11(e) of the *Charter* does not apply in respect of bail pending appeal. The term "public interest" in section 679(3) of the *Code* is not unconstitutionally vague.

R. v. Ferguson, [2008] 1 S.C.R. 96.

R. v. Gardiner, [1982] 2 S.C.R. 368. On sentence, disputed facts beyond those implicit in a guilty plea must be proved by the Crown beyond a reasonable doubt, although the strict rules of evidence, such as the rule against hearsay, do not apply.

R. v. Gladue, [1999] 1 S.C.R. 688. The Court discusses the purpose of section 718.2(e), which states that when sentencing an aboriginal, the court shall take into consideration all reasonable sanctions other than imprisonment, "with particular attention to the circumstances of aboriginal offenders."

R. v. Hunter (2001), 155 C.C.C. (3d) 225 (Ont. C.A.).

R. v. Kerr (1982), 43 A.R. 254 (Alta. C.A.). Potential immigration consequences for the accused resulting from conviction are only one factor to be considered in deciding whether to grant a discharge and are not determinative.

R. v. Knoblauch, [2000] 2 S.C.R. 780. A conditional sentence may include a requirement that the accused attend at a residential treatment centre.

R. v. L.F.W., [2000] 1 S.C.R. 149.

R. v. Ly, [1993] O.J. No. 3661 (Ontario Court of Appeal). The trial judge erred in imposing a custodial sentence, after Crown and defence provided a joint submission that a custodial sentence was not appropriate.

R. v. Lyons, [1987] 2 S.C.R. 309. The provisions for finding an accused to be a dangerous offender, and for sentencing such an offender to an indeterminate period of incarceration, are not unconstitutional.

R. v. McDonnell, [1997] 1 S.C.R. 948. The Court of Appeal should generally defer to sentences imposed by trial judges and should not lightly interfere by substituting its own views unless the sentencing judge committed an error of principle.

R. v. Morales, [1992] 3 S.C.R. 711.

R. v. Ouellette, [2009] SCC 24. Partial forfeiture of real estate property can be ordered under the *Controlled Drugs and Substances Act*.

Palmer and Palmer v. The Queen, [1980] 1 S.C.R. 759. Criteria to admit fresh evidence on appeal.

R. v. Proulx, [2000] 1 S.C.R. 61.

R. v. R.A.R., [2000] 1 S.C.R. 163.

R. v. R.N.S., [2000] 1 S.C.R. 149.

R. v. Rudyk (1975), 1 C.R. (3d) S-26 (N.S.C.A.). Pre-sentence reports should be confined to such information as the accused's background, and should not include self-serving statements by the accused about the offence. Such statements, if incorrectly included, should be ignored by the sentencing judge.

R. v. Sarson, [1996] 2 S.C.R. 223.

R. v. Shoker, [2006] 2 S.C.R. 399. A probation order cannot authorize a search and seizure of bodily substances.

R. v. Skalbania, [1997] 3 S.C.R. 995. The mental element necessary to establish guilt for misappropriation of money is a question of law, not of fact.

Canada (Director of Investigation and Research) v. Southam Inc., [1997] 1 S.C.R. 748.

R. v. Tallman (1980), 48 C.C.C. (3d) 81 (Alta. C.A.). Time spent in custody awaiting trial should be credited to an accused on sentencing following conviction. While no exact formula exists for doing so, generally the time credited should be more than the "dead time" spent awaiting trial.

CHAPTER 6: *Sentencing and Appeals*

R. v. Tan (1974), 22 C.C.C. (2d) 184 (B.C.C.A.). A previous discharge may be proven by the Crown, not as a conviction aggravating sentence, but merely as suggesting that a further discharge in the new case may be inappropriate.

R. v. Taylor (1997), 122 C.C.C. (3d) 376 (Sask. C.A.). The Court discusses sentencing circles, banishment, and principles of sentencing for aboriginal offenders. The accused was convicted of sexual assault.

R. v. Urbanovick (1985), 19 C.C.C. (3d) 43 (Man. C.A.). Pre-sentence reports present information and recommendations for the consideration of the sentencing judge, but are not evidence, and should not be used by the sentencing judge to justify reversing findings of fact made during the trial.

R. v. Warsing, [1998] 3 S.C.R. 579. The Court reaffirmed *Palmer and Palmer*, with respect to the four principles for the admission of fresh evidence on appeal. The court stated, however, that a failure to strictly meet the "due diligence" requirement is not fatal to an application to adduce fresh evidence, but rather that the failure should be considered in the context of all the circumstances, and the requirement should be relaxed if it is in the interests of justice to do so. Although it is generally impermissible to raise a new defence for the first time on appeal, on the facts of this case the accused was allowed to raise the NCRMD (not criminally responsible on account of mental disorder) defence at his appeal, and was granted a new trial where that issue could be determined.

R. v. W.(R.), [1992] 2 S.C.R. 122. Whether a jury verdict was unreasonable is a question of law, even though it requires the appellate court, to some extent, to consider and weigh the evidence. The appellate court may overturn verdicts, including those based on findings of credibility, in circumstances where it concludes that on all the evidence, the verdict was unreasonable.

R. v. Wiles, [2005] 3 S.C.R. 895.

R. v. Yebes, [1987] 2 S.C.R. 168. A determination of whether the verdict of the jury was unreasonable is a question of law on which an appeal can be taken in respect of a dissent in the Court of Appeal.

R. v. Zelensky et al. (1978), 41 C.C.C. (2d) 97 (S.C.C.). Compensation orders as part of a sentence are *intra vires* Parliament under section 91(27), the federal power over criminal law and procedure.

PART II

GATHERING EVIDENCE AND ITS ADMISSIBILITY

CHAPTER 7: *Evidence That Is Illegally or Improperly Obtained*

CHAPTER OBJECTIVES

In studying this chapter, you should develop an understanding of the following topics and concepts:

- the common law power to exclude evidence before and after the *Charter*
- the development of the exclusion of evidence as a remedy for violations of *Charter* rights
- the difference between remedies under sections 24(1) and 24(2) of the *Charter*
- the criteria for remedies and the exclusion of evidence under section 24(1)
- the criteria for the exclusion of evidence under section 24(2)

THE COMMON LAW: ILLEGALLY OBTAINED EVIDENCE

At common law, real (or physical) evidence was not excluded at trial when an accused's rights had been violated in obtaining such evidence. In one old English case the judge said, "It matters not how you get it; if you steal it even, it would be admissible as evidence" (*Leatham* 1861, quoted in Mirfield 1987–8, 434). The rationale for such a rule was that the trier of fact should have access to all relevant information prior to reaching a decision. In addition, the value or veracity of unlawfully obtained physical evidence was not reduced by the fact that it was illegally obtained (compared with, for example, confessions, discussed in Chapter 10). After reviewing the law in Canada and other countries, the Ouimet Committee concluded that it was uncertain whether a trial judge even had the discretion to exclude real evidence that was illegally obtained. The Committee provided an example where a blood sample was obtained by unlawful force but was still admissible (1969, 71–2).

The common law position on the admissibility of illegally obtained real evidence was endorsed by the majority of the Supreme Court of Canada (six to three) in 1970, in the *Wray* case. The Court recognized a rare exception to the general admissibility of illegally obtained evidence. It held that it was only evidence that was "gravely prejudicial to the accused, the admissibility of which is tenuous, and whose probative force in relation to the main issue before the Court is trifling, which can be said to operate unfairly" (17), and that could therefore be excluded. In effect, a trial judge had essentially no discretion to exclude evidence on the basis that it was illegally obtained, or because its admission would bring the administration of justice into disrepute. Discretion to exclude evidence was restricted to cases where it would be unfair to admit it, in the very restricted sense in *Wray*. The courts were accordingly very reluctant to exclude evidence under these limited conditions.

The Fruit of the Poisonous Tree in the United States

The Supreme Court of the United States developed an **exclusionary rule** for all illegally obtained evidence (sometimes referred to as the **fruit of the poisonous tree** doctrine). That Court has since created exceptions to the general rule of mandatory exclusion of improperly obtained evidence, allowing for more flexibility in the admission of evidence. In 1984, the U.S. Supreme Court developed an exception if police were acting in good faith when they violated the accused's rights: "evidence obtained by the police acting in good faith on a search warrant that was issued by a neutral and detached magistrate, but that is ultimately found to be invalid, may be admitted and used at trial"

(*Leon* and *Sheppard*, quoted in del Carmen 1995, 65–6). In *Krull* (1987), the good faith exception was extended to include situations where the police relied on a statute that was later found to be unconstitutional (del Carmen 1995, 68). In *Nix v. Williams,* the Court developed the **inevitable discovery** exception, whereby "evidence is admissible if the police can prove that they would inevitably have discovered the evidence anyway by lawful means, regardless of their illegal action" (del Carmen 1995, 69). Also see Hails (2005 Chapter 10).

The Canadian *Bill of Rights*

The *Bill of Rights,* which came into force in Canada in 1960, did not contain any enforcement provisions. This was not that unusual, in that many countries had (and still have) rights and freedoms in constitutional documents without enforcement provisions; in their absence, the courts develop their own means of enforcing rights. The Supreme Court of Canada, by comparison to the Supreme Court of the United States, was reluctant to develop enforcement procedures under the *Bill of Rights*, which was not a constitutional document, but rather was ordinary legislation that had achieved quasi-constitutional status (Laskin, J. in *Hogan*). The Court took a major step in declaring legislation to be of no force and effect under the *Bill of Rights* in the *Drybones* case; however, that was exceptional. The Court was not prepared to use the *Bill of Rights* to exclude evidence at trial, even if it was obtained in violation of the accused's rights. For example, in *Hogan,* the accused was taken to a police station for a breathalyzer test and was denied his right to counsel contrary to the *Bill of Rights*, even though his lawyer was available for consultation. Faced with choosing between providing a sample of his breath or being charged with refusing to so, Hogan obliged the police officers. Mr. Justice Ritchie, for the majority of the Supreme Court of Canada, rejected the U.S. model of excluding such evidence, because to do so would be a violation of an established rule in Canada. Mr. Justice Laskin, in dissent, believed that the violation of an accused's rights under the *Bill of Rights* called for the exclusion of the evidence if the courts were to take the violation seriously.

Suggestions for Law Reform Prior to the *Charter*

In 1969, the Ouimet Report recommended that legislation be enacted to give a trial judge the discretion to reject illegally obtained evidence. In exercising this discretion, the Committee suggested the courts should take into account whether the violation was inadvertent, the urgency of the situation, and whether the admission of the evidence would be unfair to the accused (74-5). The Committee thought that such a rule would assist in deterring the police from abusing their powers, although "the problem of unlawful arrests and illegal searches has not been as acute in Canada as in the United States" (73). In addition, "deliberate violations of the rights of the suspect may reduce respect for the entire criminal process and diminish the likelihood of the offender's rehabilitation" (74).

The Law Reform Commission of Canada, in its proposed evidence code, suggested that evidence be excluded "if it was obtained under such circumstances that its use in the proceedings would tend to bring the administration of justice into disrepute" (1975, 22). The McDonald Royal Commission (1981), which investigated activities of the Royal Canadian Mounted Police, also recommended that judges be given the discretion to exclude evidence that was illegally obtained.

One major dissent from the recommendation that judges be allowed to exclude illegally obtained evidence was voiced by the Federal/Provincial Task Force on Uniform Rules of Evidence (Uniform Law Conference of Canada 1982). The Task Force took the view that police could be more effectively disciplined directly, and that an accused should not be allowed to benefit by the exclusion of evidence (231). It also suggested that it would be very difficult to determine what evidence should be excluded, and that such rules would introduce greater uncertainty in the criminal justice arena. The Task Force was, however, prepared to allow the court to exclude evidence in circumstances similar to those discussed by the Supreme Court of Canada in *Wray*. Section 22 of their proposed *Uniform Evidence Act* stated that "the court may exclude evidence the admissibility of which is tenuous, the probative force of which is trifling in relation to the main issue and the admission of which would be gravely prejudicial to a party" (549).

SECTION 24 OF THE *CHARTER*

In the Introduction, we examined the constitutional framework of the law of criminal procedure and evidence, and the impact of the *Canadian Charter of Rights and Freedoms* on the criminal justice system. The *Constitution* of Canada, of which the *Charter* is a part, is the supreme law of Canada, as is stated in section 52 of the *Constitution Act*. The rights and freedoms guaranteed under the *Charter* are subject only to such "reasonable limits prescribed by law as can be demonstrably justified in a free and democratic society" (section 1). If a law violates our rights under the *Charter*, the government will have to defend it before the courts, establishing that the infringement is demonstrably justified in a free and democratic society. Examples of this process were discussed in detail in Part I of this text.

There may be circumstances where an accused's rights are violated by an agent of the state, and the violation is not condoned or prescribed by law, such that section 1 is not applicable. In such cases, the accused may seek a remedy under section 24 of the *Charter*, which has had a major impact on the gathering and admissibility of evidence.

Historical Development of Section 24

The initial drafts of the *Charter* in 1978–79 did not contain a specific section that dealt with the exclusion of illegally obtained evidence. Rather, the original proposals provided that any individual could apply to a court for a remedy to enforce his or her rights under the *Charter*. Following a redrafting of the provisions in 1980, some of the provinces became concerned that the courts might develop an American-style exclusionary rule, and so a section was added that seemed to adopt the Supreme Court of Canada's decision in *Wray*. Strong opposition to this approach from civil liberties advocates and the Canadian Bar Association resulted in the subsequent deletion of this draft section. Later, a forerunner of the present section 24(2) was added, after further submissions from the Progressive Conservative and New Democratic parties (McLellan and Elman 1983, 206–8).

Section 24(1) Remedies

Section 24(1) of the *Charter* now reads:

> anyone whose rights or freedoms, as guaranteed by this Charter, have been infringed or denied may apply to a court of competent jurisdiction to obtain such remedy as the court considers appropriate and just in the circumstances.

Who Qualifies for Remedies, and Under What Circumstances?

The word "anyone" is not defined in the *Charter*. It was likely intended to include corporate entities (McLellan and Elman 1983, 208–9), and the courts have allowed corporations to apply for remedies under the *Charter* (*Big M Drug Mart* and *Wholesale Travel Group Inc.*). The section is limited, however, to "anyone whose rights or freedoms…have been infringed or denied." This limitation was confirmed by the Supreme Court of Canada in *Edwards*, where it found that the police had not violated the accused's rights under section 8 of the *Charter* when they entered his girlfriend's apartment and found the drugs he kept there. The accused had no remedy under the *Charter* because *his* rights were not violated (discussed further in Chapter 8).

What is a Court of Competent Jurisdiction?

A preliminary inquiry court is not a court of competent jurisdiction under section 24, and therefore cannot provide remedies or exclude evidence under the *Charter* (see Chapter 4). Pottow (2000) suggests that the interpretation of "court of competent jurisdiction" has been one of the most contentious issues in section 24. Does a provincial judge have to have statutory authority over a remedy in order for it to grant the remedy for a *Charter* violation? For example, a provincial court judge does not have authority to award costs against the Crown. Applying this limitation to *Charter* remedies is a restrictive approach to remedies. Pottow suggests that a flexible approach would allow the provincial court judge, who has jurisdiction over the accused and the Crown, to grant remedies such as costs under section 24 (2000, 463-67). This issue has yet to be resolved by the Supreme Court of Canada.

What Remedies Can the Court Order?

Under section 24(1) of the *Charter,* the judge will decide what remedy is "appropriate and just in the circumstances." The courts have provided a number of different types of remedies; however, the court cannot exclude evidence under section 24(1) if section 24(2) applies—that is, if evidence was obtained in a manner that violated the accused's rights under the *Charter* (*Therens*). In *White*, the Supreme Court of Canada approved the use of section 24(1) to exclude evidence where the evidence was not obtained in a manner that violated the accused's rights, but where the admission of the evidence would nevertheless violate the accused's rights. White, who was involved in an accident, was required by provincial legislation to report it to the police. Although requiring her to report the accident was not a violation of White's rights under the *Charter* (therefore, there was no evidence obtained in a manner that violated her rights), the admission of her statements against her in criminal proceedings would have violated her right against self-incrimination under section 7 of the *Charter*. The trial judge was therefore justified in excluding her statements under section 24(1) of the *Charter*. According to the Supreme Court of Canada, the trial judge could also have excluded the evidence under the common law duty to exclude evidence that would render the trial unfair (*White* para. 89). This last proposition is somewhat controversial, as Plaxton (2003) suggests that if it were taken literally, judges would no longer be restricted to section 24(2) in their decision whether to admit or exclude evidence. However, the Supreme Court of Canada has stated that absent a *Charter* violation, "judges have a discretion at common law to exclude evidence obtained in circumstances such that it would result in unfairness if the evidence was admitted at trial, or if the prejudicial effect of admitting the evidence outweighs its probative value" (*Buhay* para. 40).

Judges will often enter a judicial stay of proceedings under section 24(1) of the *Charter,* if an accused's right to be tried within a reasonable time has been violated under section 11(b) (see Chapter 1). The courts may adjourn a case where the Crown has failed to disclose evidence (a violation of the accused's rights under section 7 of the *Charter*, if it affects the accused's ability to make full answer and defence), so that the accused can have an opportunity to review the evidence and exercise his or her right to make full answer and defence (see Chapter 4). According to the Supreme Court of Canada in *Bjelland*, evidence that the Crown fails to disclose in a timely manner, but that was otherwise gathered without violating the accused's rights under the *Charter*, should only be excluded at trial as a remedy in exceptional circumstances: "(a) where the late disclosure renders the trial process unfair and this unfairness cannot be remedied through an adjournment and disclosure order or (b) where exclusion is necessary to maintain the integrity of the justice system" (2009 para. 24). Since the exclusion of evidence "impairs the truth-seeking function of trials," exclusion of evidence under section 24(1) is not appropriate if some other remedy can be fashioned so as to not infringe on fairness to the accused or the integrity of the justice system (para. 24).

Additionally, illegally seized goods might be ordered returned under section 24(1) (*Lagiorgia*), and the court might order costs or damages paid to the aggrieved person. In *Leduc*, the Ontario Court of Appeal, in overturning a stay of proceedings and an award of costs against the Crown, stated that costs against the Crown should not be awarded routinely, but rather should be limited to "circumstances of a marked and unacceptable departure from the reasonable standards expected of the prosecution" (para 158). The Supreme Court of Canada dismissed Leduc's application for leave to appeal.

Section 24(2)

Section 24(2) of the *Charter* provides for the exclusion of evidence, but requires more than the mere fact of a violation of one's rights (*Therens*). Section 24(2) reads:

> Where, in proceedings under subsection (1), a court concludes that evidence was **obtained in a manner** which infringed or denied any rights or freedoms guaranteed by the *Charter*, the evidence shall be excluded if it is established that, having regard to all the circumstances, the admission of it in the proceedings would **bring the administration of justice into disrepute** (emphasis added).

Section 24(2) is a compromise between the traditional common law approach that illegally obtained evidence is admissible if it is relevant, and the historical approach in the United States, where all illegally obtained evidence is inadmissible (as discussed above). The compromise in section 24(2) is that evidence obtained in a manner that violates an accused's rights is excluded only if its admission would bring the administration of justice into disrepute. The French version of the *Charter* has resulted in the word "would" being interpreted as "could," so that the evidence shall be excluded if its admission "could bring the administration of justice into disrepute" (*Collins* 287).

The Rationale for Applying Section 24(2)

Paciocco (1989–90) discusses three possible rationales for excluding evidence under section 24(2). First, the exclusion of evidence is remedial. The only way to provide a remedy to a person is to deny the prosecution the use of the tainted evidence (332). The second reason for excluding evidence is to deter agents of the state (usually police officers) from violating an accused's rights under the *Charter* (332). The third rationale is the "imperative of judicial integrity," which incorporates the "twin goals of enabling the judiciary to avoid the taint of partnership in official lawlessness and of assuring the people…[that] the government would not profit from its lawless behaviour, thus minimizing the risk of seriously undermining popular trust in government" (Brennan, J. of the United States Supreme Court, as quoted in Paciocco 1989–90, 333).

Although Penney (2004) suggests that deterrence should be the sole rationale for excluding evidence under section 24(2), Paciocco (338–40) concludes that the remedial and deterrence rationales have been rejected by the Supreme Court of Canada, and that what is left is the imperative of judicial integrity. The Supreme Court of Canada recently confirmed Paciocco's interpretation, stating that the inquiry under section 24(2) "is not to punish the police or to deter *Charter* breaches" (*Grant* 2009 para. 73), or to provide compensation to the accused (para. 70). The focus "must be understood in the long-term sense of maintaining the integrity of, and public confidence in, the justice system" (para. 68). This focus is societal and systemic (para. 70).

According to Paciocco, the two aspects of this third rationale are sometimes incompatible. Are judges interested in focusing on their own integrity, therefore "avoiding the taint of partnership" with the police, or are they interested in "maintaining popular trust in the courts" (340)? In the former case, the judges' role might be that of educator of the public, while in the latter, they may follow public opinion. Mr. Justice Lamer has acknowledged that if the courts were to respond to the values of the majority, less evidence would be excluded under section 24(2) (*Collins,* as summarized by Paciocco 341). In addition, Paciocco writes:

> it is noteworthy that in developing the two most significant recent exceptions to the American exclusionary rule, the inevitable discovery exception and the qualified good faith exception, majority members of the court spoke about the need to maintain respect for the administration of justice while the dissenters focussed desperately on the need for the judiciary to maintain its own integrity. The dissenters did not want public opinion watering down the exclusionary rule (341).

The inquiry under section 24(2) is objective. According to the Supreme Court of Canada, it "asks whether a reasonable person, informed of all relevant circumstances and the values underlying the *Charter*, would conclude that the admission of the evidence would bring the administration of justice into disrepute" (*Grant* para. 68).

Who Has What Onus?

The standard of proof under section 24(2) is the civil test, the balance of probabilities. The onus of proof is on the party asserting that her or his rights or freedoms have been violated under the *Charter*. In the case of a criminal charge, the onus is on the accused to show that (1) one of his or her rights or freedoms under the *Charter* has been violated; (2) evidence was obtained in a manner that violated that right or freedom; and (3) the admission of the evidence obtained by reason of the infringement could bring the administration of justice into disrepute.

"Obtained in a Manner"

For section 24(2) to apply, evidence must have been obtained in a manner that infringed or denied the accused's rights under the *Charter*. What exactly does this mean? In the *Therens* case, the Supreme Court of Canada rejected a **causal connection** test. The *Charter* violation does not have to be the cause of obtaining the evidence. According to LeDain, J., it is sufficient if the *Charter* violation preceded or occurred while the evidence was being gathered. The Supreme Court of Canada reconsidered this approach in *Strachan*, where Mr. Justice Dickson again rejected a causal connection test and suggested that judges should focus "on the entire chain of events during which the *Charter* violation occurred and the evidence that was obtained" (498).

In *Goldhart*, the Supreme Court of Canada confirmed its earlier decisions in *Therens* and *Strachan*, rejecting strict causal analysis; however, it decided that the courts must look at the remoteness of both the temporal and causal connection between the evidence and the *Charter* breach to determine whether evidence was "obtained in a manner" that infringed or denied the accused's rights under the *Charter*. In *Goldhart,* the connection between a witness who became a born-again Christian following an illegal search of a grow operation, and the illegal search itself, was too remote to engage this section of the *Charter*.

Box 7.1 Public Opinion and the Reputation of the Administration of Justice

On May 19, 1995, the Supreme Court of Canada ordered a new trial for Terrence Burlingham, who was convicted in the 1984 murders of two young women from Cranbrook, British Columbia. Each woman had been sexually assaulted and shot twice in the head at close range with a .410 shotgun. On appeal, the Supreme Court of Canada ruled that the gun and the fact that Burlingham led the police to its hiding place underneath the frozen Kootenay River were inadmissible. In dealing with the role of public opinion under section 24(2), Mr. Justice Sopinka, in *Burlingham*, wrote:

> Not surprisingly, commentators no less than the public differ as to the appropriate approach to the exclusion of evidence associated with a violation of a Charter right.... While Professor Paciocco favours an approach that would be less exclusionary and, in his opinion, more in tune with the views of the average Canadian, Steven Penney, in his comprehensive article at p. 810, argues that by focusing on trial fairness, as opposed to the criminal justice system as a whole, we "render individual Canadians more susceptible to invasions of their constitutional rights."
>
> Both Professor Paciocco and my colleague are of the view that the approach we have taken is out of step with the public mood. Quite apart from the admonitions of Lamer J. (as he then was) in *Collins*, at pp. 281–82, that individual rights are not to be submitted to an adjudication by the majority, there is no accurate assessment of public opinion. Adjusting the approach to Charter rights based on public opinion surveys is fraught with difficulties. This can be illustrated by reference to the empirical study to which my colleague refers by Bryant, Gold, Stevenson and Northrup, "Public Attitudes Toward the Exclusion of Evidence: Section 24(2) of the Canadian Charter of Rights and Freedoms" (1990), 69 Can. Bar Rev. 1. It purported to show "a significant gap between public opinion and judicial opinion" regarding the application of the *Collins* factors. After publication of that study, a further study by the same authors, "Public Support for the Exclusion of Unconstitutionally Obtained Evidence" (1990), 1 S.C.L.R. (2d) 555, concluded at p. 557 that "taking into account some of the ambiguity in the case law, the gap between public and judicial opinion may not be that substantial over a broad range of cases" (paras. 139-140).

Bringing "the Administration of Justice into Disrepute"

Whether the admission of evidence would bring the administration of justice into disrepute is a question of law, and must be decided by the judge, not by the jury. A ***voir dire*** is held, in the absence of the jury, to determine whether the evidence is admissible.

While section 24(2) raises a question of law and is therefore subject to appeal, the Supreme Court of Canada has on several occasions stated that it will not normally review such decisions. In *Duguay*, the majority wrote:

> It is not the proper function of this Court, though it has the jurisdiction to do so, absent some
> apparent error as to the applicable principles or rules of law, or absent a finding that is unreasonable,
> to review findings of the courts below under s. 24(2) of the *Charter* and substitute its opinion of the
> matter for that arrived at by the Court of Appeal (para. 2).

Similarly, it is not for courts of appeal to substitute their opinion for that of the trial judge, "absent an apparent error as to the applicable principles or rules of law or an unreasonable finding" (*Mann* para. 59). So long as the trial judge's decision sets out the appropriate principles and is not unreasonable, it will not be overturned. Paciocco (1989–90) criticizes this approach because it allows for two opposing decisions to be correct, and puts far too much discretion into the hands of the trial judge.

What will bring the administration of justice into disrepute? What type of evidence will the court consider? The Supreme Court of Canada has made it clear that "disrepute" is to be evaluated from the perspective of a reasonable person, "dispassionate and fully appraised of the circumstances of the case" (*Collins* 266). It has rejected the use of public opinion polls, holding that judges are better equipped to determine this issue. Lamer, J. discussed this in *Collins:*

> The concept of disrepute necessarily involves some element of community views, and the
> determination of disrepute thus requires the judge to refer to what he conceives to be the views of the
> community at large. This does not mean that evidence of the public's perception of the repute of the
> administration of justice, which Professor Gibson suggested could be presented in the form of public
> opinion polls will be determinative of the issue. … [As suggested by Professor Gibson], "the ultimate
> determination must be with the courts, because they provide what is often the only effective shelter
> for individuals and unpopular minorities from the shifting winds of public passion" (281).

Bryant *et al.* (1990) have provided evidence, through a survey, that the public would be less likely to exclude evidence in some circumstances than the courts (also see discussion in Box 7.1). As Madame Justice Southin of the British Columbia Court of Appeal observed in *Evans,* "If reports in the press are any guide, there is a substantial body of the citizenry who have a very low opinion of the administration of justice generally and not because they believe that evidence is being admitted which they would exclude" (137).

The Grant Framework

In 2009, the Supreme Court of Canada replaced its earlier test for excluding evidence under section 24(2) of the *Charter*, as developed in its decisions in *Collins* and *Stillman,* with the *Grant* framework. The majority summarized the inquiry under section 24(2):

> When faced with an application for exclusion under s. 24(2), a court must assess and balance the effect of
> admitting the evidence on society's confidence in the justice system having regard to: (1) the seriousness
> of the *Charter*-infringing state conduct (admission may send the message the justice system condones
> serious state misconduct), (2) the impact of the breach on the *Charter*-protected interests of the accused
> (admission may send the message that individual rights count for little), and (3) society's interest in the
> adjudication of the case on its merits. The court's role on a s. 24(2) application is to balance the
> assessments under each of these lines of inquiry to determine whether, considering all the circumstances,
> admission of the evidence would bring the administration of justice into disrepute (*Grant* para. 71).

The spectrum of the seriousness of a *Charter* violation can range from "inadvertent or minor" to "wilful or reckless." The latter "will inevitably have a negative effect on the public confidence in the rule of law, and risk bringing the administration of justice into disrepute" (para. 74). Extenuating circumstances and good faith will weaken the seriousness of *Charter* violation; however, ignorance, wilful blindness, and deliberate violations tend to support the exclusion of the evidence (para. 75).

The second factor considers the impact of the *Charter* violation on the accused's rights. The impact can range from "fleeting and technical to profoundly intrusive" (para. 76). The more serious the impact, "the greater the risk that admission of the evidence may signal to the public that *Charter* rights, however high-sounding, are of little actual value to the citizen, breeding public cynicism and bringing the administration of justice into disrepute" (para. 76).

The third factor examines society's interest in having criminal charges adjudicated on their merits. According to the majority in *Grant*, this involves analyzing "whether the truth-seeking function of the criminal trial process would be better served by admission of the evidence, or by its exclusion" (para. 79). Although not a determining factor, the reliability of the evidence is "an important factor" (para. 81) that works both ways. On the one hand, admitting unreliable evidence "serves neither the accused's interest in a fair trial nor the public interest in uncovering the truth." On the other hand, "exclusion of relevant and reliable evidence may undermine the truth-seeking function of the justice system and render the trial unfair from the public perspective, thus bringing the administration of justice into disrepute" (para. 81). It is possible that the exclusion of evidence may extract "too great a toll on the truth-seeking goal of the criminal trial" (para. 82). The third factor also examines "the importance of the evidence to the prosecution's case" (para. 83); however, the section 24(2) analysis operates "independently of the type of crime" for which the accused is charged (para. 84).

In *Burlingham*, Mr. Justice Iacobucci of the Supreme Court of Canada wrote:

> …we should never lose sight of the fact that even a person accused of the most heinous crimes, and no matter the likelihood that he or she actually committed those crimes, is entitled to the full protection of the *Charter*. Short-cutting or short-circuiting those rights affects not only the accused, but also the entire reputation of the criminal justice system. It must be emphasized that the goals of preserving the integrity of the criminal justice system as well as promoting the decency of investigatory techniques are of fundamental importance in applying s. 24(2) (para. 50).

The *Grant* framework is now used in determining applications to exclude evidence under section 24(2) of the *Charter*, if the evidence was obtained in a manner that violated an accused's rights under the *Charter*.

Box 7.2 Violating the Rights of Obnoxious Accused

In *Tremblay* it was held that the police violated the accused's rights by not giving him a reasonable opportunity to contact a lawyer before being required to provide a sample of his breath in an impaired driving investigation. It was held, however, that the admission of the evidence of the results of that breath test would not bring the administration into disrepute, given that "from the moment the accused was intercepted on the road to the moment he was asked to give the first sample of his breath his behaviour was violent, vulgar and obnoxious" (567).

SUMMARY

At common law, illegally obtained evidence was generally admissible. The introduction of the *Charter of Rights and Freedoms* provided courts the power to grant remedies for the breach of constitutional rights, including the power to exclude otherwise admissible evidence. Under section 24(2) of the *Charter,* evidence must be excluded if it was "obtained in a manner which infringed or denied" a *Charter* right, and if "having regard to all the circumstances," its admission could "bring the administration of justice into disrepute." These criteria must be established by the party asserting the breach, on a balance of probabilities. Factors to be considered under section 24(2) include the seriousness of the *Charter* violation, the impact of the breach on the accused's *Charter*-protected interests, and society's interest in adjudicating the allegations on their merits. Under certain circumstances evidence can be excluded under section 24(1) of the *Charter* or the common law.

QUESTIONS TO CONSIDER

(1)   Under what circumstances would a court in Canada have excluded evidence under the common law before the introduction of the *Charter*? What was the definition of "unfairness" under the common law?

(2)   What types of remedies can the court order under section 24(1)?

(3)   Under what circumstances can evidence be excluded under section 24(1) of the *Charter*?

(4)   What are the major rationales for excluding evidence under section 24(2) of the *Charter*? Which rationale is the appropriate one, according to the Supreme Court of Canada?

(5)   What is the "fruit of the poisonous tree" doctrine? How does it differ from the law under section 24 of the *Charter*?

(6)   What are the two components of the imperative of judicial integrity? How might they conflict with one another?

(7)   How could social science research assist in developing or elaborating on the rationales used to exclude evidence under section 24(2)? What are some of the issues that would have to be addressed before using public opinion polls?

(8)   Should the court be required to consider social science research when decided whether to exclude evidence under section 24(2)? Why or why not?

(9)   What does "obtained in a manner" mean under section 24(2) of the *Charter*?

(10)   Is the application of section 24(2) of the *Charter* a question of law or a question of fact? What are the implications of this?

(11)   What is the *Grant* framework? What are the three factors the court will consider? Create a fact pattern question and then apply the *Grant* framework to your question.

BIBLIOGRAPHY

Boyle, Christine, Marilyn T. MacCrimmon, and Dianne Martin. *The Law of Evidence: Fact Finding, Fairness, and Advocacy.* Toronto: Emond Montgomery Publications Limited, 1999.

Chapter 11 contains a section on the role of the *Charter* and the exclusion of evidence under section 24(2).

Brewer, Carol A. "*Stillman* and Section 24(2): Much To-Do about Nothing." (1997) 2 *Canadian Criminal Law Review* 239.

Bryant, Alan W., Sidney N. Lederman and Michelle K. Fuerst. *Sopinka, Lederman & Bryant – The Law of Evidence in Canada* 3rd ed. Canada: LexisNexis Canada, 2009.
        Chapter 9, "Illegally Obtained Evidence," discusses the common law rule for excluding evidence and section 24 of the *Charter.*

Bryant, Alan W., Marc Gold, H. Michael Stevenson, and David Northrup. "Public Attitudes Toward the Exclusion of Evidence: Section 24(2) of the Canadian Charter of Rights and Freedoms." (1990) 69 *Canadian Bar Review* 1.

Code, Michael. "American Cadillacs or Canadian Compacts: What is the Correct Criminal Procedure for S. 24 Applications Under the *Charter*?" (1990–1) 33 *Criminal Law Quarterly* 298 (Part I), 407 (Part II).

del Carmen, Rolando V. *Criminal Procedure, Law and Practice,* 3rd ed. Belmont, California: Wadsworth Publishing Company, 1995.

Delisle, Ron Don Stuart, and David M. Tanovich. *Evidence: Principles and Problems*, 8th ed. Toronto: Carswell, 2007.
        Chapter 3 discusses the discretion to exclude evidence under the common law and section 24 of the *Charter.*

Gorham, Nathan J. S. "Eight Plus Twenty-Four Two Equals Zero-Point Five." (2003) 6 *Criminal Reports* (6th) 257.

Hails, Judy. *Criminal Evidence.* 5th ed. Belmont, CA: Thomson Wadsworth, 2005.

Hill, S. Casey, David M. Tanovich and Louis P. Strezos. *McWilliams' Canadian Criminal Evidence.* Aurora, ON: Canada Law Book Limited (available online on Criminal Spectrum).
        Chapter 16, "Illegally Obtained Evidence" discusses the exclusion of evidence under section 24(2) of the *Charter*. Also see Chapter 33 (Charter Applications).

Jull, Kenneth. "Exclusion of Evidence and the Beast of Burden." (1987–8) 30 *Criminal Law Quarterly* 178.

Law Reform Commission of Canada. *Report on Evidence*. Ottawa: Law Reform Commission of Canada, 1975.

Mahoney, Richard. "Problems with the Current Approach to s. 24(2) of the Charter: An Inevitable Discovery." (1999) 42(4) *Criminal Law Quarterly* 443.

McDonald, David C. *Freedom and Security Under the Law.* Second Report of the Commission of Inquiry Concerning Certain Activities of the Royal Canadian Mounted Police. Ottawa: The Commission, 1981.

McLellan, A. Anne, and Bruce P. Elman. "The Enforcement of the *Canadian Charter of Rights and Freedoms*: An Analysis of Section 24." (1983) 21(2) *Alberta Law Review* 205.

Mirfield, Peter. "The Early Jurisprudence of Judicial Disrepute." (1987–8) 30 *Criminal Law Quarterly* 434.

Mitchell, Gerard. "Section 24(2) Circumstances." (1993) 35 *Criminal Law Quarterly* 433.

Mitchell, Gerard E. "The Supreme Court of Canada and the Exclusion of Evidence in Criminal Cases Under Section 24 of the *Charter.*" (1987–8) 30 *Criminal Law Quarterly* 165.

Ouimet, Roger (Chair). *Report of the Canadian Committee on Corrections, Towards Unity: Criminal Justice and Corrections.* Ottawa: Information Canada, 1969.

Paciocco, David M. "*Stillman*, Disproportion and the Fair Trial Dichotomy Under Section 24(2)." (1997) 2 *Canadian Criminal Law Review* 163.

Paciocco, David M. "The Judicial Repeal of S. 24(2) and the Development of the Canadian Exclusionary Rule." (1989–90) 32 *Criminal Law Quarterly* 326.

Penney. Steven. "Taking Deterrence Seriously: Excluding Unconstitutionally Obtained Evidence Under Section 24(2) of the Charter." (2004) 49 *McGill Law Journal* 105.

Plaxton, Michael C. "Who Needs Section 24(2)? Or: Common Law Sleight-of-Hand." (2003) 10 *Criminal Reports* (6th) 236.

Pottow, J.A.E. "Constitutional Remedies in the Criminal Context: A Unified Approach to Section 24." (2000) 44 *Criminal Law Quarterly* 459.

Pottow, J.A.E. "Constitutional Remedies in the Criminal Context: A Unified Approach to Section 24 (Part II)." (2001a) 44 *Criminal Law Quarterly* 34.

Pottow, J.A.E. "Constitutional Remedies in the Criminal Context: A Unified Approach to Section 24 (Part III)." (2001b) 44 *Criminal Law Quarterly* 223.

Stuart, Don. "Eight Plus Twenty-Four Two Equals Zero." (1998) 13 *Criminal Reports* (5th) 50.

Stuart, Don, Ron Delisle, and Tim Quigley. *Learning Canadian Criminal Procedure*, 9th ed. Toronto: Carswell, 2008.
    Chapter 5 discusses remedies under section 24 of the *Charter*.

Uniform Law Conference of Canada. *Report of the Federal/Provincial Task Force on Uniform Rules of Evidence.* Toronto: Carswell, 1982.
    Chapter 16, "Illegally or Improperly Obtained Evidence," discusses the law in Canada prior to the introduction of the *Charter (Wray)* and the law as it existed in England, Australia, New Zealand, Scotland, Ireland, and the United States. After examining several proposals for reform, the Task Force recommends that no changes be made to the (then) existing law (i.e., they endorsed the approach by the Supreme Court of Canada in *Wray)*.

CASES

R. v. Big M Drug Mart Ltd., [1985] 1 S.C.R. 295. A corporation can apply for a *Charter* remedy under section 24(1).

R. v. Bjelland, 2009 SCC 38. The Court sets out the criteria for excluding evidence under section 24(1) of the *Charter* for late disclosure by the Crown.

R. v. Buhay, [2003] 1 S.C.R. 63. There is a reasonable expectation of privacy in the contents of a rented locker, and therefore the police need a search warrant to search it. On the facts, the security guards were not acting as agents of the state.

R. v. Burlingham, [1995] 2 S.C.R. 206. Real derivative evidence (the gun used in the killings) can be conscripted evidence and therefore excluded under section 24(2) of the *Charter*.

R. v. Collins, [1987] 1 S.C.R. 265.

R. v. Drybones, [1970] S.C.R. 282.

R. v. Duguay, [1989] 1 S.C.R. 93. The Supreme Court will not normally review lower court findings under section 24(2) of the *Charter,* unless there has been some error in law or principle.

R. v. Edwards, [1996] 1 S.C.R. 128. The Court sets out the factors that should be considered in determining whether a person who does not rent an apartment has a reasonable expectation of privacy in it, under section 8 of the *Charter*.

R. v. Evans (1994), 93 C.C.C. (3d) 130 (B.C.C.A.). An example of illegally obtained but pre-existing real evidence being admitted based on an assessment of good faith, its effect on the fairness of the trial, and the effect of its exclusion on the reputation of the administration of justice.

CHAPTER 7: *Evidence That Is Illegally or Improperly Obtained*

R. v. Feeney, [1997] 2 S.C.R. 13. The Court examines several items of evidence and finds that some should be excluded under section 24(2) of the *Charter* on the basis that they were conscripted and affected the fairness of the trial. Some of the others are excluded on the remaining tests under section 24(2).

R. v. Genest, [1989] 1 S.C.R. 59. In decisions concerning the exclusion of evidence under section 24(2) of the *Charter*, the courts must consider three groups of factors: the effect of the admission of the evidence on the fairness of the trial, the seriousness of the violation of rights and any justification for it, and the relative effects of admitting versus excluding the evidence.

R. v. Goldhart, [1996] 2 S.C.R. 463. The Court states the test for whether "evidence has been obtained in a manner which infringed or denied any rights..." as including both a temporal and a causal component. Both should be examined in terms of remoteness.

R. v. Grant, 2009 SCC 32. The Supreme Court of Canada established a new framework for determining whether evidence should be excluded under section 24(2) of the *Charter*, replacing the *Collins/Stillman* test.

R. v. Hogan (1975), 18 C.C.C. (2d) 65 (S.C.C.). The court rejected the argument that the conviction for impaired driving should be quashed because the breathalyzer sample was obtained after the accused was denied his right to counsel under the *Bill of Rights*.

R. v. Lagiorgia (1988), 35 C.C.C. (3d) 445 (F.C.A.); leave to appeal refused. The appropriate remedy under section 24(1) for an unreasonable search for and seizure of items that are not illicit as such is to order the return of the items to their rightful owner.

R. v. Leduc, [2003] O.J. No. 2974 (Ont. C.A.); leave to appeal to S.C.C. dismissed [2003] S.C.C.A. No. 411.
Charges should not be stayed for prosecutorial misconduct unless "(1) the prejudice caused by the abuse in question will be manifested, perpetuated or aggravated through the conduct of the trial, or by its outcomes; and (2) no other remedy is reasonably capable of removing the prejudice . . ." (para. 142). Costs against the Crown should be limited to "circumstances of a marked and unacceptable departure from the reasonable standards expected of the prosecution" (para 158).

R. v. Mann, [2004] 3 S.C.R. 59.

R. v. Stillman, [1997] 1 S.C.R. 607.

R. v. Strachan, [1988] 2 S.C.R. 980. A strict causal connection is not absolutely required between a *Charter* violation and the obtaining of evidence, before the evidence can be excluded under section 24(2). In determining whether the admission of evidence would bring the administration of justice into disrepute, the court must examine three groups of factors: the fairness of the trial, the seriousness of the violation, and the effect of the exclusion of the evidence.

R. v. Therens, [1985] 1 S.C.R. 613. Evidence obtained as a result of a *Charter* breach should only be excluded (if at all) pursuant to s. 24(2).

R. v. Tremblay, [1987] 2 S.C.R. 435.

R. v. White, [1999] 2 S.C.R. 417. The use in a subsequent criminal trial of evidence gathered by compulsion under provincial statute in quasi-criminal proceedings violates an accused's rights under section 7 (right against self-incrimination) and can be excluded under section 24(1) of the *Charter*.

R. v. Wholesale Travel Group Inc., [1991] 3 S.C.R. 154. A corporation cannot be convicted of an offence that violates section 7 of the *Charter*, even though a corporation is not protected by section 7.

R. v. Wray, [1971] 4 S.C.R. 272 . Illegally obtained evidence is generally admissible, unless it is gravely prejudicial, of little probative value, and its admission would be unfair.

CHAPTER 8: *Search and Seizure*

CHAPTER OBJECTIVES

In studying this chapter, you should develop an understanding of the following topics and concepts:

- the meaning of privacy under section 8 of the *Charter*
- when and how section 8 of the *Charter* applies
- the requirements for a constitutionally valid search
- the requirements for a search warrant under section 487 of the *Code*
- the requirements of other provisions that allow for search warrants
- the validity of "perimeter" and similar searches
- common law powers of search and seizure

SECTION 8 OF THE CHARTER

Section 8 of the *Charter* states that "everyone has the right to be secure against **unreasonable search or seizure**." Before the *Charter*, the common law was quite strict in its protection of private property (especially homes) from the invasion of state agents looking for evidence. Police officers who entered private property without proper authority, or without the consent of the owners, were considered to be trespassers. The common law exception, which permitted entry to arrest a suspect, was restricted following the decision in *Feeney*, and entry warrants were added to the *Criminal Code* (discussed in Chapter 2).

The protection against unreasonable searches under section 8 of the *Charter* was first considered by the Supreme Court of Canada in *Hunter et al. v. Southam Inc.* The case challenged a provision in the *Combines Investigation Act* (now the *Competition Act*), which allowed the Director of Investigation and Research (at that time, Mr. Hunter), or any representative authorized by the Director, to enter "any premises on which the Director believes there may be evidence relevant to the matter being inquired into and…[to] examine any thing on the premises and…[to] copy or take away" any documents. Before conducting the search, the Director had to apply to a member of the Restrictive Trade Practices Commission (RTPC) for a certificate authorizing the search. The RTPC at that time was both an adjudicative and an investigative body. In addition to making decisions on cases that came before it, the RTPC also had investigative powers that allowed it to gather evidence, if it thought that the Director's investigation was not sufficient, and the power to order the production of documents. This arrangement was challenged in *Hunter*, under section 8 of the *Charter*.

According to the Supreme Court in *Hunter*, section 8 is concerned with preventing unjustified state intrusions into the **privacy** of individuals. A system authorizing such intrusions must be examined to see if it is acceptable. The Court decided that the Combines legislation violated section 8 of the *Charter*, and was therefore of no force and effect. (Note that the section has since been replaced by a section that allows the Director, or his authorized representative, to apply to the Federal Court for a search warrant.)

In reaching his decision, Mr. Justice Dickson said that the protection under section 8 was much broader than under the common law, in that it was concerned with (perhaps among other things) the right to privacy, and when that right must give way to government interests such as law enforcement. Section 8 is aimed at preventing unjustified state intrusions into the privacy of individuals. In this context, the purpose of the *Charter* is to constrain government action that violates *Charter* rights and freedoms. The *Charter* does not authorize reasonable searches; it prohibits unreasonable searches.

Dickson provided a framework within which to evaluate the constitutionality of legislation authorizing search warrants. First, where it is feasible to obtain **prior authorization** (i.e., a **search warrant**), a search warrant is required before a search can be considered reasonable. A warrantless search is presumed to be unreasonable. This is a rebuttable presumption; the Crown might be able to establish that a warrantless search was not unreasonable, given the circumstances (e.g., to prevent the destruction of evidence). The purpose in obtaining prior authorization is so that the interests of privacy can be assessed in light of the conflicting interests of the state, and so that the right to privacy is breached only where appropriate standards are met (*Hunter* 110).

Second, the prior authorization must be by **a person who is capable of acting judicially** (not necessarily a judge). The person has to be able to act in a neutral and impartial manner, which cannot be done by a person or body that has "significant investigatory powers" (*Hunter* 112). The RTPC had a number of investigative powers, which prevented it from acting judicially (that is, in an impartial manner) in authorizing others to engage in investigative procedures.

Third, there must be sufficient evidence for the person to make a **judicial decision**. The fact that the search *may* uncover evidence of a crime is not sufficient; mere suspicion is not enough. There must be "a credibly based probability," such as is required by section 487 of the *Criminal Code*, which requires **"reasonable grounds to believe"** (*Hunter* 114–5). There may be exceptions, but in cases like *Hunter,* the minimum standard is "reasonable grounds, established upon oath, to believe that an offence has been committed and that there is evidence to be found at the place of the search" (*Hunter* 115).

After finding that the Combines legislation violated the *Charter*, the next step in *Hunter* would normally have been to decide whether the violation was demonstrably justified in a free and democratic society, under section 1 of the *Charter*. The government did not argue that the legislation authorizing an unreasonable search was justified under section 1, and so the Court did not need to consider the relationship between section 8 and section 1. In fact, if legislation authorizes "unreasonable searches and seizures," it may not make sense to consider whether such legislation is demonstrably justified in a free and democratic society. Section 8 seems to be self-contained with respect to section 1, although the Supreme Court of Canada has left this open. The framework developed by Dickson in *Hunter v. Southam* can, of course, be used to decide the constitutional validity of other legislative provisions authorizing searches and seizures.

Section 8 also requires that the person granting the search warrant have the **discretion** to decide whether or not to issue a warrant. For example, the search warrant provisions under the *Income Tax Act* previously stated that "the judge shall issue the warrant referred to in [an earlier section] where he is satisfied that there are reasonable grounds to believe that...etc." The Supreme Court of Canada in *Baron* held that there must be a "residual discretion in the judicial officer who issues the warrant." This is a fundamental aspect of section 8 of the *Charter,* since the decision to issue a warrant is based on balancing the right to privacy, and freedom from state intrusions into that privacy, against the interests of law enforcement.

Since section 8 of the *Charter* protects a person's reasonable expectation in privacy, it is important to establish whether such privacy in fact exists. This analysis requires an assessment of the "totality of the circumstances" (*Edwards* para. 45; *Tessling* para. 19; *Patrick* para. 26), no matter "whether the claim involves aspects of personal privacy, territorial privacy, . . . informational privacy" or a combination of them (*Patrick* para. 26). Although the assessment "requires close attention to context," the Court has given an analytical framework for establishing a reasonable expectation of privacy, that must be adjusted to the circumstances (*Patrick* para. 26). For example, in *Patrick*, the Court approved the following framework used by the trial judge (who relied on the analysis in

Tessling), in determining that the accused did not have a privacy interest in garbage which had been set on the ground just inside the property line for pickup:

1. What was the nature or subject matter of the evidence gathered by the police?
2. Did the appellant have a direct interest in the contents?
3. Did the appellant have a *subjective* expectation of privacy in the informational content of the garbage?
4. If so, was the expectation *objectively* reasonable? In this respect, regard must be had to [the totality of the circumstances]:

> a. the place where the alleged "search" occurred; in particular, did the police trespass on the appellant's property and, if so, what is the impact of such a finding on the privacy analysis?
> b. whether the informational content of the subject matter was in public view;
> c. whether the informational content of the subject matter had been abandoned;
> d. whether such information was already in the hands of third parties; if so, was it subject to an obligation of confidentiality?
> e. whether the police technique was intrusive in relation to the privacy interest;
> f. whether the use of this evidence gathering technique was itself objectively unreasonable;
> g. whether the informational content exposed any intimate details of the appellant's lifestyle, or information of a biographic nature (para. 27).

The Court's conclusion that the accused did not retain his privacy interests in his garbage is not without critics. See Kaiser, who suggests that *Patrick* "thrashes privacy rights" (2009).

In *Edwards*, the Supreme Court of Canada found that the accused had no reasonable expectation of privacy in his girlfriend's apartment, where the police found the drugs he kept there. The totality of the circumstances included:

> (i) presence at the time of the search;
> (ii) possession or control of the property or place searched;
> (iii) ownership of the property or place;
> (iv) historical use of the property or item;
> (v) the ability to regulate access, including the right to admit or exclude others from the place;
> (vi) the existence of a subjective expectation of privacy; and
> (vii) the objective reasonableness of the expectation (para. 45).

Absent a privacy interest, there can be no violation of rights under section 8 of the *Charter*. In *Belnavis*, Mr. Justice Cory, for the majority of the Court, decided that a passenger in a motor vehicle, who was unrelated to the driver, had no reasonable expectation of privacy regarding garbage bags filled with stolen property. Both *Edwards* and *Belnavis* have been criticized for dramatically narrowing the protection against unreasonable search and seizure (Hendel and Sankoff 1996; Schwartz 1997). In *Buhay*, the Supreme Court of Canada found that there was a reasonable expectation of privacy in lockers at a bus depot, and also commented that people had a reasonable expectation of privacy in hotel rooms.

In *M.(R.M.)*, the Supreme Court of Canada found that a 13-year-old student had a reasonable expectation of privacy in relation to searches of his body by school authorities; however, the reasonable expectation was diminished because he was in a school environment. Mr. Justice Cory summarized the rules regarding such searches:

> (1) A warrant is not essential in order to conduct a search of a student by a school authority.
> (2) The school authority must have reasonable grounds to believe that there has been a breach of school regulations or discipline and that a search of a student would reveal evidence of that breach.

(3) School authorities will be in the best position to assess information given to them and relate it to the situation existing in their school. Courts should recognize the preferred position of school authorities to determine if reasonable grounds existed for the search.

(4) The following may constitute reasonable grounds in this context: information received from one student considered to be credible, information received from more than one student, a teacher's or principal's own observations, or any combination of these pieces of information which the relevant authority considers to be credible. The compelling nature of the information and the credibility of these or other sources must be assessed by the school authority in the context of the circumstances existing at the particular school (para. 50).

The decision was criticized for limiting the rights of school children (MacKay 1997; Stuart 1999). The Court was more supportive of children's rights in *A.M.,* where a student was found to have a reasonable expectation of privacy in his knapsack left unattended in the gymnasium. However, in *A.M.* the police were doing a walk-around with a police sniffer dog.

Technological developments continue to challenge our understanding of a reasonable expectation of privacy. In *Tessling*, the police flew over the accused's property using a Forward Looking Infra-Red (FLIR) camera, which detected an unusual amount of heat radiating from his house. Although they could not determine the source of the heat, they used this information, and information from two informants that the accused was growing marijuana, to obtain a search warrant. Was the use of this camera to detect heat a search? The Supreme Court of Canada found that the technology was not sufficiently advanced to be considered a search. Considering the "totality of the circumstances" (see para. 32 in *Tessling* for the *Edwards*' test, modified to fit the circumstances), it found that Tessling did not have a reasonable expectation of privacy in the heat emanating from his house. Pomerance (2005) suggests that this result is not surprising given the Court's decision in *Plant* (discussed below), but that the Court could have provided more guidance on what is and is not protected under section 8 of the *Charter*. See Coughlan and Gorbet (2005) for a criticism of *Tessling*. Advances in technology may, however, change the Court's evaluation of whether section 8 is engaged in these circumstances.

Stringham (2005) suggests that there are at least two ways to examine if a reasonable expectation of privacy exists. A societal expectation or social convention approach, which is used in the United States, requires the court to determine what the public actually expects in terms of privacy. The Supreme Court of Canada has referred to this approach as descriptive (*Tessling* para. 42; *Patrick* para. 14). On the other hand, a normative approach asks not whether we expect privacy in certain circumstances, but whether we should be entitled to expect privacy in the circumstances because we live in a free and democratic society. Social expectation of surveillance does not necessarily mean that it should be allowed in our society without judicial authorization. For example, even if we come to expect a security camera on every street or in every public washroom, this societal expectation does not necessarily mean that we have no entitlement to privacy in a public washroom (normative approach). In 2005, Stringham suggested that the Supreme Court of Canada had moved toward using the social convention approach. However, in both *Tessling* (para. 42) and *Patrick* (para. 14), the Court confirmed that its approach is normative. As Binnie, J. explained in *Patrick*, the privacy analysis under section 8 is "laden with value judgments which are made from the independent perspective of the reasonable and informed person who is concerned about the long-term consequences of government action for the protection of privacy" (para. 14).

If the accused establishes a reasonable expectation of privacy, the court will then determine if the search was reasonable in the circumstances. A search will be reasonable if it is authorized by statute or the common law, if the authorizing law is reasonable, and if the search is conducted in a reasonable manner (*Collins* para. 23).

Excluding Evidence Under Section 24(2)

If the accused establishes that evidence was obtained in a manner that violated his or her rights under section 8 of the *Charter*, the judge will consider whether the evidence should be excluded under section 24(2) of the *Charter*, using the *Grant* (2009) framework discussed in Chapter 7. In an earlier decision (*Stillman*), the Supreme Court of Canada had decided that conscriptive evidence, which included bodily evidence that the accused was compelled to participate in producing, would render a trial unfair and should be excluded. The Court in *Grant* rejected what had, since *Stillman*, become "a near-automatic exclusionary rule for bodily evidence obtained contrary to the *Charter*" (para. 100). The Court replaced this "one-size-fits-all conscriptive test" (para. 103) with "a flexible test based on all the circumstances," as required under section 24(2), to determine the effect of admitting the evidence on the repute of the justice system. Such an analysis involves examining the seriousness of the *Charter* violation, the seriousness of the breach on the accused's rights, and the public interest in having the case adjudicated on its merits (para. 107-108).

In examining the impact of the seriousness of the *Charter* violation (during a search) on the repute of the justice system, the court will examine, for example, whether the behaviour of the police was "deliberate and egregious" or "committed in good faith" (para. 108). With regard to the impact of the breach of the accused's rights on the repute of the justice system, the court will examine "the degree to which the search and seizure intruded upon the privacy, bodily integrity and human dignity of the accused" (para. 109). Forcibly taking blood samples or dental impressions is not as serious as taking fingerprints (para. 109). The public interest in having a case adjudicated on its merits favours the admission of bodily samples as they are generally reliable (para. 110). However, "where an intrusion on bodily integrity is deliberately inflicted and the impact on the accused's privacy, bodily integrity and dignity is high, bodily evidence will be excluded, notwithstanding its relevance and reliability" (para. 111). A similar analysis can be done for non-bodily physical evidence. The court will examine all of the circumstances in light of the three factors in the *Grant* framework (para. 112-115).

SEARCH BY WARRANT—*CRIMINAL CODE*

Section 487 Warrants

Section 487 of the *Criminal Code* states that a justice "who is satisfied by information upon oath in Form 1, that there are reasonable grounds to believe that there is in a building, receptacle or place" anything that will afford evidence respecting the commission of an offence, may issue a search warrant (see Form 5) to search therein, and to seize such items and bring them before a justice. Warrants can be issued under section 487(1)(c) for anticipated offences, but only if the offence anticipated is an offence against the person for which a person may be arrested without a warrant (see Chapter 2). For example, if a police officer has reasonable grounds to believe that a person is in possession of poison for the purpose of putting it in someone's food, the police officer could obtain a warrant to seize the poison rather than having to wait until the offence was carried out. Sections 487(2.1) and (2.2) set out what a person authorized to search a computer may do, and the corresponding duties on a person in possession of a computer.

A justice presented with information upon oath in Form 1 must be satisfied "that there are **reasonable grounds** to believe...." Where legislation uses the phrase **"reasonable and probable grounds,"** the phrase has the same meaning as "reasonable grounds." Each phrase imports the standard required under section 8 of the *Charter*–"credibly based probability" (*Baron*, 531–2).

Justices must exercise their discretion in deciding whether the information under oath provided to them meets the requirements of section 487. Further, they must exercise their discretion in a judicial manner, and cannot just "rubber stamp" such applications.

The information provided to the justice for consideration is usually provided by a police officer, who provides sufficient evidence in a document known as an "Information to Obtain a Search Warrant" (or, somewhat confusingly, an "Information"), such that the justice can make a judicial decision (neutral and impartial). The **Information to Obtain** has to describe the offence that is suspected to have been committed, so that the justice knows the nature of the alleged offence (the warrant that is issued also includes this, so that the officers with the warrant and any person they show it to can ascertain the nature of the alleged offence). The Information to Obtain and the warrant must also specify the items to be seized, in respect of which there must be reasonable grounds to believe they will afford evidence of the commission of an offence. The test is whether the description is sufficient to permit the officers executing the warrant to identify the objects, and to link them to the offence described in the warrant. The warrant does not have to name the alleged offender unless the name is known.

A further confusion in terminology arises; the police officer who swears the Information to Obtain (Form 1) is known as the **Informant.** A person who provides confidential information to a police officer is also known as an informant (at least by the police—defence counsel have several less complimentary names for such persons). Although it is permissible for the police officer ("Informant" who swears the Information to Obtain) to rely on information from someone else (including a **confidential informant**), even though it is hearsay, the police officer will have to include the surrounding circumstances, so that the justice can assess the veracity of the information and be satisfied that reasonable grounds exist. The level of verification needed for reasonable grounds will depend on (1) whether the information was compelling, (2) whether the confidential informant or other third party was credible, and (3) whether the police had other information to corroborate the information from the confidential informant (*Debot*). In *Richard*, the court found that compelling information from a credible informant was sufficient, without a lot of detail about the informant, who feared for his life. It is also possible for the police to rely on an anonymous tip, if the tip is compelling, and is corroborated by other information (*Plant*). However, an anonymous tip by itself is not sufficient (*Evans*).

The location to be searched must be specified in the Information to Obtain, and in the warrant. Section 487 states that a warrant can be issued for any building, receptacle, or place, and there are numerous cases that discuss what is or is not included in that description. A search warrant cannot be issued under this section to seize bodily substances from the accused, although there are other legislative provisions that allow for this under certain circumstances (as discussed below). In *Laporte*, a search warrant to remove a bullet lodged in the accused was quashed, the Court holding that section 487 does not allow for surgical searches of a person's body. In addition, section 487 cannot be used to install a videotape in a hotel room to capture evidence of criminal activities (*Wong*), although the federal government has since added a section to the *Criminal Code* that would allow for such recordings today (discussed below).

Several cases have concluded that section 487 of the *Code* does not violate section 8 of the *Charter*, in that it is reasonable and does not, on its face, provide for unreasonable searches. Even if section 487 is valid, it does not mean that all search warrants under section 487 are valid, or that searches carried out under valid 487 warrants are reasonable, but only that the section itself complies with section 8. If a warrant is bad (contains false information) or is executed in an unreasonable manner such that section 8 of the *Charter* is violated, the court can consider whether the evidence so obtained is admissible under section 24(2) of the *Charter*. For example, in *Genest* (discussed later) the Supreme Court of Canada found that the search warrant was invalid (it resembled a "fishing licence, not a search warrant"), and that the search was carried out in an unreasonable manner.

DNA and Bodily Impression Warrants: Sections 487.04 to 487.091

Prior to amendments to the *Criminal Code* in July 1995, allowing for the taking of bodily substances for the purpose of DNA analysis, police used the scavenger method—looking for discarded facial tissues or cigarette butts to obtain material for DNA analysis. Another method was simply to invite hundreds of people to prove themselves innocent by submitting to a DNA test through voluntary "blooding lotteries" or a "DNA Dragnet." For example, a blooding lottery was conducted in Vermilion, Alberta, where the RCMP believed three separate rapes were committed by the same man, who was probably a local resident (Plischke, 1995, A2). Despite the collection of DNA from 400 men in the area, the crime remained unsolved (Rusnell 2001, A6).

The introduction of legislation allowing DNA search warrants did not put an end to these lotteries. In 1999, the police in Sudbury, Ontario, collected DNA from more than 400 men in their investigation of the stabbing death of Renee Sweeney (Stevenson 1999, A11). In 2003, the Toronto police successfully flushed out the person who sexually assaulted and killed Holly Jones through a DNA Dragnet. Michael Joseph Briere refused to give a DNA sample because he was concerned that "Big Brother is watching us" (*Briere* para. 121), and he resisted subsequent pressure to provide one. Following surveillance, the police picked up a discarded pop can and matched his DNA to that found on the victim. The facts as presented at the sentencing decision illustrate how individuals can be indirectly pressured to provide DNA samples, even where)as in this case), there is nothing in their past or behaviour at the time (other than the refusal to provide DNA) to ever put them on a list of suspects. DNA Dragnets, which continue today, are criticized for forcing individuals to provide DNA samples that they should in fact be entitled not to. Such "voluntary" collections of DNA and their results must be destroyed if there is not a match to the crime under investigation (section 487.09(3)). Although this section is thought to provide better privacy protection, it also puts greater pressure on individuals to provide voluntary DNA samples. See Rondinelli (2003) for a commentary on Dragnets.

Sections 487.04 to 487.09 provide a code of procedure to be followed in the taking of DNA samples for certain designated offences, their analysis, and the retention or destruction of the information. The list of designated offences was expanded in 2008. The offences are categorized based on whether DNA samples are mandatory or discretionary. Section 487.05 allows a Provincial Court judge (not a justice) to issue a warrant to a peace officer to obtain bodily substances under

---

Box 8.1 Avoiding DNA Results

John Schneeberger, a 38-year-old Saskatchewan doctor, sliced open his arm and implanted a 15-centimetre tube filled with a patient's blood, in an attempt to foil a DNA analysis that would link him to the drugging and sexual assaults of a teenager and a 23-year-old woman. He was sentenced to six years in jail. In August 2003, his Canadian citizenship was revoked for concealing the fact that he was under investigation for these offences at the time he applied for citizenship *Schneeberger*). He was deported to South Africa (Kyle and Switzer 2004, A1).

Box 8.2 DNA Databank

Amendments to the *Criminal Code* in 2000 allowed the RCMP to create a National DNA Data Bank. The NDDB is used to link crimes where there are no suspects, identify or eliminate suspects, and identify possible serial offenders. As of 2008, the NDDB contained 128,124 profiles of individuals convicted of offences, and 40,947 profiles gathered from crime scenes (National DNA Data Bank 2008).

---

certain conditions. The sample must be taken by one of the investigative procedures in section 487.06 (see Pomerance 1995 for a discussion of the sections). The use of DNA is not without controversy (see Brodsky 1994; Federico 1990–1; Holmgren 2008; McDonald 1998; Walsh 1991–2 for a discussion of some of the issues raised by the use of DNA evidence). However, the Supreme Court of Canada has found that these warrant provisions do not violate the *Charter* (*S.A.B.*) (see Stratas 2004 for a discussion). In 2006, a majority of the Court found that section 487.055(1), which allows for the taking of DNA samples from individuals imprisoned before the legislation came into force, did not violate the *Charter* (*Rodgers*).

Other Warrants Under the *Criminal Code*

Sections have been added to the *Criminal Code* as a result of police creativity in investigating offences, the development of technology, and the judicial interpretation of the *Charter*. In *Wong*, the Supreme Court of Canada held that hidden video surveillance of a hotel room, where a group of people gathered to gamble, violated section 8 of the *Charter*. In coming to this conclusion, Mr. Justice LaForest commented that the expectation of privacy in our society is in sharp contrast to the picture painted by George Orwell in his novel *1984* (47). Without judicial authorization, agents of the state cannot have unrestricted discretion to conduct covert video surveillance of citizens. The police had consulted the Crown, which had correctly concluded that the wiretap provisions of the *Criminal Code* could not be used to authorize video surveillance. LaForest, J. stated that it was up to Parliament, not the courts, "to widen the possibility of encroachments...on personal liberties" (57).

In 1993, the federal government responded by adding a general warrant provision to the *Criminal Code*:

> 487.01(1) A provincial court judge, a judge of a superior court jurisdiction or a judge as defined in section 552 may issue a warrant in writing authorizing a peace officer to, subject to this section, use any device or investigative technique or procedure or do any thing described in the warrant that would, if not authorized, constitute an unreasonable search or seizure in respect of a person or a person's property if
>> (a) the judge is satisfied by information on oath in writing that there are reasonable grounds to believe that an offence against this or any other Act of Parliament has been or will be committed and that information concerning the offence will be obtained through the use of the technique, procedure or device or doing of the thing;
>> (b) the judge is satisfied that it is in the best interests of the administration of justice to issue the warrant; and

> (c) there is no other provision in this or any other Act of Parliament that would provide for a warrant, authorization or order permitting the technique, procedure or device to be used or the thing to be done.
>
> (2) Nothing in subsection (1) shall be construed as to permit interference with the bodily integrity of any person.

Subsection 3 states that the warrant shall contain terms and conditions to ensure that the search and seizure is reasonable, and subsection 4 states that if the warrant authorizes the use of a camera or similar electronic device to record activities in circumstances where the persons have a **reasonable expectation of privacy**, the warrant shall contain conditions to ensure privacy as far as possible. The section is quite broad, in that it contemplates a warrant that may authorize the peace officer to "do any thing" (see discussion by Coughlan 2003 and Watt 2008). There is also some question whether the general warrant section can be used simply because another warrant section does not allow for more invasive searches (see Coughlan 2009 and *Ha*).

Warrants to search by camera or similar electronic devices are only available for the offences listed in section 183, and several of the provisions regarding the interception of private communications (see Chapter 9) apply to these warrants. If the warrant authorizes covert entry into a place, the warrant shall require "notice of the entry and search be given within any time after the execution of the warrant that the judge considers reasonable in the circumstances" (subsection 5.1).

Another case that inspired the federal government to codify the law of search and seizure was *Wise*, where the police, without judicial authorization, installed a "beeper" in the accused's car in order to track his location (this was done while the car was in police custody pursuant to a warrant to gather evidence in a homicide investigation). The police used the beeper to maintain visual surveillance on the suspect, which resulted in observing the accused's car near the scene of a crime (the destruction of a $2 million Bell Canada communication tower). The trial judge found that the accused's rights under section 8 of the *Charter* had been violated, and excluded all evidence obtained directly or indirectly from the use of the beeper. The accused was consequently acquitted, but the Ontario Court of Appeal ordered a new trial. At the Supreme Court of Canada, the Crown conceded that the installation of the beeper violated the accused's rights under section 8, even though the use of such a beeper had been held to be neither a search nor a seizure under the Fourth Amendment in the United States (*Wise* 1992, 219). In a four to three decision, the Supreme Court decided that even though the accused's rights under section 8 were violated, the evidence should not be excluded under section 24(2). All the judges held that it would be better if such surveillance were dealt with under legislation. As a result of this decision, Parliament enacted section 492.1(1) in 1993 which allows a justice to issue a warrant to install and maintain a tracking device if "there are reasonable grounds to suspect that an offence . . . has been or will be committed . . ." The warrant is valid for 60 days, with the possibility of extension by further warrants. Note that the section allows for a warrant based on reasonable suspicion and does not require reasonable grounds. Does it meet the requirements in *Hunter*?

In a similar fashion (that is, "on reasonable grounds to suspect" an offence has taken place), section 492.2 allows for a warrant to "install, maintain and remove a number recorder in relation to any telephone or telephone line," and to monitor the number recorder. A **dial number recorder** can record the telephone numbers called from a phone and their locations. Again, the warrant lasts for 60 days, with the provision for further warrants. In *Nguyen* the British Columbia Supreme Court found that the section was unconstitutional and should be read to require "reasonable grounds to believe"; however, a Quebec Superior Court found that the section did not violate the *Charter* (*Cody and Langille*). See Penney (2008) for commentary on these sections.

Box 8.3 What Happened to Wise?

In 1996, the Ontario Court of Justice (General Division) stayed the charges against Wise. Chadwick, J. stated, "Given the compendium of factors which I have already reviewed, this is clearly a case where the community's sense of decency and fair play and the integrity of our judicial process requires that a stay of proceedings be granted. When one considers the harm already inflicted on Mr. Wise as a result of the prosecutorial misconduct, coupled with the personal prejudice that would be portrayed as a result of a third trial, there is no other remedy short of stay of proceedings which could be considered as 'just and appropriate in the circumstances.'" (*Wise*, 1996 para. 55).

In May 1997, section 487.092 was added to allow a Justice to issue a warrant to a peace officer to obtain "any handprint, fingerprint, footprint, foot impression, teeth impression or other print or impression of the body of any part of the body in respect of a person," where there "reasonable grounds to believe that an offence...has been committed," and "that it is interests of the administration of justice to issue the warrant."

A number of sections in the *Criminal Code* allow for search warrants for specific offences: sections 164 (crime comics, obscenity and child pornography), 199 (gaming offences, common bawdy house), 256 (blood samples in drinking and driving offences), 320 (hate propaganda), 395 (valuable minerals), and 462.32 (proceeds of crime). Some sections provide for entry to make arrests (entry warrants are discussed in Chapter 2).

Provisions Regarding Search Warrants

Section 487.1 allows for warrants under sections 256 or 487 to be issued over the telephone or by other means of telecommunication (**telewarrants**). The section states that "where a peace officer believes that an indictable offence has been committed and that it would be impractical to appear personally before a justice to make application for a warrant," the peace officer may apply for a warrant by means of a telephone or other means of telecommunication. Telewarrants are requested by information on oath, and the section stipulates how the information is later to be reduced to a record.

The *Criminal Code* specifies how search warrants are to be executed. Section 488 states that a warrant under section 487 or 487.1 shall be executed by day, unless the justice authorizes execution by night. Section 489 allows everyone who executes a warrant to "seize, in addition to the things mentioned in the warrant, anything that the person believes on reasonable grounds has been obtained by [or]...has been used in the commission of an offence." Section 492 allows for the seizure of explosives under a section 487 or 487.1 warrant, if the person executing the warrant suspects that the explosive is "intended to be used for an unlawful purpose." Section 489 covers seizures under the *Criminal Code* and other federal legislation, and specifies what police officers are to do with seized property. The officer shall return the property seized under certain circumstances, or bring it before a justice to have matter dealt with under section 490.

Section 29(1) of the *Criminal Code* states that everyone who executes a warrant should have it with them "where it is feasible to do so," and should produce it upon request. This legislative requirement has to be read in conjunction with the rights of a person upon arrest or detention (see section 10 of the *Charter*).

Production Orders

Effective September 15, 2004, sections 487.011 to 497.017 allow police officers to get **production orders** from a justice or judge, to compel a person (other than one under investigation) to produce documents or to prepare documents from existing data. Failure to comply with a production order is a summary conviction offence, subject to a fine not exceeding $250,000, six months in prison, or both (section 487.017). The costs of production orders under section 487.012 are absorbed by the company required to produce the information as a cost of doing business (*Tele-Mobile Co.*).

SEARCHING A LAWYER'S OFFICE

Section 488.1 deals with the procedures to determine if material that is the subject of a search of a lawyer's office is protected by solicitor–client privilege (privilege is discussed in more detail in Chapter 12). In *Lavallee*, the majority of the Supreme Court of Canada stated that it is clear that section 488.1 "was never intended to supersede the common law principles pertaining to the issuance of [warrants] in the law office context, as discussed by Lamer J. in *Descôteaux*,. . .but merely with the manner in which they are carried out" (para. 22). They held that the legislation violated section 8 of the *Charter*, in that it compromised solicitor-client privilege. The legislation was therefore of no force and effect under section 52 of the Constitution Act (para. 3). The Court stated that it would be best for Parliament to remedy this problem, but provided some guidelines to govern the search authorization process, and the manner in which searches must be conducted in order to protect solicitor-client privilege (para. 49). The Court's guidelines include:

> 2 Before searching a law office, the investigative authorities must satisfy the issuing justice that there exists no other reasonable alternative to the search.
>
> . . .
>
> 5 Every effort must be made to contact the lawyer and the client at the time of the execution of the search warrant. Where the lawyer or the client cannot be contacted, a representative of the Bar should be allowed to oversee the sealing and seizure of documents.
>
> . . .
>
> 8 The Attorney General may make submissions on the issue of privilege, but should not be permitted to inspect the documents beforehand. The prosecuting authority can only inspect the documents if and when it is determined by a judge that the documents are not privileged.
>
> . . .
>
> 10 Where documents are found to be privileged, they are to be returned immediately to the holder of the privilege, or to a person designated by the court (para. 49).

SEARCH BY WARRANT—NON-*CODE* PROVISIONS

Legislation other than the *Criminal Code* provides for search warrants. For example, section 11(1) of the *Controlled Drugs and Substances Act* authorizes a justice to issue a search warrant for a controlled substance or precursor (both defined in Schedules to the Act), anything they might be contained in, any offence-related property, or "any thing that will afford evidence in respect of an offence" under the *Act*. Section 11(5) allows a peace officer who is executing the warrant to "search any person found in the place set out in the warrant," if the peace officer has reasonable grounds to believe the person has any thing set out in the warrant. Section 11(7) allows a peace officer to conduct searches without warrants "if the conditions for obtaining a warrant exist but by reason of exigent circumstances it would be impracticable to obtain one."

Section 12 allows peace officers to "enlist such assistance as the officer deems necessary," and to "use as much force as is necessary in the circumstances." This does not mean that there are no limitations to what police officers can do. In *Genest*, several police officers conducted a "full-scale search," in which they broke down the door of the accused's house without advance warning. While the police did have a search warrant, the Supreme Court of Canada found that it was defective, in that it did not name a police officer who was to be responsible for the search, and it did not specify the items to be searched for and seized. Mr. Justice Dickson stated that the Justice had "issued a fishing licence, not a search warrant" (405). In executing a search warrant, the manner of the search must be reasonable. "The greater the departure from the standards of behaviour required by the common law and the *Charter*, the heavier the onus on the police to show why they thought it necessary to use force in the process of an arrest or a search" (*Genest*, 408). The Court in *Genest* excluded the seized evidence under section 24(2) of the *Charter*.

WARRANTLESS SEARCHES TO OBTAIN INFORMATION TO SUPPORT AN APPLICATION FOR A SEARCH WARRANT

In *Kokesch*, police officers, who were suspicious that the accused was growing marijuana, searched the yard and smelled the odour of marijuana coming from a house vent. Based on that evidence, they obtained a warrant to search the house. The Supreme Court of Canada found that section 8 was violated by the warrantless search of the yard (the **perimeter search**); therefore the subsequent warrant based on it was invalid. The police officers had only been suspicious in *Kokesch*, and did not have reasonable and probable grounds to believe there would be a narcotic present before conducting the perimeter search (the legislation existing at the time would have allowed for a warrantless search, had the police officers had reasonable grounds to believe such an offence was taking place). The evidence from the search was accordingly excluded under section 24(2) of the *Charter*.

In *Plant*, the police received a tip from an unknown informant, examined electrical consumption (which was four times higher than normal for the size of the house under investigation), and conducted a warrantless perimeter search (sniffing at and examining a basement exhaust vent) (para. 4). Again, the Supreme Court of Canada made it clear that warrantless perimeter searches violate section 8 of the *Charter* unless some type of exigent circumstances exist (para. 15). In deciding whether obtaining information from the utility company violated the accused's rights under section 8, Mr. Justice Sopinka (for the majority of the Court) cited *Hunter* for the proposition that the purpose of section 8 is to protect people's privacy

---

Box 8.4 Perimeter Search

The classic statement of the law respecting the search of dwellings is that "the house of everyone is to him as his castle and fortress" (*Semayne*). In a case just subsequent to the trial decision in *Kokesch*, the Crown was trying to distinguish that decision by laying a better factual basis to counter the *Charter* argument of the defence. In explaining his decision to conduct a perimeter search to sniff for the odour of growing marijuana, Detective Stu Gillette of the Vancouver Police Department testified: "Well, Your Honour, the way I saw it, just because a man's home is his castle doesn't mean you can't swim in his moat!" The evidence was excluded.

---

against unwarranted state intrusion, by balancing the individual's reasonable expectation of privacy against the state's interest in law enforcement. Since section 8 protects people, not property, there is no need for a person to have a proprietary interest in something for his or her privacy to be invaded. When it comes to utility-use computer records, a number of factors have to be considered:

> the nature of the information itself, the nature of the relationship between the party releasing the information and the party claiming its confidentiality, the place where the information was obtained, the manner in which it was obtained, and the seriousness of the crime being investigated. [These factors]…allow for the balancing of the societal interests in protecting individual dignity, integrity and autonomy with effective law enforcement (para. 19).

To be protected by section 8, the information must be of a "personal and confidential nature," that might reflect "intimate details of the lifestyle and personal choices of the individual" (para. 20). Sopinka decided that the information did not reveal intimate lifestyle details, the relationship between the electrical company and the accused was not a confidential one, the search was not an intrusive or high-handed search, and the seriousness of the offence outweighed the privacy interest claimed by the accused (para. 23). Madame Justice McLachlin (in the minority) differed as to whether there was a reasonable expectation of privacy regarding electrical records. Although it was a borderline issue, she was of the view that the police ought to obtain a warrant before gathering such information. She wrote, "a reasonable person looking at these facts would conclude that the records should be used only for the purpose for which they were made…and not divulged to strangers without proper legal authorization" (para. 41). She felt that such records do reveal intimate details of a lifestyle—for example, the fact that the persons are growing marijuana (para. 42).

In turning to whether the evidence should be excluded under section 24(2), based on the finding that the warrantless perimeter search violated section 8 and was an integral part of the search (i.e., there was sufficient temporal connection), Sopinka, J. considered the seriousness of the offence and the necessity of the evidence to a conviction. He concluded that the exclusion of the evidence would have a more negative impact on the repute of the administration of justice than its admission, and the evidence was admissible.

After the *Kokesch* decision from the Supreme Court of Canada (which determined that warrantless perimeter searches violated section 8 of the *Charter*), the police started to use **knock-ons** and other investigative techniques in situations where they suspected an indoor marijuana growing operation, but had insufficient information to obtain a search warrant. A knock-on is simply that; the police approach the suspicious house and knock on the door. If there is an overwhelming smell of marijuana, the suspect who answers the door is arrested, and the police use the additional evidence of odour to then obtain a search warrant.

Is a knock-on different from a warrantless perimeter search, in which the police enter the property and sniff at vents or open windows? In *Evans,* the police had received a tip from Crime Stoppers that the tipster ("very experienced with drugs") had gone to the front door of Evans's house and had been overwhelmed by the smell of growing marijuana. After observing Evans's car in the driveway, three police officers attended at the front door, and Evans identified himself. The police officers detected the odour of marijuana and arrested Evans. While securing the residence, the police officers observed marijuana plants in the basement, and obtained a warrant to search the house. The trial judge decided that the knock-on was not a search, or if it was a search, it did not violate section 8. All three judges of the British Columbia Court of Appeal found that the knock-on constituted a search, and violated section 8 of the *Charter*. The Supreme Court of Canada agreed; however it also decided that the evidence was properly admitted under section 24(2). Again, this put the police on notice that any similar subsequent violations would result in the exclusion of evidence.

SEARCHING MOTOR VEHICLES

As discussed in Chapter 2, random spot checks of motor vehicles, authorized by provincial legislation, violate section 9 of the *Charter* (the right not to be arbitrarily detained). However, the Supreme Court of Canada has found such violations to be demonstrably justified under section 1. Police officers at these random spot checks are entitled to ask questions about sobriety and the mechanical condition of the vehicle, and to require the driver to produce a driver's licence, proof of ownership, and insurance (*Hufsky* and *Thomsen*, discussed in Chapter 2).

What happens if the police officer, without any grounds, goes beyond these acceptable questions? This happened in *Mellenthin*, where the accused was stopped at a random road-side check, and the police officer improperly asked him what was in the gym bag on the front seat beside him. Through a series of questions by the police officer and responses by Mellenthin, the police officer established reasonable grounds to believe Mellenthin was in possession of marijuana. The officer subsequently demanded that Mellenthin hand over the gym bag. This search violated section 8 of the *Charter*. Mr. Justice Cory was of the view that, "as a result of [the] detention, it can reasonably be inferred that the [driver] felt compelled to respond to questions put to him by the police officer" (487). It is up to the Crown to show that there was informed consent to the search—was the driver aware of his rights to refuse to respond to the questions? Cory, J. agreed with the trial judge that the driver in this case felt compelled to answer the questions, and that therefore there was no consent. According to Cory, J. "unless there are reasonable…grounds for conducting the search, or drugs, alcohol or weapons are in plain view in the interior of the vehicle, the evidence flowing from such a search should not be admitted" (491). To admit such evidence would adversely affect the fairness of the trial. See *Harrison* for another case in which the Supreme Court of Canada excluded drugs that were discovered after an unfounded search of a motor vehicle. Also see the discussion below on Search As An Incident to Arrest.

SEARCH WITHOUT WARRANT—LEGISLATION

In 1983, the Law Reform Commission of Canada reported that there were 82 federal enactments that allowed for warrantless seizures. Most of these provisions have now been amended. Section 462, which could face constitutional challenge, still allows for a warrantless seizure of counterfeit money, and materials or tools used to produce it. Section 117.02 of the *Code* allows warrantless searches of persons, vehicles, or places other than dwelling-houses, for weapons or ammunition, where the police officer has grounds to obtain a warrant but it would not be practical to obtain a warrant because of exigent circumstances. Warrantless searches or entries to private dwellings are only allowed in limited exigent circumstances (for example, see section 529.3 of the *Criminal Code* as discussed in Chapter 2, and section 11(7) of the *Controlled Drugs and Substances Act*).

Section 254(3), which requires a person to provide samples of breath, and in some cases blood, allows for searches and seizures without a warrant. However, the requirement "that reasonable and probable grounds exist is not only a statutory but a constitutional requirement as a precondition to a lawful search and seizure under s. 8 of the Canadian Charter of Rights and Freedoms" (*Shepherd* para. 13). Given the warrantless search under this section, it is up to the Crown to show that the search is reasonable (para.15). In the case of *Shepherd*, the Crown had established both subjective and objective grounds for the search (para. 23).

There are many provincial statutes and other federal statutes that allow for search and seizure without authorization. Although these sections can be challenged as violating section 8 of the *Charter*, the courts might find that they are reasonable under the circumstances.

SEARCH WITHOUT WARRANT—COMMON LAW

The Law Reform Commission of Canada (1984b) has written about the different historical roots of the search warrant and the warrantless search. Search warrants developed in the context of protecting property; warrantless searches developed in the context of searching suspected criminals. Theoretically, property had (and perhaps still has) much greater legal protection against searches than people. This section examines some of the common law powers of search.

Search by Sniffer Police Dogs

In a split decision, the Supreme Court of Canada in *Kang-Brown* elaborated on the common law power of police to use sniffer dogs. Five of the nine judges found that the police had a common law power to use sniffer dogs; four required reasonable suspicion (para. 90) and one required a general suspicion. Such searches can be done without a search warrant (*Kang-Brown* para. 26). In *A.M.*, where the Supreme Court of Canada found that a student had a reasonable expectation of privacy in his knapsack, the Court also found that the random search by a police dog was not authorized by law and was unreasonable. For comments on the use of sniffer-dogs and the two cases, see Bailey (2009), Coughlan (2008), Davis-Baron (2007), Jochelson (2009), Kerr and McGill (2007), Marks (2007), Quigley (2008), Shapiro (2008), Stuart (2008), Tanovich (2008).

Search During Investigative Detention

In Chapter 2 we saw that the Supreme Court of Canada in *Mann* found that the police are allowed to detain persons for investigative purposes, if they have "reasonable grounds to detain," based on the officer's reasonable suspicion. Another issue addressed in that case was the extent to which the police could search a person held in an investigative detention. According to Iacobucci, J. for the majority, there is no automatic right to search as an incident to investigative detention; rather, the search must "be reasonably necessary in light of the totality of the circumstances" (para. 40). That is, "where a police officer has reasonable grounds to believe that his or her safety or that of others is at risk, the officer may engage in a protective pat-down search of the detained individual" (para. 45). Furthermore, "investigative detention should be brief in duration and does not impose an obligation on the detained individual to answer questions posed by the police" (para. 45).

According to Latimer (2007), the Court in *Mann* required the police to have reasonable grounds to believe that their safety was in issue before they were allowed to conduct a protective search. However, in *Clayton*, where it was alleged that firearms were involved, the Court implied that the threshold test to conduct a protective search appears to be reasonable suspicion. The investigative detention search powers are therefore different from the powers following an arrest.

Search As an Incident to Arrest

In *Cloutier v. Langlois*, a lawyer swore a private information against two police officers, alleging assault. The police officers had stopped the lawyer because he made a right turn from the centre lane, contrary to a municipal bylaw. They asked for his identification, and the discussion became somewhat heated. There was a warrant of committal outstanding for the lawyer for unpaid traffic fines, and the police arrested him. Before putting him in the police car, the officers frisked him (hands on the hood of the car, legs spread, and a pat down, just as you see on television). At the assault trial, the charges against the police officers were dismissed. The trial judge found that the police officers had reasonable and probable grounds for doing a frisk search; the lawyer was abusive, therefore, the police officers had a concern for their safety. The Quebec Court of Appeal overturned the trial decision and entered a verdict of guilty against the police officers, holding that the search was unlawful and technically an assault.

On further appeal, Madame Justice L'Heureux-Dubé, for the Supreme Court of Canada, surveyed the history of the common law power to search as an incident to arrest, a major question being whether police officers require reasonable grounds to conduct a **search incidental to arrest.** The common law in England developed to the point where it appeared that police officers did not require grounds for such a search; however, in 1984, England's Parliament codified the search of lawfully arrested persons. A person may be searched if "the constable has reasonable grounds for believing that the arrested person may present a danger to himself or others...or [has] reasonable grounds for believing that the person has anything on him which he might use to escape or which might be evidence" (*Cloutier* 173). In the United States, there is an unfettered power to search incidental to arrest, and reasonable grounds are not required for the search (*Cloutier* 174).

Madame Justice L'Heureux-Dubé concluded that police officers do not need reasonable grounds to conduct a search incidental to arrest (however, they first need reasonable grounds to make the arrest; see arrest powers discussed in Chapter 2). She recited several general principles regarding these searches. First, there is no duty to conduct a search, and the police officer has the discretion whether or not to conduct a search. Second, the search must have a valid object, such as the discovery of evidence or a weapon that might endanger the safety of the police, the public, or the accused, or something that might facilitate escape. The police cannot search to intimidate or harass. Third, the search cannot be conducted in an abusive fashion, and the physical and psychological constraint used has to be proportionate to the objectives sought (186). She noted that the search in *Cloutier* was not an intrusive frisk.

The Supreme Court of Canada revisited the power of search incidental to arrest in *Caslake*. Following Caslake's arrest for possession of marijuana for the purpose of trafficking, the police towed his car to the RCMP detachment and searched it six hours later for the sole purpose of conducting an inventory (required by RCMP policy). The police discovered cocaine. Chief Justice Lamer stated the three prerequisites to a reasonable search under section 8: "In order to be reasonable, a search must be authorized by law, the law itself must be reasonable, and the search must be carried out in a reasonable manner" (para. 10). Warrantless searches are *prima facie* unreasonable, and the Crown must show that they are reasonable. A search incident to arrest must be related to the purpose of the arrest (protecting the police, the evidence, and so on). In this case, the "inventory search" was not related to the purpose of the arrest, and was therefore without lawful authority (paras. 28-29). However, the evidence was admitted under section 24(2) of the *Charter*. With regard to searching vehicles, Lamer, C.J. stated that vehicles can be the legitimate target of a search incidental to arrest, provided they are conducted "to achieve some valid purpose connected to the arrest"–that is, "ensuring the safety of the police and public, the protection of evidence from destruction at the hands of the arrestee or others, and the discovery of evidence which can be used at the arrestee's trial," as set out by L'Heureux-Dubé J. in *Cloutier* (para. 19). Therefore, "the right to search a car incident to arrest and the scope of that search will depend on a number of factors, including the basis for the arrest, the location of the motor vehicle in relation to the place of the arrest, and other relevant circumstances" (para. 23). For example, in *Belnavis*, the Supreme Court of Canada decided that an arrest for outstanding traffic tickets could not be used as authority to search the trunk of a car. The question is whether it is reasonable to conduct such a search.

As suggested in *Caslake*, the right to search as an incident to arrest is not limited to searches conducted immediately on arrest. The six-hour delay was not by itself problematic in that case. The *Miller* case (one of the cases Justice L'Heureux-Dubé referred to in *Cloutier* without criticism), considered evidence (a bandage around the accused's head) seized with the assistance of a doctor, 18 hours after the arrest. The Ontario Court of Appeal held that it was a valid seizure incidental to arrest, since if a seizure can be made on the spot, it can be made later when the person is in custody. Thus, if a police officer arrests someone and wants the suspect's clothes for evidence, there is no requirement that the suspect strip on the spot.

Some questions have been raised regarding the intensity of a search. In the *Morrison* case, referred to in *Cloutier* by Justice L'Heureux-Dubé, the Ontario Court of Appeal approved a three-minute strip-search of the accused, who had been taken into custody on charges of theft and stolen property. However, in *Flintoff*, the Ontario Court of Appeal excluded a breath sample because an accused was strip-searched following his arrest (see discussion by Sankoff 1998). Numerous restrictions are imposed by the courts on some of the more intrusive searches, which are only justified if there are reasonable grounds for believing that there is evidence to be had. The courts will not allow some searches at all. Some more intrusive searches are now allowed by warrant (for example, the sections allowing for bodily substances for DNA analysis, discussed above).

In *Golden*, the Supreme Court of Canada examined the law surrounding strip searches as an incident to arrest. Iacobucci and Arbour, J.J. for the 5-4 majority, addressed three questions: "(1) Was the search authorized by law? (2) Is the law itself reasonable? (3) Was the search conducted in a reasonable manner?" (para. 44). They found that the common law power to search incidental to arrest includes the power to strip search, defined as "the removal or rearrangement of some or all of the clothing of a person so as to permit a visual inspection of a person's private areas, namely genitals, buttocks, breasts (in the case of a female), or undergarments" (para. 47). Once reasonable grounds exist to make an arrest, an accused may be searched as an incident to that arrest. However, if the search involves a strip search, the police must have "reasonable and probable grounds justifying the strip search in addition to reasonable and probable grounds justifying the arrest" (para. 99). Strip searches should be conducted at the police station, although they can be conducted in the field "where there is a demonstrated necessity and urgency to search for weapons or objects that could be used to threaten the safety of the accused, the arresting officers or other individuals" (para. 102). The majority decided that the common law was reasonable, but the search in these circumstances was unreasonable, and violated section 8. Given the circumstances of the case, the majority refused to conduct a section 24(2) analysis, and entered an acquittal. The majority also suggested that legislative guidance from Parliament "as to when and how strip searches should be conducted would be of assistance to the police and to the courts" (para. 103) (see Gottardi 2002 for a discussion of the case).

Plain View

Another common law search-related power is that a police officer who is lawfully on any premises (for example, by invitation) may seize things in **plain view** as evidence of an offence. The plain-view doctrine is not a power to search as such, but rather a power to seize items in plain view. So, for example, if a police officer is invited into a house to investigate or prevent an assault, she or he can seize any items that are in plain view if they constitute evidence of an offence, even though that offence may have nothing to do with the assault. Section 489(2) of the *Code* allow polices officers who are executing a warrant to seize evidence of an offence that is not mentioned in the search warrant, without the requirement of plain view, if they have reasonable grounds to do so.

Box 8.5 Searching for Swallowed Drugs

Based on undisclosed and unspecified confidential information, the accused in *Greffe* was suspected of importing heroin. He was arrested at the Calgary airport for unpaid traffic tickets and was strip-searched. He was informed that a doctor would conduct a body-cavity search and was advised of his right to counsel. He was taken to a hospital, where a rectal examination was conducted with the use of a sigmoidoscope. A condom containing 40 grams of heroin was removed from the accused's lower bowel, with the use of "Kelly grasping forceps." He was then arrested and warned in respect of importing a narcotic. Lamer, J., for the majority, ruled that the evidence, which was conceded to have been obtained through breaches of sections 8, 10(a), and 10(b), should be excluded under section 24(2) of the *Charter*. He said, "this court cannot condone rectal searches incident to an arrest for outstanding traffic warrants" (799).

In *Monney*, the Supreme Court of Canada approved a "bedpan vigil" in a "drug loo facility," under section 98 of the *Customs Act*, as a means of "searching" for heroin swallowed by the accused.

By Consent

At common law, a person or a place may also be searched by consent. Consent must be voluntary and can be withdrawn. The Supreme Court of Canada has dealt with the question of consent in light of section 8 of the *Charter* in several cases involving search and seizure.

In *Borden*, the police officers obtained the consent of the accused to take a sample of his blood for the purpose of DNA testing. Borden was charged with sexually assaulting a woman who was able to identify him. He was arrested for the offence, told of his rights, and detained. The police were also investigating a second offence, the violent sexual assault of a 69-year-old woman who could not identify her assailant. In an effort to get Borden's consent to providing a blood sample for the purpose of DNA analysis, the police officers consulted with Crown counsel and came up with the following consent form:

> I, Josh Randall Borden…do hereby give my consent to the New Glasgow Police Department to take a sample of my blood for the purposes relating to their investigations.

Borden was not aware that he was being investigated for the second sexual assault, and the police did not inform him of this, nor of his right to counsel in light of this second offence. The trial judge found that the *Charter* violation under section 8 was technical, and concluded that the admission of the evidence would not bring the administration of justice into disrepute. The accused was convicted of the second sexual assault and sentenced to six years. The Nova Scotia Court of Appeal found that both section 8 and section 10(b) of the *Charter* had been violated, and held that the evidence should have been excluded under section 24(2) of the *Charter*. They set aside the conviction. On further appeal, the Supreme Court of Canada found that Borden's rights under both sections 8 and 10(b) had been violated, and that the Nova Scotia Court of Appeal did not err in concluding that the evidence should have been excluded under section 24(2). The trial judge had found only a technical breach, whereas the Court of Appeal was correct in finding this was more than a technical breach.

With regard to consent, Mr. Justice Iacobucci wrote, "In the absence of a statutory scheme whereby the police can demand a blood sample in cases such as these (a scheme that may raise Charter concerns), the police require the true consent of an accused" (422). Consent, he reasoned, requires the waiver of right:

> In order for a waiver of the right to be secure against unreasonable seizure to be effective, the person purporting to consent must be possessed of the requisite informational foundation for a true relinquishment of the right. A right to choose requires not only the volition to prefer one option over another, but also sufficient available information to make the preference meaningful (417).

However in *Arp*, the Supreme Court of Canada did not think it was unfair or illegal for the police to retain evidence obtained by consent in one investigation, and to use it in a subsequent investigation, where the second investigation was unanticipated at the time the consent was obtained. In this case, neither Arp nor the police had put any limits on the consent to provide bodily samples. See Luther (2008) for some concerns over the use of consent searches by the police.

SUMMARY

Section 8 of the *Charter* guarantees everyone the right "to be secure against unreasonable search or seizure." This right is aimed at protecting the privacy interests of individuals. For a search or seizure to be reasonable, it must generally be conducted under the authority of prior authorization, where feasible. This authorization must be by a person who is capable of acting in a judicial capacity, and who in fact acts judicially in exercising a discretion to grant or refuse the authorization. The standard for issuing such authorization is "a credibly based probability," sufficiently expressed by legislative requirements of "reasonable grounds" or "reasonable and probable grounds."

Search warrants are commonly obtained under section 487 *Criminal Code* to search for evidence in respect of the commission of an offence. Such warrants require the applicant (the "Informant") to have sworn to facts (in the "Information to Obtain") constituting "reasonable grounds to believe" that there is evidence in a certain location relating to a certain offence. The *Code* contains other sections under which search warrants may be granted, which may cover such evidence-gathering techniques as video surveillance, tracking devices, and dial number recorders. Warrants should normally be present when they are executed, to be available for production to persons at the location in question. Other legislative provisions exist respecting authorization to search and seize.

Unauthorized perimeter searches, or the use of techniques such as knock-ons to gain evidence to support applications for search warrants for more intrusive searches, are not permissible. However, depending on the circumstances of the case, evidence ultimately obtained through the execution of the resultant search warrant may still be admissible, if the search warrant could have been obtained without the initial unauthorized searches.

In addition to statutory powers of search and seizure, valid common law powers remain. The power to search a person incidental to their arrest has survived, so long as the search follows a valid exercise of discretion, for a valid object, and is conducted in a reasonable manner. Searches may be justified as incidental to arrest even if they occur some time removed from the actual arrest. The police also retain the common law power to seize items under the "plain-view" doctrine, which allows someone who is lawfully on premises to seize items that are in plain view as evidence of an offence. A person may consent to an otherwise unreasonable or unjustified search, although such consent must constitute an informed waiver of rights under section 8.

QUESTIONS TO CONSIDER

(1)   What is the court concerned with protecting under section 8 of the *Charter*?

(2)   Discuss the three criteria given by the Supreme Court of Canada in *Hunter* for evaluating authorizations to search under section 8 of the *Charter*.

(3)   In *Edwards*, the Supreme Court of Canada stated that "a reasonable expectation of privacy is to be determined on the basis of the totality of the circumstances." The Court again examined the issue in *Patrick*. What are the factors the court will use in assessing the circumstances? Provide an example of a situation in which an accused may not have a reasonable expectation of privacy.

(4)   What rules did the Supreme Court of Canada provide in the *M. (R.M.)* case regarding searches in schools?

(5)   A reasonable expectation of privacy can be determined through a societal expectation approach or a normative approach. What is the difference between these two approaches, and what would be the advantages and disadvantages of each? Provide an example of each approach. Which approach does the Supreme Court of Canada prefer?

(6)   What three factors will the court consider in determining whether there are sufficient grounds to issue a search warrant, when police are relying on information from a third party.

(7)   What rules apply to a search incidental to arrest?

(8)   Do police officers in a) Canada, b) England, and c) the USA need reasonable grounds to search a person incidental to arrest?

(9)   Give an example of circumstances under which the police could lawfully search a motor vehicle without a warrant.

(10)   Identify and discuss three circumstances under which the police can seize evidence without a warrant.

(11)   Create a fact pattern question that would require an analysis of section 8 and section 24(2) of the *Charter*. Answer your question.

(12)   Read section 462(2) of the *Criminal Code*. How might a person subjected to such a seizure argue that the machines used to make counterfeit money seized under the section should not be admissible at the person's trial for possession of such machines?

(13)   What is the plain view doctrine?

BIBLIOGRAPHY

Bailey, Jane. "Across the Rubicon and into the Apennines: Privacy and Common Law Police Powers after A.M. and Kang-Brown." (2009) 55 *Criminal Law Quarterly* 239.

Boilard, Jean-Guy. *Guide to Criminal Evidence*. Cowansville, Quebec: Les Editions Yvon Blais Inc., 1999, looseleaf. Chapter 6 contains a section on search.

Brodsky, Daniel J. "Telewarrants."(1986–7) 29 *Criminal Law Quarterly* 345.

Brodsky, G. Gregg. "DNA: The Technology of the Future is Here." (1994) 36 *Criminal Law Quarterly* 10.

Coughlan, Steve. "General Warrants at the Crossroads: Limit or Licence?" (2003) 10 *Criminal Reports* (6th) 269.

Coughlan, Steve. "Improving Privacy Protection, But By How Much?" (2008) 55 *Criminal Reports* (6th) 394.

Coughlan, Steve and Marc S. Gorbet. "Nothing Plus Nothing Equals...Something? A Proposal for FLIR Warrants on Reasonable Suspicion." (2005) 23 *Criminal Reports* (6th) 239.

Coughlan, Stephen G. "*R. v. Ha*: Upholding General Warrants without Asking the Right Questions." (2009) 65 *Criminal Reports* (6th) 41.

Davis-Barron, Sherri. "The Lawful Use of Drug Detector Dogs." (2007) 52 *Criminal Law Quarterly* 345.

Fontana, James A. and M. David Keeshan. *The Law of Search and Seizure in Canada.* 7th ed. Markham, Toronto: LexisNexis Canada, 2007 (looseleaf).
    The text covers material from the nature of search warrants, their issuance, and execution, and search warrants under various federal and provincial statutes. The impact of the *Charter* on the law of search and seizure is also discussed.

Federico, Ricardo G. "The Genetic Witness: DNA Evidence and Canada's Criminal Law." (1990–1) 33 *Criminal Law Quarterly* 204.

Freidland, M.L., and Kent Roach. *Criminal Law and Procedure: Cases and Materials,* 8th ed. Toronto: Emond Montgomery Publications Limited, 1997.
    Chapter 2, Part III, ("Search and Seizure") covers section 8 of the *Charter*, reasonable expectation of privacy, judicial authorization of search and seizure, the effect of obtaining a search warrant, and exclusion of evidence under section 24 of the *Charter*.

Gottardi, Eric V. "The *Golden* Rules: Raising the Bar Regarding Strip Searches Incident to Arrest."(2002) 47 *Criminal Reports* (5th) 48.

Hendel, Ursula, and Peter Sankoff. "*R. v. Edwards*: When Two Wrongs Might Make A Right." (1996) 45 *Criminal Reports* (4th) 330.

Holmgren, Janne. *DNA Evidence: Judge and Jury Challenges, Judge and Jury Interpretations, Perceptions, and Understanding of DNA Evidence.* Saarbrücken: VDM Verlag Dr. Müller, 2008.

Jochelson, Richard. "Multidimensional Analysis as a Window into Activism Scholarship: Searching for Meaning with Sniffer Dogs." (2009) 24 *Canadian Journal of Law and Society* 231.

Kaiser, Archibald. "*Patrick*: Protecting Canadians' Privacy Interest in Garbage; 'a step too far' for the Supreme Court." (2009) 64 *Criminal Reports* (6th) 30.

Kerr, Ian, and Jena McGill. "Emanations, Snoop Dogs and Reasonable Expectation of Privacy." (2007) 52 *Criminal Law Quarterly* 392.

Kyle, Anne and Tim Switzer. "He's left the Country: Former Kipling Doctor sent to South Africa." (22 July 2004) *Regina Leader Post* A1.

Latimer, Scott. "The expanded scope of search incident to investigative detention." (2007) 48 *Criminal Reports* (6th) 201.

Law Reform Commission of Canada. *Writs of Assistance and Telewarrants.* Report #19. Ottawa, 1983.

Law Reform Commission of Canada. *Investigative Tests.* Working Paper #34. Ottawa, 1984a.
    The Commission discusses investigative tests that require passive acquiescence, and divides them into those of simple body inspection (e.g., lineups, fingerprinting, dental impressions), those that require the removal of external substances (e.g., hair examination, fingernail scrapings), those which require removal of internal substances (e.g., blood tests, semen samples), and those in which substances are administered (narcoanalysis and other drugs). It discusses tests that require active participation such as performance tests, polygraph examination, and hypnosis. Uses of the test results as evidence, the law regarding the administration of the investigation tests, and issues and guidelines for the statutory regulation of investigative tests are examined.

CHAPTER 8: *Search and Seizure*

Law Reform Commission of Canada. *Search and Seizure.* Report #24. Ottawa, 1984b.

Law Reform Commission of Canada. *Recodifying Criminal Procedure.* Report #33. Ottawa, 1991.

Luther, Glen. "Consent Search and Reasonable Expectation of Privacy: Twin Barriers to the Reasonable Protection of Privacy in Canada." (2008) 41 *UBC Law Review* 1.

McDonald, Trevor R. "Genetic Justice: DNA Evidence and the Criminal Law." (1998) 26(1) *Manitoba Law Journal* 1.

MacKay, A. Wayne. "Don't Mind Me, I'm From the RCMP: *R. v. M.(M.R.)*—Another Brick in the Wall Between Students and Their Rights." (1997) 7 *Criminal Reports* (5th) 24.

MacKinnon, William. "Do We Throw Our Privacy Rights Out With the Trash? The Alberta Court of Appeal's Decision in *R. v. Patrick*." (2008) 46 *Alberta Law Review* 225.

Hill, S. Casey, David M. Tanovich and Louis P. Strezos. *McWilliams' Canadian Criminal Evidence.* Aurora, ON: Canada Law Book Limited (available online on Criminal Spectrum).

Marks, Amber. "Drug Detection Dogs and the Growth of Olfactory Surveillance: Beyond the Rule of Law?" (2007) 4 *Surveillance and Society* 257.

National DNA Data Bank. 2007-2008 Annual Report (Ottawa, 2008); www.nddb-bndg.org; accessed October 12, 2009.

Penney, Steven. "Updating Canada's Communications Surveillance Laws: Privacy and Security in the Digital Age." (2008) 12 *Canadian Criminal Law Review* 12:115.

Plischke, Helen. "Widespread DNA Tests Used for Tracking Rapist," *Vancouver Sun*, June 14, 1996, A6.

Plischke, Helen. "Blooding Lottery Has Only a Looser: Police Hope Mass DNA Testing Will Expose Rapist." *Vancouver Sun,* May 13, 1995, A2.

Pomerance, Renee M. "Bill C-104: A Practical Guide to the New DNA Warrants." (1995) 39 *Criminal Reports* (4th) 224.

Pomerance, Renee M. "Shedding Light on the Nature of Heat: Defining Privacy in the wake of *R. v. Tessling*." (2005) 23 *Criminal Reports* (6th) 229.

Quigley, Tim. *Procedure in Canadian Criminal Law,* 2nd ed. Toronto: Carswell, 2006.

Quigley, Tim. "Welcome Charter scrutiny of dog sniffer use: Time for Parliament to act." (2008) 55 *Criminal Reports* (6th) 376.

Rondinelli, Vincenzo. "The DNA Dragnet: A Modern Day Salem Witch Hunt?" (2003) 10 *Criminal Reports* (6th) 16.

Rusnell, Charles. "Trail Cold in Search for Vermilion Rapist." (2001 December 31) *Edmonton Journal* A6.

Salhany, Roger E. *Canadian Criminal Procedure*, 6th ed. Toronto: Canada Law Book Inc., (available online on Criminal Spectrum).
> Chapter 3 discusses search on arrest and seizure of property, including a consideration of section 8 of the *Charter*, *Code* provisions governing searches and search warrants and related topics.

Sankoff, Peter. "Routine Strip-searches and the *Charter*: Addressing Conceptual Problems of Right and Remedy." (1998) 16 *Criminal Reports* (5th) 266.

Schwartz, David J. "*Edwards* and *Belnavis*: Front and Rear Door Exceptions to the Right to be Secure from Unreasonable Search and Seizure." (1997) 10 *Criminal Reports* (5th) 100.

Shapiro, Jonathan. "Narcotics Dogs and the Search for Illegality: American Law in Canadian Courts." (2007) 43 *Criminal Reports* (6th) 299.

Shapiro, Jonathan. "Confusion and Dangers in Lowering the *Hunter* Standards." (2008) 55 *Criminal Reports* (6th) 396.

Stratas, David. "*R. v. B. (S.A.)* and the Right Against Self-Incrimination: A Confusing Change of Direction." (2004) 14 *Criminal Reports* (6th) 227.

Stringham, J. A. Q. "Reasonable Expectations Reconsidered: A Return to the Search for a Normative Core for Section 8?" (2005) 23 *Criminal Reports* (6th) 245.

Stevenson, Mark. "400 Men Submit DNA Samples to Prove They are Innocent of Murder." *National Post,* December 2, 1999, A11.

Stuart, Don. "Reducing *Charter* Rights of School Children." (1999) 20 *Criminal Reports* (5th) 230.

Stuart, Don. "Revitalising Section 8: Individualised Reasonable Suspicion is a Sound Compromise for Routine Dog Sniffer Use." (2008) 55 *Criminal Reports* (6th) 376.

Stuart, Don, Ron Delisle, and Tim Quigley. *Learning Canadian Criminal Procedure*, 9th ed. Toronto: Carswell, 2008.
    Chapter 2 deals with search and seizure.

Tanovich, David M. "The Constitutionality of Searches Incident to Vehicle Stops." (1993) 35 *Criminal Law Quarterly* 323.

Tanovich, David M. "A Powerful Blow Against Police Use of Drug Courier Profiles." (2008) 55 *Criminal Reports* (6th) 379.

Walsh, John J. "The Population Genetics of Forensic DNA Typing: 'Could It Have Been Someone Else?'" (1991–2) 34 *Criminal Law Quarterly* 469.

Watt, Daniel. "General Warrants Take the Wrong Path: Challenging the Constitutionality of Section 487.01 of the Code." (2008) 12 *Canadian Criminal Law Review* 297.

CASES

R. v. A.M., [2008] 1 S.C.R. 569. The police cannot conduct random or full scale searches of schools to look for drugs; they require reasonable suspicion for specific searches.

R. v. Arp, [1998] 3 S.C.R. 339.

The Queen in Right of Canada et al. v. Baron et al. (1993), 78 C.C.C. (3d) 510 (S.C.C.). The words "reasonable grounds" create the same standards as "reasonable and probable grounds," in respect of legislation authorizing the issuance of search warrants. Such legislation is invalid, however, if it deprives the judicial authority the residual discretion to refuse to issue a search warrant on the satisfaction of specified conditions.

R. v. Belnavis, [1997] 3 S.C.R. 341. The Court followed its decision in *Edwards* and determined that a passenger in a vehicle did not have a reasonable expectation of privacy under the circumstances.

R. v. Borden, [1994] 3 S.C.R. 145. The Court clarified the meaning of consent when accused provide evidence to the police (in this case, blood for the purpose of DNA analysis).

R. v. Briere, [2004] O.J. No. 5611 (Ont. S.C.).

R. v. Buhay, [2003] 1 S.C.R. 631.

R. v. Caslake, [1998] 1 S.C.R. 51.

R. v. Clayton, [2007] 2 S.C.R. 725.

Cloutier v. Langlois, [1990] 1 S.C.R. 158. A search authorized by a reasonable law and conducted in a reasonable manner will be acceptable under the *Charter*. The common law power of search incidental to arrest to secure or preserve evidence, ensure the safety of the officer or the suspect, and prevent escape is lawful if the search is conducted reasonably. Such a search is not mandatory, and must be for a valid purpose, such as those listed.

R. v. Cody and Langille, [2004] Q.J. No. 14164. Section 492.2 (dial number recorders) does not violate section 8 of the *Charter*.

R. v. Collins, [1987] 1 S.C.R. 265. Suspicion is an insufficient basis for a search. Where the validity of a search is constitutionally challenged, the Crown is entitled to support it by adducing evidence as to the searching officer's belief and the bases for it, even if such evidence is hearsay. Hearsay evidence is capable of supporting a reasonable belief, for example, that the accused is in possession of a narcotic.

R. v. Debot, [1989] 2 S.C.R. 1140. The reasonable grounds that police officers must have before conducting a warrantless search should address the following questions: "First, was the information predicting the commission of a criminal offence compelling? Second, where a 'tip' originating from a source outside the police, was that source credible? Finally, was the information corroborated by police investigation prior to making the decision to conduct the search?" (1143).

Descoteaux et al. v. Mierzwinski and A.G. Quebec et al. (1982), 70 C.C.C. (2d) 385 (S.C.C.). Solicitor–client privilege includes communications with those who assist lawyers professionally, so long as those communications are made to them in their professional capacity. This includes information exchanged before retainer respecting the legal problem, respecting payment, and so forth. A justice normally has no jurisdiction to authorize a search for information that is privileged. Exceptions exist if the communications are criminal or are made with a view to committing a crime.

R. v. Edwards, [1996] 1 S.C.R. 128. The Court set out the factors that should be considered in determining whether a person who does not rent an apartment has a reasonable expectation of privacy in it, under section 8 of the *Charter*.

R. v. Evans and Evans, [1996] 1 S.C.R. 8. Police officers knocking on the front door of a house to determine if they can smell marijuana are conducting an illegal search under section 8 of the *Charter*. An anonymous tip is not by itself sufficient information to obtain a search warrant.

R. v. Feeney, [1997] 2 S.C.R. 13.

R. v. Flintoff, [1998] O.J. No. 2337 (Ont. C.A.).

Genest v. The Queen (1989), 45 C.C.C. (3d) 385 (S.C.C.). Although the importance of the evidence to obtaining a conviction is an important factor in applications to exclude such evidence under section 24(2) of the *Charter,* the result does not automatically flow from that consideration. Detailed discussion of the factors to be considered, as related to the fairness of the trial, the seriousness of the breach, and the balancing of the effect of excluding versus admitting the evidence.

R. v. Golden, [2001] 3 S.C.R. 679.

R. v. Grant, 2009 SCC 32.

R. v. Greffe, [1990] 1 S.C.R. 755.

R. v. Ha, [2009] O.J. No. 1693; application for leave to appeal to the Supreme Court of Canada dismissed (without reasons), November 12, 2009; [2009] S.C.C.A. No. 295.

R. v. Harrison, 2009 SCC 34.

Hunter et al. v. Southam Inc. (1984), 14 C.C.C. (3d) 97 (S.C.C.). Consideration of the constitutional validity of legislation authorizing searches involves a balancing of the government's interest in intruding on an individual's privacy with the public's interest in being free from state interference. Generally, prior authorization by a person capable of acting in a judicial manner will be a minimum requirement for valid search and seizure provisions. Generally, the provisions must require a test or standard of "credibly based probability."

R. v. Kang-Brown, [2008] 1 S.C.R. 456. Police have a common law power to use drug-sniffing dogs to search for drugs. Four justices required reasonable suspicion for such searches, and on required only a generalized suspicion.

R. v. Kokesch, [1990] 3 S.C.R. 3.

R. v. Laporte (1972), 8 C.C.C. (3d) 343 (Que. Q.B.). A Justice has no jurisdiction to issue a warrant to seize bullets from the body of an accused, and the powers of search incident to arrest similarly do no authorize the surgical removal of such bullets for evidence.

Lavallee, Rackel & Heintz v. Canada (Attorney General); White, Ottenheimer & Baker v. Canada (Attorney General); R. v. Fink, [2002] 3 S.C.R. 209. Section 488.1 of the *Criminal Code* violates sections 7 and 8 of the *Charter* because it does not adequately protect solicitor–client privilege.

R. v. Mann, [2004] 3 S.C.R. 59.

R. v. M.(R.M.), [1998] 3 S.C.R. 393.

R. v. Mellenthin, [1992] 3 S.C.R. 615. Mellenthin was randomly stopped and asked questions about the contents of a bag in his car. The narcotics that were found were not admissible, as the questioning was improper (there were no reasonable grounds), and he could not be conscripted to provide evidence against himself. A random spot check constitutes a detention which is justified, but does not permit baseless general questioning or searches.

R. v. Miller (1987), 38 C.C.C. (3d) 252 (Ont. C.A.). Although there is no authority to issue a search warrant to seize a bandage worn by the accused, it may be validly seized incident to arrest, as evidence to connect the accused to the crime.

R. v. Monney, [1999] 1 S.C.R. 652. A "bedpan vigil" in a "drug loo facility" is a legitimate means of searching for heroin swallowed by the accused, under section 98 of the *Customs Act*.

R. v. Morrison (1987), 20 O.A.C. 230.

R. v. Nguyen, [2004] B.C.J. No. 247. Section 492.2 (dial number recorders) is unconstitutional, and should be read to require reasonable grounds to believe that an offence has been committed, and so on.

R. v. Patrick, [2009] 1 S.C.R. 579. The Court examined the expectation of privacy regarding garbage left for pickup.

R. v. Plant, [1993] 3 S.C.R. 281. The accessing by police of computer data relating to the accused's hydro-electric consumption does not constitute a search. Where a search warrant is obtained in part based on information from an impermissible and unauthorized perimeter search, it will still be valid if it can be supported on the remaining information.

R. v. Richard (1995), 99 C.C.C. (3d) 441 (N.S.C.A.); appeal allowed, [1996] 1 S.C.R. 896. The police had compelling information from a reliable informant that was sufficient to obtain a search warrant. It was not practical to obtain corroboration without putting the informant's life in jeopardy.

R. v. Rodgers, [2006] 1 S.C.R. 554. Section 487.055(1) of the *Code* does not violate the *Charter*.

R. v. S.A.B., [2003] 2 S.C.R. 678. DNA warrant sections of the *Criminal Code* (487.04 to 487.09) do not violate the *Charter*.

The Minister of Citizenship and Immigration v. Schneeberger, [2004] 1 F.C.R. 280; application for leave to appeal to the Supreme Court of Canada dismissed with costs; [2004] S.C.C.A. No. 2.

Semayne's Case (1604), 5 Co. Rep. 91a, 77 E.R. 194 (K.B.).

CHAPTER 8: *Search and Seizure*

R. v. Shepherd, 2009 SCC 35.

R. v. Stillman, [1997] 1 S.C.R. 607.

Tele-Mobile Co. v. Ontario, [2008] 1 S.C.R. 305.

R. v. Tessling, [2004] 3 S.C.R. 432.

R. v. Wise (1992), 70 C.C.C. (3d) 193 (S.C.C.). The installation and monitoring of a radio tracking device in the accused's car constitutes an unreasonable search and seizure. In the circumstances of this case, however, the admission of the evidence obtained as a result would not bring the administration of justice into disrepute.

R. v. Wise, [1996] O.J. No. 571.

R. v. Wong, [1990] 3 S.C.R. 36. The surreptitious (i.e., secret) video surveillance of an accused may amount to a search and seizure, in circumstances where a reasonable expectation of privacy existed.

CHAPTER 9: *Electronic Surveillance and the Interception of Private Communications*

CHAPTER OBJECTIVES

In studying this chapter, you should develop an understanding of the following topics and concepts:

- the nature of private communications
- the illegality of intercepting private communications, and other means to protect private communications
- the legislative scheme whereby the interception of private communications may be lawful
- the procedure for testing the admissibility of intercepted private communications at trial

HISTORY OF ELECTRONIC SURVEILLANCE

At English common law, it was a public nuisance offence to eavesdrop. However, the offence was not recognized in Canada when the issue first arose in 1918, and the Canadian courts did not recognize a common law right to privacy in this area (MacDonald 1987, 142–3). In addition, there was no need to establish that wiretap evidence was legally obtained to enter it at trial, since, as we have seen, most illegally obtained evidence was admissible as long as it was relevant. In the mid- to late 1960s, a series of events raised the level of concern over the use of wiretap equipment by the police and private individuals.

In 1966, the Ontario Court of Appeal, disturbed by the behaviour of the police, halved a $10 000 fine for a person convicted of keeping a common betting house. The police had searched a house with a valid warrant for evidence that the accused was keeping a common betting house, and while in the house they planted a listening device. The wiretap evidence was admitted at trial, and the accused was convicted (*Steinberg;* see discussion in Beck 1968, 643–4).

In 1967, a public inquiry was held in British Columbia into allegations that rival unions were bugging one another's conventions. The inquiry found other examples of the use of electronic surveillance: car sales firms were bugging booths (so they could determine how much people were prepared to pay for a car), partners and spouses were hiring private detectives to record conversations, dance studios and health clubs were listening in on client conversations to improve their sales pitch, and so on (Sargent Report 1967, and Beck 1968, 645–6).

In 1968, British Columbia created a tort, allowing the recovery of civil damages (without proof of specific damage) from anyone who invaded another's privacy. Electronic surveillance was included within the scope of the tort. Other provinces followed suit in the late 1960s and early 1970s. This legislation was limited in application to private individuals; police acting in the course of their duty were, and still are, exempt (MacDonald 1987, 143–4).

A major controversy over the use of electronic surveillance by the police arose in Toronto in 1968, during an inquiry into the fitness of two magistrates to perform their duties. The Toronto police had recorded 60 telephone conversations of a man convicted of six theft charges; some of these conversations were with one of the magistrates. Much of the editorial comment in the newspapers expressed concern that the police had recorded this man's conversations over a two-month period, without any judicial authorization (Grant Report 1968, and Beck 1968, 648).

These public inquiries found that the use of wiretap equipment was far more extensive than the public had imagined. The police themselves admitted that their major use of such equipment was to assist in investigations rather than to gather evidence to be presented at trial. This meant that the police were gathering information by electronic means, but that this fact was never disclosed to other actors in the criminal justice system, the targets of the surveillance, or to the public (Beck 1968, 648–9). The wiretapping was being done without judicial authorization. The only control was the Police Commission's guidelines, which required police officers (who had reasonable grounds for conducting a wiretap) to get permission from their chief to conduct electronic surveillance. The guidelines were not always followed (MacDonald 1987, 144).

The issue was dealt with again in the Ouimet Report (1969), which recommended judicial control over electronic interceptions by law enforcement agencies, unless one of the parties agreed to the interception (85). Such interceptions would be required to be authorized by a superior court judge, and would be available only for more serious offences (86). The Committee recommended that the trial judge have discretion to exclude illegally obtained electronic surveillance after considering factors such as the deliberate or inadvertent nature of the illegality (87). In addition, the Committee recommended that Parliament create an offence of possession (without lawful excuse) of equipment capable of such interceptions. The proposals were not aimed at controlling the interception of conversations in which one of the parties to the conversation consented to the interception, since these conversations were assumed to take place without an expectation of privacy.

The House of Commons Standing Committee on Justice and Legal Affairs made several recommendations in 1970. It suggested criminalizing electronic surveillance, limiting the crimes for which authorizations could be issued, and implementing time limits, annual reporting to Parliament, and stringent controls for their use by police (MacDonald 1987, 145). A Bill on Privacy was introduced into the House of Commons in 1973, and became law on June 30, 1974. It has since been amended a number of times.

In 1986, the Law Reform Commission of Canada observed that the legislation introduced to *protect* privacy was used to a much greater extent to *invade* privacy (7). The Commission noted that such authorizations were 20 times more common in Canada than in the United States (10). Concerned with the number of such invasions, the Commission made 76 recommendations directed at making the invasion of privacy sections conform to section 8 of the *Charter*, and at imposing further restraints on the use of this very intrusive form of surveillance. A key concern to law reformers was to reach a balance between the constitutional right to privacy and the need for effective law enforcement. Many of the amendments suggested by the Law Reform Commission in their report on *Recodifying Criminal Procedure* (1991) took into account decisions of the Supreme Court of Canada that had applied the *Charter* to this area of law.

It was (and probably still is) difficult to determine exactly how much use is made of electronic surveillance, because of the possibility that it is being used illegally as an investigative tool by the police, private investigators, or anyone who wants to intercept private conversations. The increasing use of private detectives and private police in our society suggests that surveillance may occur (illegally) without the intervention of the law or the benefit of judicial authorization.

THE LEGISLATIVE SCHEME: WHAT IT COVERS

The purpose of the legislation in 1974 was, and still is, twofold: to protect privacy and to authorize the appropriate authorities to intercept some private communications. Authorization to intercept private communications by specified means is available for a limited number of offences.

Offences for Which Authorization is Available

The first draft of the legislation in 1973 defined "offence" to include any federal offence that could be prosecuted by indictment. Suggestions for change at that time included limiting electronic surveillance to crimes against national security (Law Reform Commission of Canada 1986, 3). After first reading of the draft bill, the definition of "offence" was changed to consist of a list of specific offences considered sufficiently serious to warrant the use of electronic surveillance. Since that time, offences have been added to the list in section 183.

The Meaning of "Intercept" and "Means of Interception"

The legislation is designed to protect and, with authorization, intercept private communications. **Intercept** "includes listen to, record or acquire a communication or acquire the substance, meaning or purport thereof" (section 183). In the early 1990s, a number of cellular telephone conversations by public figures were reported in the news media. In 1990, the Attorney-General of British Columbia (Bud Smith) resigned after a reporter intercepted and published some of his conversations (Cobb 1992, A6). In 1992, the conversations of a Quebec civil servant (Diane Wilhelmy), in which she criticized Premier Robert Bourassa's performance at the constitutional talks, were taped and made public (Canadian Press 1992; Cobb 1992). In England, cellular phone calls have also led to rifts between the media and the Royal Family. In Canada, they have resulted in embarrassments, resignations, and new legislation to cover cellular phone conversations.

The amendments to the legislation to prohibit the interception of cellular telephone calls were strongly opposed by the Canadian Daily Newspaper Association and the Canadian Association of Journalists; both argued before the House of Commons Committee that these provisions would infringe freedom of the press (Canadian Press 1993). The press wanted the freedom to publish anything it heard in cellular phone calls. Despite the strong opposition, section 183 was amended so that the definition of private communication now includes "radio-based telephone communication that is treated electronically or otherwise for the purpose of preventing intelligible reception by any person other than the person intended by the originator to receive it." Cellular phone calls that are not so electronically treated are dealt with under section 184.5 (discussed below).

Box 9.1 Statistics on Wiretap Applications in Canada

Between 2000 and 2004, there were 746 applications for authorizations granted. Two applications were refused. Between 2004 and 2008, there were 584 applications granted and three were refused. Some of the difference between the two reporting periods may be from a lag in reporting. Of the 584 authorizations in the latter period, 75 percent were under section 185 for audio surveillance, 18 percent were under section 487.01 for video surveillance, 6.5 percent were renewals of existing authorizations, fewer than 1 percent were for emergency audio surveillance under section 188, and there were no authorizations for emergency video surveillance under section 487.01 (Public Safety Canada 2009, 5). The vast majority of these authorizations are for offences related to controlled drugs (2009, 22).

Private Communication

A **private communication** is defined in section 183 as:

> any oral communication, or any telecommunication, that is made by an originator who is in Canada or is intended by the originator to be received by a person who is in Canada and that is made under circumstances in which it is reasonable for the originator to expect that it will not be intercepted by any person other than the person intended by the originator thereof to receive it, and includes any radio-based telephone communication that is treated electronically or otherwise for the purpose of preventing intelligible reception by any person other than the person intended by the originator to receive it.

A communication is "private" if there is a **reasonable expectation of privacy**. The circumstances surrounding the communication will assist the courts in determining whether there is a reasonable expectation of privacy. Lower courts have decided, for example, that there was no expectation of privacy when kidnappers used the phone to demand a ransom and hung up repeatedly because they thought they were being monitored (*Tam*). In addition, the father of the victim, who recorded the conversations, was not considered to be an agent of the state. In *Samson*, the Ontario Court of Appeal commented that since the accused made a statement on the telephone to the effect that there was a danger the phones were being tapped, there was some question whether the conversations really were private communications. However, an *obiter* comment by the Supreme Court of Canada in *Tessling* has suggested these conversations may still be private:

> It is one thing to say that a person who puts out the garbage has no reasonable expectation of privacy in it. It is quite another to say that someone who fears their telephone is bugged no longer has a subjective expectation of privacy and thereby forfeits the protection of s. 8. Expectation of privacy is a normative rather than a descriptive standard (para. 42).

There is no expectation of privacy where correctional officers are allowed to record communications in a correctional centre that might disclose "a threat to the management, operation, discipline or security" of the centre (*Napope*). It may be necessary, however, in such a situation, to clearly inform inmates that their conversations are being monitored (*Rodney*). In *McIsaac*, the accused was charged with a number of counts of assault causing bodily harm and uttering threats to cause death or bodily harm against his common law partner. While detained in the Prince George Regional Correctional Centre, he made 36 calls to the complainant which were passively recorded by the Centre. When the complainant recanted her story, the police obtained a search warrant to seize the telephone recordings. After reviewing the factors in *Edwards* (see Chapter 8), the judge decided that McIsaac, as an inmate in a correctional centre, had no reasonable expectation of privacy in the conversations. Upon entering the centre, inmates sign a document that states: "All oral and written incoming/outgoing communication may be monitored and taped. Visits and common areas are video taped" (para. 38). In addition biometric voice prints are obtained from all inmates who register for a smart card (para. 41).

In 1986, and again in 1991, the Law Reform Commission of Canada recommended that a communication should not cease to be private simply because a person believes it might be the subject of an authorized interception (1986, 20, and 1991, 120). This approach is in line with the normative approach referred to by the Supreme Court of Canada in a number of cases, but it would not prohibit the unauthorized interception of conversations between hostage takers and the outside world, because there is clearly no reasonable expectation of privacy in these circumstances (1986c, 18–9).

Box 9.2 Prayers are not Private Communications

In *Davie*, an accused about to undergo a polygraph examination in an arson investigation was monitored on a video surveillance system. While in the examination room alone, he slipped off the chair and onto his knees. He held his arms up and said, "Oh, God, let me get away with it just this once." A majority of the Court held that this "prayer" was not a "private communication," since the legislation refers to communications between persons, which implicitly meant "humans." Hutcheon, J.A. said:

> In my opinion, the word "person" is used in the statutes of Canada to describe someone to whom rights are granted and upon whom obligations are placed. There is no earthly authority which can grant rights or impose duties upon God. I can find no reason to think that the Parliament of Canada has attempted to do so in the enactment of sections of the *Criminal Code* dealing with the protection of privacy" (223).

The widespread use of electronic mail (e-mail, text messaging) has led to concerns about privacy on the information highway. Are these communications private like a telephone conversation, or do they fall outside the ambit of legislative protection? The *Weir* case (Box 9.3) recognized some privacy in e-mail. Penney (2008) suggests that the *Criminal Code* has not kept up with technology, and recommends amendments to clarify which authorizations need what type of supporting information (see also Pomerance 2005 on concerns over data mining).

Some governments take the position that their employees' e-mail is the property of the government (Munro 1994, B1-2), so no judicial authorization is required for the government to intercept their communications. The issue of privacy in the workplace is an important one, but beyond the scope of this book (for discussion, see: Geist, 2003, Cockfield, 2003 and 2004).

In 2004, section 184(2)(e) was added to the *Code* to allow those in possession or control of computer systems to intercept private communications to manage the performance and integrity of the system, and to protect the system from unauthorized use of a computer (section 342.1(1)), or mischief in relation to data 430(1.1)). Any evidence of a crime can be disclosed to a peace officer or prosecutor (section 184(3)).

THE LEGISLATIVE SCHEME: PROTECTING PRIVACY

Offences Designed to Protect Privacy

To protect privacy, Part VI of the *Criminal Code* (Invasion of Privacy) creates several offences. Section 184(1) creates the indictable offence of wilfully intercepting a private communication by means of an "electro-magnetic, acoustic, mechanical or other device," which is broadly defined in section 183. The maximum penalty for contravening section 184(1) is five years imprisonment. In 1993, section 184.5 was added to cover cellular telephone communications that had been found by the courts not to be included in section 183.

Section 184(2) specifies the exceptions to the offences. It is not an offence if the interception is made with consent of one of the parties, in accordance with a **judicial authorization**, or pursuant to section 184.4 (discussed below). It is permissible, under the first exception, for anyone to record their own conversations, or the conversations of others, with the consent of one of the parties. Another exception to the offence includes those who provide communication service to the public if the interception is necessary for the provision of services, to check for quality control, or to protect "rights or property directly related to providing such services" (section 184(2)(c)). Controversies over the 2004 amendment, which provides some exemptions for those in charge of a computer system, are available on Justice Canada's website (see Nevis Consulting Group Inc. 2003).

Section 191 makes it an indictable offence to possess, sell, or purchase the various types of equipment designed to intercept private communications. Subsection (2) provides a number of exceptions, including "a person in possession of such a device . . . under the direction of a police officer . . ." Sections 193 and 193.1 make it an indictable offence to use or disclose information that is intercepted, or the fact that it was intercepted. These offences have exceptions, and carry a maximum penalty of two years imprisonment.

Concerned with the use of hidden video cameras in locations where people believe they are acting in private, the federal government introduced the offence of voyeurism in 2005. Section 162 now makes it an offence to surreptitiously observe or make visual recordings of persons "in circumstances that give rise to a reasonable expectation of privacy," if the person is nude, specified parts of the body are exposed, or "the observation or recording is done for a sexual purpose."

Box 9.3 The Privacy of E-Mail?

In *Weir*, the Alberta Court of Queen's Bench found that there is a reasonable expectation of privacy regarding Internet e-mail, although it is not as great as the expectation of privacy in mail sent by Canada Post. Weir had requested that his Internet service provider (ISP) repair his e-mail box (it had exceeded its two megabytes limit and was therefore inaccessible to Weir). The ISP viewed some of the child pornography attachments in his e-mail box, and reported them to the police. In response to a request by the police, the ISP forwarded the e-mail to the police. Based on this information, the police obtained a search warrant and searched Weir's house after the e-mail box was fixed (enlarged) and Weir had downloaded the child pornography. Weir was convicted of possession of child pornography. The Alberta Court of Appeal dismissed Weir's appeal. It found that the ISP was acting as an agent of the state when it sent the e-mail to them at their request. This was a warrantless search; however, details regarding the e-mail was unnecessary to obtain the search warrant, as the police already had sufficient information to obtain a search warrant without it (see Hubbard *et. al.* 2001 and Penney 2008 for discussions of privacy issues in internet communications).

Damages

Section 194 allows a judge, under some circumstances, to order a person convicted of an offence under section 184, 184.5, 193, or 193.1 to pay an aggrieved person up to $5000 as punitive damages. An application for damages under section 194 must be made by the aggrieved person at the time the accused is sentenced.

Notification to Target

Section 196 requires that those who have been the objects of authorization under section 185 be notified of such authorization and interception within 90 days after the authorization or renewal expires. An application, or successive applications, can be made to extend this period. Each **deferral** may be for up to three years, and may be granted if a judge is of the opinion "the interests of justice warrant" such an extension. Note that section 196 does not apply to authorizations of interceptions under section 184.1, nor to authorizations under 184.2 (3), 184.3(6), or 188(2).

Reporting

Section 195 requires that the Minister of Public Safety and Emergency Preparedness prepare an annual report for Parliament, including detailed statistics about the number of applications for authorizations and renewals, general descriptions of the methods used to intercept, and so on (see Box 9.1). A less detailed annual report must be prepared for the public by the Attorney-General of each province, under section 195(5).

JUDICIAL AUTHORIZATION TO INTERCEPT

Judicial authorization by state agents to intercept a private communication by means defined in section 183 is sought by an **application** under section 185. The application is *ex parte,* meaning that no notice is given to the person whose communications are to be intercepted. It would not make sense to give the target of such an application notice of this hearing, but what other means of protecting the person's right to privacy might be considered? Would notice to a "public advocate" be of any assistance?

The application under section 185 must be signed by the Attorney-General of the province, the federal Minister of Public Safety, or their designate (unlike applications under sections 184.2 and 184.3, discussed below). The police do not make the application, but must obtain the approval of a designated prosecutor for the application (unlike the procedure followed to obtain a search warrant). The application is accompanied by an **affidavit**, usually that of a police officer, which includes the information required by section 185(1):

> (c) the facts relied upon to justify the belief that an authorization should be given together with the particulars of the offence,
> (d) the type of private communication proposed to be intercepted,
> (e) the names, addresses and occupations, if known, of all persons, the interception of whose private communications there are reasonable grounds to believe may assist the investigation of the offence, a general description of the nature and location of the place, if known, at which the private communications are proposed to be intercepted and a general description of the manner of interception proposed to be used,...
> (g) the period for which the authorization is requested, and
> (h) whether other investigative procedures have been tried and have failed or why it appears they are unlikely to succeed or that the urgency of the matter is such that it would be impractical to carry out the investigation of the offence using only other investigate procedures.

In practice, an application to intercept is presented to a judge in chambers by a designated prosecutor, usually along with the police officer who swore the affidavit. False affidavits can lead to charges of perjury and obstruction of justice, as happened following a lengthy investigation by various police forces throughout Ontario in an effort to "dismantle" Satan's Choice Motorcycle Club (Project Dismantle). Following a number of findings against a Detective Sergeant of the Ontario Provincial Police (see *Bogiatzis et al.*) that he destroyed evidence and swore false affidavits, he was charged with perjury and obstruction of justice in 2004 (Kari 2004). In 2009, the charges were withdrawn, in part, because the Crown did not want to identify a confidential source who might have raised a reasonable doubt as to the Sergeant's guilt (Powell 2009).

The application under section 186 is made to a judge of "a superior court of criminal jurisdiction," as opposed to a magistrate or a Provincial Court judge who can issue search warrants. Section 186(1) states that the judge who issues the authorization must be satisfied

> (a) that it would be in the best interests of the administration of justice to do so; and
> (b) that other investigative procedures have been tried and have failed, other investigative procedures are unlikely to succeed or the urgency of the matter is such that it would be impractical to carry out the investigation of the offence using only other investigative procedures.

Section 186(4) specifies what must be contained in the judicial authorization to intercept private communications.

Several Supreme Court of Canada decisions have interpreted the requirements of such authorizations. For example, in *Paterson et al.*, the Court upheld an Ontario Court of Appeal decision that a **basket clause,** which permitted the police to intercept the private communication of unknown persons, provided there were reasonable and probable grounds to believe the communications would assist the investigation, was invalid, because it delegated to the police the judge's function of determining who should be the **target** of interception. If the name of the targeted person is unknown, the authorization must specify the class of persons whose communications can be intercepted. Such decisions cannot be left to the police.

Section 186 does not specifically state that the judge must be satisfied that there are reasonable grounds to believe that an offence has been or is being committed. However Mr. Justice LaForest, for the majority in *Duarte*, stated that "the best interest of the administration of justice" imports a minimum requirement "that the issuing judge must be satisfied that there are reasonable and probable grounds to believe that an offence has been, or is being, committed and that the authorization sought will afford evidence of that offence" (12). The Court found that this section authorized a search or seizure, and that the legislation met all the requirements of section 8 of the *Charter*. Mr. Justice Sopinka, in *Garofoli*, noted that jurisprudence on the "reasonable and probable grounds" necessary to obtain a search warrant are also applicable to wiretap authorizations (191).

Investigative Necessity

Electronic surveillance is a highly invasive investigative tool that captures every sound that people make. When used in a home, it captures conversations and sounds indiscriminately, many irrelevant to the investigation, and many by people who are not targets of investigation. This is why section 186(1)(b) requires what is sometimes referred to as "investigative necessity" before an authorization is issued. Although enforcement officials have lobbied to remove this restriction, the Law Reform Commission in Canada recommended that it be retained, as "wiretap legislation was intended to be the tool of last resort" (1986, 34).

Prior to 2000, the Supreme Court of Canada had no definitive definition of "investigative necessity," and failed to "distinguish between a 'last resort' test and a 'no real practical alternative' test'" (*Araujo* para. 23). According to the Court in *Araujo*, section 186(1)(b) contemplates three kinds of circumstances in which wiretap authorizations may be issued. That is, where "other investigative procedures have been tried and have failed, other investigative procedures are unlikely to succeed *or* the urgency of the matter is such that it would be impractical to carry out the investigation of the offence using only other investigative procedures" (para. 27; emphasis added by the Court). The section must be applied keeping in mind an appropriate balance between "privacy interests and the realities and difficulties of law enforcement" (para. 22). The Court rejected "a pure last resort test," because it "would turn the process of authorization into a formalistic exercise that would take no account of the difficulties of police investigations targeting sophisticated crime" (para. 29). Rather, it concluded that section 186(1)(b) requires that "there must be, practically speaking, *no other reasonable alternative method of investigation*, in the circumstances of the particular criminal inquiry" (emphasis by the Court) (para. 29).

According to section 186(1.1), the necessity requirement does not apply to applications for authorization in relation to various offences involving criminal organizations, "an offence committed for the benefit of, at the direction of or in association with a criminal organization," or a terrorism offence. The exceptions to the requirement of investigative necessity, introduced in 1997 as part of the anti-criminal organization legislation, and in 2001 as part of the anti-terrorism legislation, have been challenged unsuccessfully under the *Charter* in lower courts (*Doiron, Doucet, Lucas, Pangman*). For an argument that investigative necessity should be a constitutional requirement for all wiretap authorizations, see Whitling (2002).

Interception of Communications by Lawyers

Sections 186(2) and (3) state that authorizations may not be made to intercept conversations of lawyers at their offices, residences, or at any other place ordinarily used by a lawyer for the purpose of consulting with clients, unless the judge is satisfied that there are reasonable grounds to believe that the lawyer is or is about to become a party to an offence. Section 186(3) requires the judge to include terms and conditions to protect solicitor–client privilege.

Renewals

Authorizations are made for a set period of time (not exceeding 60 days); however, the Minister of Public Safety or the Attorney-General of a province (or their designate) may apply for **renewals** under section 186(6), and such an application may be accompanied by an affidavit as set out in section 186(6). Renewals cannot exceed 60 days.

Emergency Authorizations, Section 188

Section 188(2) allows for authorizations (which are more like "pre-authorizations") in case of emergencies, where there is insufficient time to obtain an authorization under section 185. These applications may be made by police officers who are specially designated by the Attorney-General of a province or the Minister of Public Safety. The authorization can be for a period of up to 36 hours. If a subsequent authorization is based on the same circumstances as an emergency authorization, a trial judge can deem the evidence gathered under section 188 inadmissible (section 188(5)).

CONSENT INTERCEPTIONS

In 1990, the Supreme Court of Canada's decision in *Duarte* had substantial impact on the use of consent to legitimize interceptions of private communications during **participant** or **consent surveillance** (that is, surveillance where one of the parties agrees to interceptions by the police). Although consent surveillance was allowed under the *Criminal Code*, the accused challenged such surveillance as violating their rights under section 8 of the *Charter*. Mr. Justice LaForest decided that the sections themselves did not infringe or deny rights under section 8, but their use by the state, even with the consent of the originator or intended recipient, but without prior authorization, did infringe section 8. Secret electronic surveillance of an individual by an agency of the state, he held, constitutes unreasonable search and seizure.

In an earlier case, the Supreme Court of Canada had ruled that section 8 "must be held to embrace all existing means by which agencies of the state can electronically intrude on the privacy of the individual, and any means which technology places at the disposal of law enforcement authorities in the future" (*Wong*, para. 8). This broad scope given to section 8 was key to the Supreme Court of Canada's decision that "participant surveillance" by agents of the state constitutes unreasonable search and seizure, unless prior authorization is obtained. LaForest concluded that it is completely unacceptable in a free and democratic society to allow agents of the state to use this form of electronic surveillance at their own discretion (para. 13). This decision was in line with the decision in *Hunter* (see Chapter 8), which required the state to obtain judicial authorization before conducting a search. Following *Duarte*, the *Criminal Code* was amended to require authorization for what was previously known as consent surveillance by the state, and to provide certain exceptions to such a requirement.

Judicially Authorized Interception by Consent, Section 184.2

Section 184.2 was added, following *Durate*, in order to allow the interception of private communications by the police in situations where one party consents, and where a designated peace officer or public officer has obtained prior authorization from a judge. These authorizations, unlike the ones under section 186, can be made by either a provincial court or a superior court judge, and can be obtained where the judge is satisfied that:

> (a) there are reasonable grounds to believe that an offence against this or any other Act of Parliament has been or will be committed;
> (b) either the originator of the private communication or the person intended by the originator to receive it has consented to the interception; and
> (c) there are reasonable grounds to believe that information concerning the offence referred to in paragraph (a) will be obtained through the interception sought (section 184.2(3)).

Thus, while authorization is now needed for consent interceptions, such authorization is much easier to get (see requirements under section 186), and can be obtained for any federal offence; it is not limited to the offences listed in section 183. In addition, section 184.3 allows applications for such authorizations by means of telecommunication, where "it would be impracticable...to appear personally before the judge."

If the courts find fault with the authorization (e.g., grounds for the authorization are insufficient), and declare the recordings inadmissible, the evidence by an undercover police officer who engaged a suspect in the conversation may still be admissible, following a section 24(2) analysis (*Fliss*).

Unauthorized Interception To Prevent Bodily Harm, Section 184.1

Section 184.1 allows an agent of the state (a police officer or "a person acting under the authority of, or in cooperation with, a peace officer") to intercept private communications with the consent of one of the parties, where there are grounds to believe there is a risk of bodily harm to the person who consented, and where the interception is made to prevent bodily harm. These interceptions are inadmissible in court, except in proceedings where "actual, attempted or threatened bodily harm is alleged." They are also inadmissible on applications for authorizations and search warrants. Section 184.1(3) requires the agent of the state to destroy the recording and any notes of it, unless the conversations suggest bodily harm has or is likely to occur. What problems might arise from the required destruction of such recordings?

Unauthorized Emergency Interceptions, Section 184.4

Section 184.4 allows peace officers to intercept private communications in situations where it is not practical to get an authorization for urgency reasons, the "peace officer believes on reasonable grounds that such an interception is immediately necessary to prevent an unlawful act that would cause serious harm to any person to property," and either party to the communication is "likely to cause the harm or is the victim, or intended victim of the harm." Some lower courts have decided that this section violates section 8 of the *Charter*, and is not saved under section 1(*Six Accused Persons*; *Brais*); however there is at least one lower court which has upheld its validity (*Riley*).

CAMERA SURVEILLANCE, SECTION 487.01, AND PART VI

In 1993, Parliament added section 487.01 in Part XV (Special Procedures and Powers) of the *Criminal Code*, to allow judges of provincial or superior courts to issue warrants permitting the use of "any device or investigative technique or procedure" that would constitute unreasonable search and seizure without a warrant (see discussion in Chapter 8). If the warrant is for surveillance by television camera or similar device, in circumstances where there is a reasonable expectation of privacy, the warrant "shall contain such terms and conditions as the judge considers advisable to ensure that the privacy of the person or of any other person is respected as much as possible" (section 487.01(4)). In addition, warrants for camera surveillance are also governed by many of the provisions of Part VI (Invasion of Privacy) of the *Code*. Section 487.01(5) limits such surveillance to offences listed in section 183. The provisions for judicially authorized consent interceptions (under section 184.2), authorized interceptions (section 186), and emergency authorizations (section 188) all apply to search warrants to record activities by camera or similar device (in circumstances where there is a reasonable expectation of privacy). For example, search warrants to record information by camera, without the consent of the parties, must meet the requirements of section 186 or section 188. These warrants must be reported to Parliament and to the public as required under section 195.

Section 487.1.01(5) also incorporates some, but not all, of the offences and remedies under Part VI. It is an offence to disclose the content of a private communication without consent (section 193), and a judge can award damages of up to $5000 to the aggrieved person at the time of sentencing (section 194). The section does not incorporate section 184, so it is not an offence to record activities by camera or similar device, even if there is a reasonable expectation of privacy in the circumstances. However, see the earlier reference to section 162, creating an offence of voyeurism.

THE ADMISSION OF ELECTRONIC SURVEILLANCE AS EVIDENCE

Generally, if the proper procedures under the *Criminal Code* for the interception of private communications are followed, the evidence will be admissible. However, if the evidence was obtained in a manner that violated the accused's rights under section 8 of the *Charter*, the court will determine the admissibility of the evidence under section 24(2), in accordance with the *Grant* framework (see Chapter 7).

Notice—Section 189(5)

The 1993 amendments require the party seeking to introduce evidence of private intercepted communications to give reasonable notice to the other side, along with:

> (a) a transcript of the private communication where it will be adduced in the form of a recording, or a statement setting out full particulars of the private communication, where evidence of the private communication will be given *viva voce*; and
> (b) a statement respecting the time, place and date of the private communication and the parties thereto, if known (section 189(5)).

Section 189(6) provides that privileged information that is intercepted remains privileged (see Chapter 12) and inadmissible, unless the person enjoying the privilege consents to its admission.

Box 9.4 Surveillance of Care Providers

An increasing number of people rely on professional surveillance services to install video surveillance in their home to monitor childcare workers and care providers for the elderly. Should such surveillance be allowed? In *Jamieson*, the parents of a 19 month old child with medical problems were concerned about bruises on their child. They had a video camera installed in their home that showed the nursing assistant "shaking, punching and slapping the child, hitting her with a book, twisting and manipulating limbs in an abnormal manner, picking the child up by her hair or legs, rubbing a cloth in the child's face after cleaning her genital area and holding her face down in the crib" (para. 35). The court found that the parents were not acting as state agents, and held that the surveillance was not an invasion of privacy under section 8. In sentencing the accused to five years imprisonment, an automatic section 109 weapon prohibition, and an order to provide a DNA sample, the judge commented, the abuse was "repulsive had it been inflicted by a non-professional. By a nurse, it is nothing short of torture" (para. 12).

Box 9.5 Reasonable Expectation of Privacy on the Street

Bank and store cameras (closed circuit TV) are often used to identify suspects in crimes they capture in progress. Do more widely panning street cameras invade our privacy and violate our rights under the *Charter*? Some municipal governments use them (Kelowna, Sudbury, London and Sherbrooke), while others are thinking about using them. When the Privacy Commissioner of Canada sought a declaration from the Supreme Court of British Columbia that the RCMP video surveillance in Kelowna violates people's rights under the *Charter* (sections 2(d), 6, 7 and 8), and is in breach of the United Nations' *Universal Declaration of Human Rights and the International Covenant on Civil and Political Rights*, the court decided that the Privacy Commissioner did not have the capacity to commence the lawsuit (*Canada Privacy Commissioner*). See Hubbard *et al.* 2004 for a discussion of the privacy issue involved in the use of these cameras.

REVIEW OF AUTHORIZATIONS

In 1993 (following a number of Supreme Court of Canada decisions, including *Garofoli* and *Dersch*), section 187 was enacted to codify the accused's access to the sealed packet of documents related to a wiretap authorization (including the affidavit filed in support of the application). Section 187(1.4) allows a trial judge, whether in superior court or provincial court, to open the packet. The Crown has the discretion to first delete any part of the document that the prosecutor "believes would be prejudicial to the public interest," including any part that the prosecutor believes could:

(a) compromise the identity of any confidential informant;
(b) compromise the nature and extent of ongoing investigations;
(c) endanger persons engaged in particular intelligence-gathering techniques and thereby prejudice future investigation in which similar techniques would be used; or
(d) prejudice the interests of innocent persons.

Subsection 187(7) allows the accused to ask the trial judge to review the Crown's editing. The judge can order the information disclosed if he or she believes that disclosure is "required in order for the accused to make full answer and defence and for which the provision of a judicial summary would not be sufficient." The edited packet is accessible to the accused under the right to full disclosure (*Pires* para. 25).

In 1990, Mr. Justice Sopinka (in *Garofoli*) wrote that "the law with respect to testing the admissibility of wiretap evidence is a procedural quagmire," (182) in that there were at least four different applications that the accused could make (182-183). According to the Supreme Court of Canada in *Pires*, "the Court in *Garofoli* consolidated these hearings, wiping away much of the complexity created by the earlier litigation, by using the overriding constitutional nature of the challenge to the admissibility of the evidence as the relevant framework of analysis" (para. 7). Now such challenges are decided on the basis of whether they violate section 8 of the *Charter*. According to the Court in *Pires*, "Without substantive compliance with the statutory regime, the wiretap is illegal and, given the consonance between the statutory provisions and the constitutional requirements, also unconstitutional" (para. 8). It follows that when an accused challenges a wiretap authorization, the decision on "whether the interception constitutes an unreasonable search or seizure . . . . involves an inquiry into whether the statutory preconditions have been met" (para. 8). If it is decided that the wiretap authorization violates section 8 of the *Charter*, the reviewing judge then decides whether the evidence should be excluded under section 24(2) of the *Charter* (para. 8).

The decision in *Garofoli*, which the Court was asked to revisit in *Pires*, involved the circumstances under which an accused would be allowed to cross-examine an affiant on the affidavit filed in support of the wiretap authorization. The affiant, usually the police officer, swears an information in support of the wiretap authorization. The Court in *Pires* confirmed that the accused must obtain leave of the court to cross-examine the affiant, by showing that the "cross-examination is necessary to enable the accused to make full answer and defence" (para. 10). The threshold test requires the accused to "show a reasonable likelihood that cross-examination of the affiant will elicit testimony of probative value to the issue for consideration by the reviewing judge" (*Pires* para. 3). The cross-examination should then "be confined to questions directed to the issue for consideration by the court" (para. 10). If the police officer relies on an informant, such a person, who is not a witness in the proceedings, cannot be cross-examined "unless the accused brings himself within the 'innocence at stake' exception" (para. 10). For a discussion of issues of excision and amplification, see Luther 2006.

SUMMARY

A private communication is one in respect of which there exists a reasonable expectation of privacy. Contrary to the common law position, it is generally illegal to intercept a private communication in Canada, except with the consent of one of the parties, or under the authority of a judicial authorization. Further, such evidence is generally inadmissible, unless it is lawfully obtained.

The legislative scheme for authorizing the interception of private communications allows the provincial Attorneys-General, the federal Minister of Public Safety, or their specifically designated agents to apply to a superior court judge, in respect of certain serious offences. The application is accompanied by evidence in the form of a sworn affidavit, which must include specific information, along with sufficient information to satisfy the judge that there are reasonable and probable grounds to believe that the authorization sought would be in the best interests of justice, and that other investigative procedures have been tried and failed or are unlikely to succeed, or that they are impractical because of urgency. Authorizations may be for periods of up to 60 days, and on application, may be renewed for periods not exceeding 60 days each. Special provisions deal with the interception of the private communications of lawyers, and with emergency authorizations.

As a result of judicial decisions in the early 1990s, sections were added to the *Code* to expressly permit persons such as undercover police officers to wear body-packs, or to otherwise have their conversations monitored for their own protection. Information obtained through such interceptions is not admissible at trial (except under limited circumstances), nor can it be used to support an application for a judicial authorization to intercept private communications. Other provisions allow for a more expedient form of application to a superior or provincial court judge, for an authorization to intercept private communications in situations where one party consents. Information gathered through this method can be used in court. The legislation also permits judicial authorizations to use other devices or techniques that would otherwise constitute an unreasonable search or seizure. This provision would cover, for example, video surveillance.

The legislative changes, and the case law that prompted them, have removed some of the complex aspects of determining the admissibility of intercepted private communications at trial. The situation is now considered entirely under sections 8 and 24(2) of the *Charter*. Aside from compliance with the search and seizure provisions, other prerequisites for the use of wiretap or other intercepted communications in court involve notice requirements. The *Criminal Code* sets out the circumstances under which the accused is entitled to examine the material filed in support of the application for authorization to intercept the private communications being tendered by the Crown. A procedure exists for the material to be edited first, to protect the identity of informants, ongoing investigations, the integrity of investigational techniques, and the interests of innocent persons.

Persons who have been the subject of a judicial authorization for the interception of private communications must be notified of that fact, although the time limit for doing so may be deferred on application to the court, if it is in the interests of justice to do so. Further, the Minister of Public Safety must report to Parliament the details of interception activity during each year, and must make public a similar, less detailed report.

QUESTIONS TO CONSIDER

(1)   What redress does a person have against someone who illegally intercepts a private communication?

(2)   Is it an offence to record your telephone conversations with a friend of yours without judicial authorization? Why or why not?

(3)   Is it an offence for a police officer to record telephone conversations with you without judicial authorization?

(4)   Is it an offence if you put a secret recorder on your telephone to record all the conversations of your roommate?

(5)   In an application under section 186 for authorization to intercept a private communication, what must the judge be satisfied of? What exception is there?

(6)   Does section 186 require the judge to have reasonable grounds to believe that an offence has been committed?

(7)   Joey's grandfather, Joseph, likes to visit the local coffee shop and sit in the corner booth, reading his newspaper. In fact, what he really does is listen to the conversations of those around him by turning up his hearing aid, so that he can hear at least twice as well as most normal people in the coffee shop. Joseph then uses this information to entertain his friends at the Home for the Elderly, where he lives. Joey is concerned that his grandfather might get into trouble for doing this. Explain to Joey the law governing his grandfather, and the type of "trouble" his grandfather might find himself in.

(8)   Sally and Susan are two students attending Simon Fraser University. They regularly "chat" by e-mail, and more recently found it a useful method of setting up meeting times for Sally to buy stolen goods from Susan. Campus security becomes aware of this, and passes the information on to the RCMP. The RCMP ask the University for permission to monitor and record e-mail exchanges between the Sally and Susan. Armed with six e-mail messages that implicate the students in buying and selling stolen property, the RCMP obtain a search warrant and search their apartments. At their trial, their defence counsel wants the evidence

obtained by the search warrant excluded on the basis that the interception of their e-mail messages violated the *Criminal Code* and their rights under the *Charter*. Make an argument on their behalf. What response would you expect from the Crown?

(9)   Under what circumstances can an accused cross-examine an affiant who swears an information to obtain a wiretap authorization?

BIBLIOGRAPHY

Beck, Stanley M. "Electronic Surveillance and the Administration of Criminal Justice." (1968) 46 *Canadian Bar Review* 643.

Boilard, Jean-Guy. *Guide to Criminal Evidence.* Cowansville, Quebec: Les Editions Yvon Blais Inc., 1999, with updates. Cases on monitoring are discussed in Chapter 6, Part B.

Canadian Press. "New Law in Works to Protect Privacy of Cellular Phones." *Vancouver Sun,* December 5, 1992, A5.

Canadian Press. "Journalists Fight Proposed Law on Cellular Phone Conversations." *Vancouver Sun,* March 25, 1993, A18.

Cobb, Chris. "Cellular Eavesdropping Poses Difficult Questions." *Vancouver Sun,* September 19, 1992, A6.

Cockfield, Arthur J. "Who Watches the Watchers? A Law and Technology Perspective on Government and Private Sector Surveillance." (2003) 29 *Queen's Law Journal* 364.

Cockfield, Arthur J. "The State of Privacy Laws and Privacy-Encroaching Technologies after September 11: A Two-Year Report Card on the Canadian Government." (2003-2004) 1 *University of Ottawa Law & Technology Journal* 325.

Frankel, S. David. "The Impact of the *Charter* on Electronic Surveillance." (1993) 51(1) *Advocate* 45.

Geist, Michael. "Computer and E-Mail Workplace Surveillance in Canada: The Shift From Reasonable Expectation of Privacy to Reasonable Surveillance." (2003) 82 *Canadian Bar Review* 152.

Grant, Campbell. *Inquiry Re Magistrate Frederick J. Bannon and Magistrate George W. Gardhouse.* Unpublished report, Toronto, 1968.

Hill, S. Casey, David M. Tanovich and Louis P. Strezos. *McWilliams' Canadian Criminal Evidence.* Aurora, ON: Canada Law Book Limited (available online on Criminal Spectrum).

Hubbard, Robert W., Peter M. Brauti, and Scott K. Fenton. *Wiretapping and Other Electronic Surveillance: Law and Procedure.* Aurora, Ont.: Canada Law Book, 2000.

Hubbard, Robert W., Peter DeFreitas and Susan Magotiaux. "The Internet–Expectations of Privacy in a New Context." (2001) 45 *Criminal Law Quarterly* 170.

Hubbard, Robert W., Susan Magotiaux, and Matthew Sullivan. "The State Use of Closed Circuit TV: Is There a Reasonable Expectation of Privacy." (2004) 49 *Criminal Law Quarterly* 222.

Kari, Shannon. "OPP officer lied to get wiretaps." (6 July 2004) *National Post* A7.

Law Reform Commission of Canada. *Electronic Surveillance.* Working Paper #47. Ottawa, 1986.

Law Reform Commission of Canada. *Recodifying Criminal Procedure.* Report #33. Ottawa, 1991.

Luther, Glen. "Of Excision, Amplification and Standing: Making Sense of the Law of Evidence in the Context of Challenges to Warranted Searches." (2006) 11 *Canadian Criminal Law Review* 1.

MacDonald, Norman. "Electronic Surveillance in Crime Detection: An Analysis of Canadian Wiretapping Law." (1987) 10 *Dalhousie Law Journal* 141.

Munro, Harold. "Ministry Admits to E-mail Tapping." *Vancouver Sun,* October 12, 1994, B1.

Nevis Consulting Group Inc. *Lawful Access Submission Summary Report,* 2003; www.justice.gc.ca/en/cons/la_al/summary/las_report_042803_e.pdf; accessed 7 October 2005.

Ouimet, Roger (Chair). *Report of the Canadian Committee on Corrections, Towards Unity: Criminal Justice and Corrections.* Ottawa: Information Canada, 1969.
    At pages 80–8, the Committee discusses advancing technology and the use of electronic surveillance. In 1969, the only legislation remotely related to this area was directed at damage to and interference with telephone use. The Committee confined its brief discussion to the use of wiretapping and electronic surveillance by police officers and criminals. The extent of such use was unknown, however the Committee noted that it was "used extensively in the investigation of certain kinds of suspected criminal activity" (82) and made recommendations for new legislation governing its use by police officers and preventing its use for a criminal purpose.

Penney, Steven. "Updating Canada's Communications Surveillance Laws: Privacy and Security in the Digital Age." (2008) 12 *Canadian Criminal Law Review* 12:115.

Pomerance, Renee M. "Redefining Privacy in the Face of New Technologies: Data Mining and the Threat to the 'Inviolate Personality.'" (2005) 9 *Canadian Criminal Law Review* 273.

Powell, Betsy. "Case against officer withdrawn; Crown ends perjury trial of retired OPP wiretap specialist after almost five-year prosecution." (29 January 2009) *Toronto Star* GT4.

Public Safety Canada. *Annual Report on the Use of Electronic Surveillance-2008* (Ottawa, Public Safety Canada, 2009).

Sargent, R.A. *British Columbia Report of the Inquiry Into Invasion of Privacy* (1967).

Stuart, Don, Ron Delisle, and Tim Quigley. *Learning Canadian Criminal Procedure,* 9th ed. Toronto: Carswell, 2008.
    Chapter 2 discusses the legislation governing electronic surveillance and the exclusion of evidence.

Whitling, N. J. "Wiretapping, Investigative Necessity, and the *Charter.*" (2002) 46 *Criminal Law Quarterly* 89.

CASES

R. v. Araujo, [2000] 2 S.C.R. 992.

R. v. Bogiatzis, Christodoulou, Cusato and Churchill, [2003] O.J. No. 3335 (Ont. S.C.).

R. c. Brais, [2009] J.Q. no 2487.

Canada (Privacy Commissioner) v. Canada (Attorney General), [2003] B.C.J. No. 1344 (B.C.S.C).

R. v. Davie (1980), 54 C.C.C. (2d) 216 (B.C.C.A.).

Dersch v. Canada (Attorney General), [1990] 2 S.C.R. 1505. Upon asserting that the admissibility of wiretap evidence is being challenged and that access is required to make full answer and defence, the accused shall have access to the contents of the sealed packet of material supporting the application for the authorization.

R. v. Doiron, [2004] N.B.J. No. 208 (N.B.Q.B.); affirmed, [2007] N.B.J. No. 189; application for leave to appeal refused, [2007] S.C.C.A. No. 413.

R. v. Doucet, [2003] J.Q. No. 18497 (Q.C.S.C.).

R. v. Duarte (1990), 53 C.C.C. (3d) 1 (S.C.C.). The interception of private communications through participant surveillance is still a search or seizure under section 8, and constitutes a breach of that right unless there has been prior judicial authorization.

R. v. Fliss, [2002] 1 S.C.R. 535. Even if an authorization is subsequently declared invalid, and the illegally recorded conversations are not inadmissible under section 24(2) of the *Charter*, the police officer may still give evidence about his recollection of the conversations. The inadmissible transcripts could also be used by the police officer to refresh his memory.

R. v. Garofoli (1990), 60 C.C.C. (3d) 161 (S.C.C.). An accused has a right to the contents of the sealed packet of material filed in support of the application for the authorization. The court has the power to edit the affidavit before providing it to the accused, to protect ongoing investigations, police investigative techniques, the identities of informants, and the interests of innocent third parties. Procedure discussed. Since wiretap is a search, the constitutional requirements of section 8 of the *Charter* apply.

R. v. Jamieson, [2004] O.J. No. 1780 (the *voir dire* on the admissibility of the video recording and other evidence), and [2005] O.J. No. 2495 for the sentencing decision.

R. v. Lucas, [2009] O.J. No. 2250.

R. v. McIsaac, [2005] B.C.J. No. 946.

R. v. Napope, [1992] B.C.J. No. 1330 (B.C.S.C.). Correctional centre rules authorized the monitoring of inmates' conversations for security purposes, and therefore there was no expectation of privacy, and the conversations were not private communications. Interception did not constitute unreasonable search and seizure under section 8 of the *Charter*. The senior corrections officer who directed the taping had formed a belief that the communications might disclosure "a threat to the management, operation, discipline or security" of the centre under the rules. *Rodney* distinguished.

R. v. Pangman, [2000] M.J. No. 300 (Man. Q.B.).

R. v. Paterson et al. (1987), 39 C.C.C. (3d) 575 (S.C.C.). Basket clauses discussed.

R. v. Pires; R. v. Lising, [2005] 3 S.C.R. 343.

R. v. Riley, [2008] O.J. No. 2887.

R. v. Rodney (1984), 12 C.C.C. (3d) 195 (B.C.S.C.). The Crown must establish that it was reasonable for an accused to expect that telephone communications would be intercepted while confined in a correctional centre, before such communications are admissible. In this case, the interception was ruled inadmissible. The same evidence was also ruled inadmissible at a subsequent re-trial (1991), 65 C.C.C.(3d) 304.

R. v. Samson (1983), 9 C.C.C. (3d) 194 (Ont. C.A.). Basket clauses permitting the interception of the private communications of "any other persons" unnamed, at specified locations, are valid.

R. v. Six Accused Persons, [2008] B.C.J. No. 293.

R. v. Steinberg, [1967] 1 O.R. 733 (Ont. C.A.). Police executing a search warrant on an accused's home left a hidden tape recorder. The accused's statements recorded thereon were not inadmissible as contrary to section 1(a) (the right to enjoy property) or section 2(d) (freedom from self-incrimination) of the *Bill of Rights*.

R. v. Tam (1993), 80 C.C.C. (3d) 476 (B.C.S.C.). Kidnappers were using the phone to demand a ransom, and hung up repeatedly because they thought they were being monitored. The court found that there was no expectation of privacy, and therefore the conversations were not private.

R. v. Tessling, [2004] 3 S.C.R. 432.

R. v. Weir, [1998] A.J. No. 155 (Alta. Q.B.); [2001] A.J. No. 869 (Alta. C.A.)

R. v. Wong, [1990] 3 S.C.R. 36. The legislation that existed at this time did not allow for electronic surveillance authorizations to be used to conduct video surveillance. Note, in 1993, the federal government added section 487.01 (see Chapter 8).

CHAPTER 10: *Admissions and Confessions*

CHAPTER OBJECTIVES

In studying this chapter, you should develop an understanding of the following topics and concepts:

- the principle against self-incrimination
- the nature of and distinction between admissions and confessions
- the problems associated with jail-house informants
- the meaning of "person in authority," for the purposes of the confession rule
- the criteria for a confession to be admissible in evidence
- the impact of *Charter* rights to counsel and silence on the admissibility of confessions
- issues with undercover police officers eliciting confessions (Mr. Big)
- the nature of the privilege against testimonial self-incrimination

THE PRINCIPLE AGAINST SELF-INCRIMINATION

According to the Supreme Court of Canada, the principle against self-incrimination is a "general organizing principle of criminal law," and an "overarching principle within our criminal justice system" (*Singh* para. 21). It is now the source from which a number of common law and *Charter* rights emanate, such as the confessions rule and the right to remain silent. The principle is also "embodied in several of the more specific procedural protections such as, for example, the right to counsel in s. 10(*b*), the right to non-compellability in s. 11(*c*), and the right to use immunity set out in s.13. The *Charter* also provides residual protection to the principle through s. 7" (*Singh* para. 21). This overarching principle means that unless required to do so by law, citizens do not need to respond to questions and requests from police officers (para. 27).

ADMISSIONS

An **admission** is a statement by a person that is adverse to the person's interests. In the context of the criminal justice system, an admission can be formal or informal.

Formal Admissions

A **formal admission** (also referred to as a **solemn admission**, or a **judicial admission**) is one that is made in legal proceedings, and it relieves the Crown of the burden of proving a certain fact (or facts) in a case. A formal admission can be made orally or in writing. For example, a guilty plea is a formal admission of all of the elements of the alleged offence.

Under the common law, an accused could not admit facts; the Crown had to prove all the elements of the offence (Uniform Law Conference of Canada 1982, 34). Section 655 of the *Criminal Code* changed the common law, and now an accused may formally admit any facts alleged by the Crown. For example, an accused may admit to the fact of killing someone, because the only issue is that of sanity, or level of intent. The Crown need not accept the admission of facts, and can call evidence on issues even if facts are admitted by the accused. This may happen, for example, in a jury trial, where the defence admits certain facts hoping to lessen the impact on the jury of evidence given by upset witnesses. It is also possible to admit the voluntariness of a statement made to a person in authority (discussed below under "Confessions"), even though such an issue is to be decided by the trial judge and not by the jury.

Pursuant to section 657 of the *Criminal Code*, anything the accused says at the preliminary inquiry under section 541 can be led as evidence against him or her at trial. Such a statement by the accused is not under oath. Presumably, section 13 of the *Charter* (the right of witnesses not to have their incriminating testimony used against them at any other proceeding) does not apply to these statements because they are not testimony, and the preliminary hearing is not an "other proceeding" (*Yakeleya*). However, the Supreme Court of Canada in *Dubois* left this question open. There are limitations on the use of testimony given by a witness in one proceeding as evidence against that person in another proceeding. These rules are discussed in greater detail below, under "The Privilege Against Testimonial Self-Incrimination."

Informal Admissions

Admissions not made in judicial proceedings are referred to as **informal admissions** (or **unsolemn admissions**). There is some debate as to the basis on which such admissions are admissible in evidence—whether they are an exception to the hearsay rule (discussed in Chapter 12) or are not hearsay at all (see discussion McWilliams 1999, 14–9). In Canada, to a large extent, admissions are treated as hearsay (an out-of-court statement made by someone other than the witness, tendered to prove the truth of what the statement says), and their admissibility is based on their treatment as an exception to the hearsay rule. In other words, although hearsay is inadmissible and admissions are hearsay, admissions are an exception to the exclusionary rule.

In practical terms, if a person admits to a friend or stranger that he or she robbed a bank (an admission), that friend or stranger can be called at the person's trial to testify that the person admitted to robbing the bank. It is important to remember that once an admission is introduced as evidence, it is only one factor to be considered and weighed by the trier of fact (the judge, or the jury in the case of a jury trial) in determining guilt or innocence. All the circumstances surrounding the admission can be examined to determine how much weight to give to the admission. The accused can always take the stand and explain away the admission ("I was only joking when I said I robbed the bank"). Admissions can also be made by way of gesture or conduct, or by adopting or accepting a statement made by someone else, even if that statement is hearsay (*Streu*).

CONFESSIONS

A **confession** is an informal admission made to a **person in authority.** It is admissible as evidence against the accused only if the Crown can show, beyond a reasonable doubt, that it was voluntary (the **confession rule**), and only if it is not excluded under section 24(2) of the *Charter*. Since the burden is on the Crown to show that a confession is voluntary so that the confession can be admitted into evidence, and the burden is on the accused to show that a right under the *Charter* has been violated so that the confession ought to be excluded under section 24(2), the Ontario Court of Appeal has suggested that there should be one *voir dire* on voluntariness, and a second, separate *voir dire* on section 24(2) of the *Charter* (*Voss*).

Despite recommendations by the Law Reform Commission of Canada (1984a, 1984b) that the law governing admissions and confessions be codified, the area is still governed by the common law. There is, for example, no federal legislation governing the police questioning of suspects.

A Historical Note and Rationale

The present rule regarding the use or admissibility of confessions as evidence against an accused developed at a time when the accused was not allowed to testify on his or her own behalf, and when torture was a commonly used method of extracting the truth from the accused. According to one source, torture was the only way to ensure that a person was telling the truth; the theory was that if the accused persisted in a story under torture, it must be true.

The English courts began excluding confessions obtained under threats or promises at a time when they were biased toward acquitting people, to compensate for the numerous offences punishable by death. The general rule, which began in 1783 and subsequently made its way into Canadian law, was stated by the English Appeal Court in the case of *Ibrahim v. The King*:

> It has long been established as a positive rule of English criminal law, that no statement by an accused is admissible in evidence against him unless it is shewn [*sic*] by the prosecution to have been a voluntary statement, in the sense that it has not been obtained from him either by fear of prejudice or hope of advantage exercised or held out by the person in authority (quoted in the Uniform Law Conference Report 1982, 173).

The Confessions Rule—Voluntariness

The requirement that a confession be voluntary to be admissible was adopted by the Supreme Court of Canada in 1921 (*Prosko*). For a statement made to a person in authority to be admissible as evidence, the Crown must show beyond a reasonable doubt that the confession was voluntary. Whether a confession is voluntary is a question of fact, or a question of mixed fact and law, to be determined by the judge in a *voir dire* (*Oickle* para. 22). Whether the judge believes the statement is not relevant to the issue of whether it was voluntary. If the statement is voluntary and admissible, the trier of fact (judge or jury) then decides whether to believe the statement, or how much weight to give it.

The meaning of "voluntary" is often assessed in light of the purpose of the rule. Historically, the courts were concerned with the trustworthiness of the statement, such that a statement was involuntary and not admissible if it was obtained either by **fear of prejudice** or **hope of advantage** exercised or held out by a person in authority (see discussion in *Oickle* para. 13). The focus, therefore, was on the police officer's behaviour. If the police officer held out neither fear of prejudice nor hope of advantage, the statement was voluntary. This interpretation is quite narrow, however. A wider interpretation is that fear of prejudice or hope of advantage are only two factors that could render a statement involuntary.

Other rationales for the rule developed beyond the concern for reliability. The accused has a right to remain silent, and the Crown must prove the case against the accused without the compulsory assistance of the accused (free choice rationale). In addition, the introduction of an involuntary statement by the accused can be considered unfair to the accused, and reflect negatively on the integrity of the criminal justice system (abuse prevention rationale) (Penney 2003; also discussed in McWilliams 1999 Chapter 15; Uniform Law Conference Report 1982, 174–5).

In *Oickle*, the Supreme Court of Canada revisited the confession rule, which is designed to protect the rights of the accused "without unduly limiting society's need to investigate and solve crimes" (para. 33). It discussed four factors that the trial judges should examine when determining whether a confession is voluntary: 1) threats or promises, 2) oppression, 3) operating mind of the suspect, and 4) other police trickery. With regard to threats or promises, the most important factor is whether the police have offered something in return for a confession. Moral inducements by themselves are not improper. The Court provided a number of examples that could create an atmosphere of oppression: "depriving the suspect of food, clothing, water, sleep, or medical attention; denying access to counsel; and excessively aggressive, intimidating questioning for a prolonged period of time" (para. 60). Fabricated evidence may, under some circumstances, constitute oppression. Examples of unacceptable police trickery, which might "shock the community," include: "a police officer pretending to be a chaplain or a legal aid lawyer, or injecting truth serum into a diabetic under the pretense that it was insulin" (para. 66). However, a confession made when confronted with evidence which is inadmissible at trial is not necessarily involuntary (para. 102).

In *Spencer*, the Supreme Court of Canada stated that promises do not necessarily render an accused's statement involuntary. Inducements "becom[e] improper only when . . . standing alone or in combination with other factors, [they] are strong enough to raise a reasonable doubt about whether the will of the subject has been overborne" (*Spencer* para. 13, quoting *Oickle*). One of the rationales for allowing admissions to be tendered in evidence is the assumption that people do not admit to things against their interest unless they are true. This assumption that people are averse to falsely admitting to crimes has existed throughout the history of the common law and remains today, despite little or no empirical evidence to support it. The Supreme Court of Canada in *Oickle* was, however, concerned about the possibility of false confessions. Are people usually aware of the adverse consequences of making admissions? Are there ever reasons for which persons might falsely confess to crimes in their own interests? (See, for example, Keenan and Brockman 2010; Leo 2009; Roach 2007, 228-35).

Operating Mind

If the courts are concerned with the trustworthiness of the statement, it is important to look beyond the behaviour of the person in authority, to examine the mental state of the accused at the time the statement was made. The decisions in *Horvath* and *Ward* by the Supreme Court of Canada in the late 1970s produced a major change in the law regarding the voluntariness of confessions. In *Ward,* the accused was charged with causing death by criminal negligence in the operation of a motor vehicle. Ward and his girlfriend, who was killed in the accident, were both found outside the vehicle. There was no evidence as to who was driving. When he was revived, Ward said he had not been driving the vehicle. Later, when he was interviewed, he said he was driving. At the *voir dire* he testified that he had no recollection at all as to who was driving at the time of the accident. In finding that there was a reasonable doubt whether the confession was voluntary, Mr. Justice Spence stated that the trial judge was entitled to consider the mental and physical state of the accused to determine (1) whether a person in his condition would be subject to hope of advantage or fear of prejudice in making the statements, when perhaps a normal person would not be, and (2) whether, due to mental and physical condition, the accused's words could really be the utterances of an **operating mind.**

In *Horvath*, a 17-year-old accused was charged with the murder of his mother. Through lengthy, skillful interrogation (described as "hot and furious" by the police officers, and as "the most skillful example of police interrogation that has ever come to my attention in 36 years as a lawyer and a judge" by the trial judge), the police managed to obtain a confession from Horvath. The interrogation resulted in the "emotional disintegration" of Horvath. A psychiatrist testified that the accused was put into a state of hypnosis by the skilled police officers. The Supreme Court of Canada ruled (four to three) that the confessions were inadmissible. Two judges relied on the emotional disintegration, and two on the hypnotic state, to conclude that the confession was not voluntary.

In *Whittle*, the Supreme Court of Canada stated that the operating mind test includes a limited mental component that "requires that the accused have sufficient cognitive capacity to understand what he is saying and what is being said," and the ability "to understand a caution that the evidence can be used against the accused" (941). However, there was no "separate awareness of the consequences test" (947). The court does not have to inquire as to whether the accused was "capable of making a good or wise choice or one that is in his or her interest" (939).

Persons in Authority

The requirement in Canada that admissions be voluntary only applies to statements made to persons in authority (i.e., to confessions). The problems associated with determining who is a person in authority do not exist in the United States, where the prosecutor has to prove the voluntariness of all statements by the accused (Uniform Law Conference Report 1982, 175). Despite the English Criminal Law Revision Committee's recommendation to adopt a rule similar to that in the United States, the Uniform Law Conference (175) and the Law Reform Commission of Canada (1984a; 1984b) both recommended that Canada retain the distinction between persons in authority and others. The Supreme Court of Canada has stated that any change in the law "should be studied by Parliament and remedied by enactment" (*Hodgson* para. 29).

In *Hodgson*, the Supreme Court of Canada elaborated on the meaning of persons in authority:

> 3. . . .Though no absolute definition of "person in authority" is necessary or desirable, it typically refers to those formally engaged in the arrest, detention, examination or prosecution of the accused. Thus, it would apply to person such as police officers and prison officials or guards. When the statement of the accused is made to a police officer or prison guard a *voir dire* should be held to determine its admissibility as a voluntary statement, unless the *voir dire* is waived by counsel for the accused.

> 4. Those persons whom the accused reasonably believes are acting on behalf of the police or prosecuting authorities and could therefore influence or control the proceedings against him or her may also be persons in authority. That question will have to be determined on a case-by-case basis.

> 5. The issue as to who is a person in authority must be resolved by considering it subjectively from the viewpoint of the accused. There must, however, be a reasonable basis for the accused's belief that the person hearing the statement was a person in authority.

> 6. The issue will not normally arise in relation to undercover police officers. This is because the issue must be approached from the viewpoint of the accused. On that basis, undercover police officers will not usually be viewed by the accused as persons in authority.

> 7. If it is contended that the recipient of the statement was a person in authority in the eyes of the accused then the defence must raise the issue with the trial judge. This is appropriate for it is only the accused who can know that the statement was made to someone regarded by the accused as a person in authority.

> 8. On the ensuing *voir dire* the accused will have the evidential burden of demonstrating that there is a valid issue for consideration. If the accused meets the burden, the Crown will then have the persuasive burden of demonstrating beyond a reasonable doubt that the receiver of the statement was not a person in authority or if it is found that he or she was a person in authority, that the statement of the accused was made voluntarily.

> 9. In extremely rare cases the evidence adduced during a trial may be such that it should alert the trial judge that the issue as to whether the receiver of a statement made by an accused was a person in authority should be explored by way of *voir dire*. In those cases, which must be extremely rare in light of the obligation of the accused to raise the issue, the trial judge must of his or her own motion direct a *voir dire*, subject, of course, to waiver of the *voir dire* by counsel for the accused.

10. The duty of the trial judge to hold a *voir dire* of his or her own motion will only arise in those rare cases where the evidence, viewed objectively, is sufficient to alert the trial judge of the need to hold a *voir dire* to determine if the receiver of the statement of the accused was, in the circumstances, a person in authority.

11. If the trial judge is satisfied that the receiver of the statement was not a person in authority but that the statement of the accused was obtained by reprehensible coercive tactics, such as violence or credible threats of violence, then a direction should be given to the jury. The jury should be instructed that if they conclude that the statement was obtained by coercion, they should be cautious about accepting it, and that little if any weight should be attached to it (*Hodgson* para. 48, and *Wells* para. 14).

In *Hodgson*, the relatives of the accused confronted him at his place of work with allegations that he had raped the girl he babysat, over the period when she was between seven and eleven years old. After his admission, the father of the girl held a knife to his back until the police arrived. At trial, the accused denied making the admission, and also stated that he did not feel threatened by the encounter. The Supreme Court of Canada found that there was no evidence that the parents of the complainant were persons in authority, as there was no evidence they had first talked to the police or were planning on making a complaint (See Presser 1998 for a criticism of this approach).

In *Wells*, the parents of the complainants had first spoken to the police, before they confronted Wells with a bread knife at his throat. Wells confessed to sexually touching three boys for a sexual purpose. Neither the Crown nor the defence requested a *voir dire*. The Court ordered a new trial, as this was one of those "rare cases" where the trial judge should have initiated a *voir dire* to determine whether the statement was made to a person in authority. Wells was convicted again at his second trial. Although the Court did not state that judges have the option of excluding such admissions under the common law, where their prejudicial effect outweighs their probative value, Penney (2004a: 294-95) suggests that they do.

In *Grandinetti*, undercover police officers who claimed that they might be able to influence the course of an investigation through their corrupt police contacts were not persons in authority when they obtained a confession. According to the Supreme Court of Canada, "the state's coercive power is not engaged" in such circumstances (para. 44).

The Relationship Between the Confession Rule and Section 7

Can an accused argue that his or her right to remain silent was violated once the Crown proves beyond a reasonable doubt that a confession is voluntary? In *Singh*, the police continued to question the detainee, who was in custody, after he talked to counsel and asserted his right to remain silent eighteen times (para. 58). The trial judge found his statements to be voluntary and therefore

---

Box 10.1 What Happened to John Horvath?

Four years after the death of his mother, John Horvath answered an advertisement to share an apartment. He stabbed his potential roommate with a knife, severed her jugular vein, and stole sixty dollars and some jewellery. She lived to identify him, and in 1981 Horvath was sentenced to life imprisonment for attempted murder (Jonas 1983, 91).

admissible. The Court of Appeal dismissed Singh's appeal. The Supreme Court of Canada addressed the issue of whether the accused could argue that his right to remain silent was violated once the judge decided his statement was voluntary. Charron, J., for the majority, wrote: "if the Crown proves voluntariness beyond a reasonable doubt, there can be no finding of a *Charter* violation of the right to silence in respect of the same statement" (para. 37). For commentaries on *Singh*, see Akhtar 2008, Ives 2007, and Walker 2009.

Derived Confessions Rule

The **derived confessions rule** addresses the question of when a second statement, which follows an inadmissible first statement, should be excluded. According to Mr. Justice Sopinka, the admissibility of the second statement depends on the "degree of connection between the two statements." In order to determine this, the court must examine factors such as: "the time span between the statements, advertence to the previous statement during questioning, the discovery of additional incriminating evidence subsequent to the first statement, the presence of the same police officers at both interrogations and other similarities between the two circumstances" (*I. (L.R.)* 526). A second confession "would be involuntary if either the tainting features which disqualified the first confession continued to be present or if the fact that the first statement was made was a substantial factor contributing to the making of the second statement" (526).

An involuntary confession cannot be used to cross-examine and challenge the credibility of an accused. This would violate the accused's right to a fair trial, and "could lead to abuse and serious injustice" (*G.(B.)* para. 33). In fact, an involuntary confession cannot be used by the Crown for any purpose (*Calder* para. 26). This rule does not apply to challenging the credibility of a witness who is not the accused, and who makes inconsistent statements (discussed in Chapter 12).

Other Derivative Evidence

If a confession is ruled inadmissible, can evidence obtained or derived from that confession still be used as evidence against the accused? Further, can those parts of the otherwise inadmissible confession confirmed by that **derivative evidence** be admitted anyway, since they are obviously trustworthy? At common law, the rule in *St. Lawrence* governed:

> Where the discovery of the fact confirms the confession—that is, where the confession must be taken to be true by reasons of the discovery of the fact—then that part of the confession that is confirmed by the discovery of the fact is admissible, but further than that no part of the confession is admissible.

The rule was applied in *Wray* where, as the result of an involuntary confession, incriminating real evidence was discovered. Wray was charged with the 1968 murder of Donald Comrie, a 20-year-old man working at a service station, and with the robbery of the service station. The police took Wray to a polygraph operator in Toronto. Following a grueling examination by the polygraph operator (who was described by the trial judge as resembling a thug), which included a series of lies to Wray that the police had a substantial amount of evidence to place him at the scene of the crime, Wray confessed to the murder, and showed the police the swamp where the gun used in the murder was hidden (Jonas 1983). The trial judge found that both the confession and Wray's involvement in the finding of the rifle were inadmissible, and he was acquitted. The Crown's appeal to the Ontario Court of Appeal was dismissed, but on further appeal, the Supreme Court of Canada ordered a new trial.

According to the majority of the Court, the confession was inadmissible, but the accused's involvement in the finding of the gun was admissible because it fell within the rule in *St. Lawrence*, that any part of the confession confirmed by the finding of the article was admissible. Thus, Wray's statement, which led the police to the site of the weapon, was admissible, even though Wray's statement that he had put it there was not admissible because it was not confirmed by the finding—anyone could have put it there. In addition, the Court held that trial judges have no discretion to exclude such evidence. Mr. Justice Martland wrote: "the admission of the evidence relevant to the issue before the Court and of substantial probative value may operate unfortunately for the accused, but not unfairly" (17).

These older cases were concerned with the reliability of the evidence obtained by an involuntary confession, and there was little (if any) discretion to exclude evidence, even if its admission would bring the administration of justice into disrepute. Today, such evidence is examined under section 24(2) of the *Charter*. The Supreme Court of Canada recently revamped the law surrounding the admissibility of derivative evidence by stating that the admission of such evidence

---

Box 10.2 What Happened at Wray's Second Trial?

Defence counsel argued that there was no evidence to go to the jury, because the fact that the accused led the police to the murder weapon was not evidence he had committed the offence. The jury convicted Wray, and he was sentenced to life imprisonment (Jonas 213).

---

must be dealt with under the *Grant* framework (discussed in Chapter 7), not simply on the basis of its discoverability, as was suggested in earlier cases such as *Stillman* and *Burlingham* (para. 120-21). In determining whether the admission of derivative evidence would bring the administration of justice into disrepute, the court will first examine the circumstances surrounding the police conduct in obtaining a statement that resulted in the discovery of real evidence. Was their behaviour deliberate and systematic in "flouting" the accused's rights, or did they act in good faith (*Grant* para. 124)? Secondly, what is the impact of the breach on the accused's *Charter*-protected interests? Although discoverability is not determinative, the fact that the evidence was "independently discoverable" will lessen the impact of the breach on the accused (para. 125). With respect to "society's interest in having the case adjudicated on its merits," the majority stated: "Since evidence in this category is real or physical, there is usually less concern as to the reliability of the evidence. Thus, the public interest in having a trial adjudicated on its merits will usually favour admission of the derivative evidence" (para. 126). However, "deliberate and egregious police conduct" may result in the evidence being excluded even if it is reliable (para. 127).

THE RIGHT TO RETAIN AND INSTRUCT COUNSEL—SECTION 10(b)

In addition to considering the voluntariness of a confession, the courts also must decide whether a person's rights to retain and instruct counsel under the *Charter* have been violated, and if so, whether a confession should be excluded under section 24(2). Section 10(b) of the *Charter* is not concerned with the probative value of evidence, but rather with ensuring that a person is made aware of his or her right to counsel when detained in circumstances that may give rise to a "significant legal consequence."

Duties on Police Officers Regarding the Right to Counsel

Several cases from the Supreme Court of Canada deal with the right to counsel under section 10(b) the *Charter*. In *Manninen*, the accused had said, "I ain't saying anything until I see my lawyer" (1238). The police continued to ask questions, including questions about the location of the knife and the gun that were used in the robbery. The accused responded, "He's lying, I only had a gun. The knife was in the tool box" (1238). The Supreme Court of Canada found that the accused's rights under section 10(b) had been violated, and that the confession should have been excluded under section 24(2). The Court established two duties on a police officer, under section 10(b): (1) the duty to give detainees reasonable opportunity to exercise their right (this includes a duty to facilitate the exercise of the right to counsel), and (2) the duty to cease asking questions or otherwise attempting to elicit the evidence from the detainees until they have had a reasonable opportunity to retain and instruct counsel (1241-42).

In *Burlingham* (see discussion in Chapter 7), the police violated both of these duties. Mr. Justice Iacobucci added that section 10(b) "specifically prohibits the police, as they did in this case, from belittling an accused's lawyer with the express goal or effect of undermining the accused's confidence in and relationship with defence counsel" (para. 14). The police also violated Burlingham's right to counsel by entering into a plea bargain with him, without first giving him the opportunity to consult with his lawyer (para. 15) (see discussion by Stuart 1995).

Informational Rights Under Section 10(b) of the *Charter*

In *Brydges*, the Supreme Court of Canada discussed **informational rights** under section 10(b) of the *Charter*. In circumstances where an accused indicates an inability to afford counsel, there is a duty on the police to provide information about the availability of legal aid and duty counsel. In *obiter* (statements in a judgment that are not essential to the decision), Mr. Justice Lamer said that information regarding legal aid and duty counsel should be part of the routine information given to detainees. Since duty counsel was available in *Brydges,* there was no need for the Court to speculate about situations in which there is no accessible duty counsel. Many provinces have, however, set up 24-hour telephone services so that detainees can have access to counsel.

In a series of cases in 1994 following *Brydges*, the Supreme Court of Canada stated that section 10(b) places a duty on police officers to provide suspects with comprehensible information about their right to counsel, so that they can make informed decisions (*Bartle* 174-5). This includes telling suspects about the availability of free duty counsel (*Cobham* 373). Although 24-hour-a-day free duty counsel is not constitutionally mandated, the Court stressed the "desirability from the point of view of fairness and administrative convenience of a system of preliminary legal advice universally available to all detainees upon request and free of charge" (*Prosper* 265). Mr. Justice Lamer commented that the *Charter* requires the police to "refrain from eliciting incriminatory evidence from the detainee until he or she has had a reasonable opportunity to reach counsel" (374–5).

Whether there is a "*Brydges* duty counsel" available to the detainee will affect the evaluation of whether the detainee has exercised "reasonable diligence" in asserting his or her right to counsel, and whether the accused has been given a reasonable opportunity to consult with counsel (*Prosper* 269). The absence of duty counsel extends the time that might be needed to exercise one's right to counsel, perhaps until the legal aid office opens the next day, in jurisdictions where there is none available outside office hours (270). Lamer, J. based these conclusions on the underlying purpose of section 10(b), "to protect the privilege against self-incrimination," and the "right to remain silent," which exist as "principles of fundamental justice" under section 7 of the *Charter* (271-2). The non-availability of counsel could, in some circumstances, affect the fairness of the trial, and the government runs the risk of the evidence being excluded under section 24(2) of the *Charter* (274).

Lamer, J. added another informational requirement in situations where a detainee has been unsuccessful in contacting counsel after asserting the right:

> At this point, the police will be required to tell the detainee of his or her right to a reasonable opportunity to contact a lawyer and of the obligation on the part of the police during this time not to take any statements or require the detainee to participate in any potentially incriminating process until he or she has had that reasonable opportunity. This additional informational requirement on police ensures that a detainee who persists in wanting to waive the right to counsel will know what it is that he or she is actually giving up (*Prosper* 274).

The burden of establishing that the detainee waived his or her right to counsel is on the Crown, and the standard is very high. "A person who waives a right must know what he or she is giving up if the waiver is to be valid" (*Prosper* 275).

Lamer, J. conceded that there might be urgent circumstances in which the police cannot "hold off" until the accused contacts counsel. He was of the view, however, that the two-hour period in which samples must taken under section 258(1)(c)(ii) did not by itself amount to urgent circumstances (278). The Crown's loss of the evidentiary presumption that the accused's alcohol level was over 80 mg at the time of the offence is "simply one of the prices which has to be paid by governments which refuse to ensure that a system of '*Brydges* duty counsel' is available to give detainees free, preliminary legal advice on an on-call 24-hour basis" (*Prosper* 275).

In applying the law to the facts, Lamer, J. decided that Prosper's rights under section 10(b) had been violated. Although there was no obligation on the government to provide free and immediate legal advice, there was a duty to provide Prosper with a reasonable opportunity to consult with counsel. Prosper's right to counsel was violated in two ways:

> First, after asserting his right to counsel and trying repeatedly to contact a lawyer, the appellant was not informed when he changed his mind and agreed to take the breathalyser test that the police had to hold off from their investigation until he had had a reasonable opportunity to contact counsel. Secondly, the police failed in fact to hold off and provide the appellant with the reasonable opportunity to contact counsel to which he was entitled under section 10(b). That is, the police failed under the circumstances…to put off administering the breathalyser tests until either the appellant contacted a Legal Aid lawyer, or was taken before a justice of the peace for a bail hearing and his situation could be assessed (*Prosper* 280).

In ruling that the evidence should be excluded under section 24(2) of the *Charter*, Lamer, J. wrote, "the breach of the appellant's right to counsel goes directly to his privilege against self-incrimination and receipt of the breathalyser evidence resulting from this breach would undermine this privilege, thereby rendering the trial process unfair" (284).

The Right to Counsel When There Is a Change in Jeopardy

In *Black*, the Supreme Court of Canada dealt with the sufficiency of a *Charter* warning in circumstances where the legal jeopardy changed while the accused was being detained. Madam Justice Wilson, for the Court, made it clear that the right to counsel is given in a particular context, and that if the charges being investigated change (in this case from attempted murder to murder), the accused is then entitled to be informed again of his or her right to counsel on the new charges (see commentary by Penney 2004a).

Waiver

In *Clarkson*, Madame Justice Wilson considered the issue of waiver by an intoxicated accused:

> [I]t is clear that the waiver of the section 10(b) right by an intoxicated accused must pass some form of "awareness of the consequences" test. Unlike the confession itself, there is no room for argument that the court in assessing such a waiver should only be concerned with the probative value of the evidence so as to restrict the test to the accused's mere comprehension of his or her own words.... [A]ny voluntary waiver in order to be valid and effective must be premised on a true appreciation of the consequences of giving up the right (219).

At a minimum, Wilson, J. decided, the police should have waited until the accused sobered up, so that she could consult with counsel, or waive her right to counsel with full awareness of the consequences of doing so (219). She concluded that the actions of the police officers were a "clear case of deliberate exploitation by the police of the opportunity to violate the [accused's] rights," and excluded the evidence under section 24(2) (220).

Bruce Duncan, Crown counsel in Calgary, suggested that the police, faced with a murder, had a duty to interview this chattering witness, who was probably the only witness to the murder. He wrote, "why should the police not take advantage of a criminal's loose tongue, as they would a criminal's careless fingerprint? What is the social utility in a rule that provides otherwise? Will it ever have any effect other than to shield the guilty from their own carelessness?" (308)

In *Black*, Madam Justice Wilson also confirmed her position in *Clarkson* on the waiver of one's right to counsel. Waiver, she held, is dependent on awareness of the consequences of waiving one's right to counsel. Black, a known alcoholic with a grade 4 education, had a high blood-alcohol level, and could not be seen to have waived her right to counsel simply by responding to the officer.

The *Grant* Analysis for Violation of the Accused's Right to Counsel

Although a violation of an accused's right was at one time thought to result in almost automatic exclusion under section 24(2) of the *Charter*, the Supreme Court of Canada in *Grant* stated there is "no absolute rule" of exclusion in these circumstances (para. 91), even though such statements "tend to be excluded under s. 24(2)" (para. 98). Rather, the trial judge must examine the *Charter* violation in light of the three factors discussed in *Grant*. The first inquiry focuses on the seriousness of the violation and its impact on the repute of the justice system (para. 93-94). Secondly, what is the impact on the suspect's protected interests? A technical breach will not have a serious impact on the accused's right (para. 96). Third, what is the impact of the public interest on having a trial on its merits? One concern is that a suspect without access to a lawyer "may make statements that are based more on a misconceived idea of how to get out of his or her predicament than on the truth" (para. 97).

THE RIGHT TO REMAIN SILENT—SECTION 7

Individuals have a right to remain silent in the face of police questioning, even if they initiate police contact and volunteer some information (*Turcotte*). The common law right to remain silent is entrenched by section 7 of the *Charter*. Since a person has a right to remain silent, this silence is usually irrelevant and inadmissible at the accused's trial (*Turcotte* para. 56). Silence may be relevant if the defence raises an issue which makes it relevant, such as seeking to show the accused's cooperation, expounding a theory of mistaken identity, or claiming an alibi (*Turcotte* para. 49-50). The right to remain silent exists in both in-custody and out-of-custody situations. There are, however, a number of ways that the police have managed to obtain statements from suspects through the use of jail-house informants and undercover agents in both in- and out- of-custody circumstances.

Jail-House Informants (In-Custody Informers)

A **jail-house informant** is a person who, while in custody with the accused, claims that the accused made an admission of guilt. Police officers who are in custody as part of an undercover operation are not included in this definition. Sherrin (1997a, 1997b) identifies some of the problems with the use of such jail-house informants. The reliability of such evidence has caused concern, because the informants are generally promised something (e.g., privileges in jail, money, reduction in sentence, stay of charges) in exchange for their evidence. Informants, who typically present evidence for the Crown, may be more believable because of their association with the Crown. They may also be good witnesses if they have experience testifying in court. Mr. Justice Cory in his Sophonow Report is more blunt; he wrote, "Jailhouse informants are polished and convincing liars" (online). Jurors may be unable to properly assess the credibility of such witnesses. Informants may abuse other inmates to obtain information with which they can bargain.

The Kaufman Commission of Inquiry into the wrongful conviction of Guy Paul Morin contains 220 pages on the evidence of two jail-house informants, and recommendations regarding the use of jail-house informants (Kaufman 1998, Chapter 3). The Commission recommended ministry guidelines and limitations on the use of jail-house informants in its 34 recommendations; however, it did not recommend an absolute prohibition on the use of jail-house informants' testimony, as was suggested by the Association in Defence of the Wrongfully Convicted (AIDWYC) (Kaufman 602).

Mr. Justice Cory, in his review of the wrongful conviction of Thomas Sophonow, was less kind to jail-house informants. He referred to them as a "festering sore" and a "uniquely evil group," who "rush to testify like vultures to rotting flesh or sharks to blood." He recommended that "as a general rule, jailhouse informants should be prohibited from testifying," and in the rare cases where they might be allowed to testify, the jury should be instructed in the clearest terms about the dangers of accepting their evidence. Failure to warn should result in a mistrial (Cory 2001, online). In *Brooks*, the Supreme Court of Canada was split on the question of whether a jury should be warned about the evidence of a jailhouse informant in a particular case, but refused to take a categorical approach to such warnings, stating that the trial judge should consider all of the circumstances in the case before making this decision. Roach (2007, 226-27, citing Bastarache, J. in *Brooks*) suggests that warnings may actually have the unintended consequences of causing greater prejudice to the accused. Sherrin (1997b) examines a number of ways of dealing with jail-house informants: (1) eliminating their testimony from trials, (2) requiring corroboration, (3) mandatory warnings to the jury, (4) full disclosure and cross-examination, (5) limiting permissible incentives, (6) providing disincentives to fabrication, (7) limiting the types of available informants, (8) limiting the role of the informant, (9) giving the trial judge discretion to exclude unreliable evidence, and (10) introducing a presumption that informants are state agents. Roach (2007, 229-30) suggests that their evidence ought to be excluded, or at the very least scrutinized for reliability.

In-Custody Admissions to Undercover Police Officers

The right to remain silent under the *Charter* extends to gathering evidence from a suspect in custody (*Hebert*). Hebert was arrested for armed robbery of the Klondike Inn in Whitehorse. After consulting counsel, he told the police that he did not want to make a statement. The police then put an undercover police officer in his cell, posing as a person who had been arrested, and Hebert made incriminating statements to the officer. The Supreme Court of Canada ruled that the admissions to the undercover police officer were not admissible. The majority decision, written by McLachlin, J., traced the right to remain silent, including the right to choose whether to make a statement to the police under section 7 of the *Charter*, to two common law doctrines: the confession rule and the privilege against self-incrimination. Both of these common law concepts depend on the notion of "the right of the individual to choose whether to make a statement to authorities or to remain silent, coupled with concern for repute and integrity of the judicial process" (34). She concluded that it would be illusory to give an accused a right to remain silent at trial (the privilege against self-incrimination), but not to extend that right to the investigatory stages.

The right to remain silent, which McLachlin, J. based on section 7 of the *Charter*, is a major source of the philosophy underlying consideration of whether confessions are voluntary. She found that section 7 "confers on the detained person the right to choose whether to speak to authorities or to remain silent" (35). An accused has a right to counsel under section 10(b), and McLachlin expressed the view that one of the most important rights counsel can advise an accused of is the "right to remain silent" (35). The suspect must be "allowed to make an informed choice about whether or not he [or she] will speak to the authorities" (35). Thus, any tricks employed to induce a suspect to make a statement would "effectively deprive the suspect of this choice" (38). The right to make an informed choice does not go beyond "the basic requirement that [the suspect] possess an operating mind" (35–6). The test is objective: "was the suspect accorded his or her right to counsel" (36). Was there any police conduct that deprived the accused of the right to speak to counsel (39)?

McLachlin, J. made four points regarding the scope of the right to remain silent when an accused is in custody. First, the police are allowed to question a suspect after the suspect has consulted with counsel (*Hebert* 41). Second, the right to silence in this context applies only after detention (this was later confirmed by the Supreme Court of Canada in *McIntyre*). In *Hebert*, McLachlin suggested that pre-detention tricks to get a suspect to confess would not violate this right to silence. Wilson and Sopinka, JJ. (dissenting on this point) expressed the view that the right should arise "when the coercive power of the state is brought to bear upon the citizen…and this could well predate detention" (6). Third, the right to silence does not affect voluntary statements made to fellow cellmates. The violation occurs when the police act to "subvert the suspect's constitutional right to choose not to make a statement" (41). Finally, there is a difference between undercover agents who merely observe a suspect and hear an inculpatory statement without doing anything to violate the suspect's right to remain silent, and undercover police officers who actively elicit information in violation of the claimed right to silence.

The Supreme Court of Canada in *Broyles* provided guidance to police officers trying to elicit information from in-custody suspects. Broyles was charged with killing his grandmother. As part of the investigation, the police officers arranged for a friend of Broyles's to visit him in custody. The friend wore a body-pack to record his conversations with the accused. These facts raised two questions that had not been in issue in the *Hebert* case. First, was this friend an agent of the state? Second, did the friend elicit the statements?

Mr. Justice Iacobucci, for the Court, stated that the purpose of the right to remain silent under section 7, in these circumstances, is to "prevent the use of state power to subvert the right of an accused to choose whether or not to speak to authorities" (*Broyles* 607). The purpose is not to prevent the accused from incriminating himself or herself, but to "limit the use of the coercive power of the state to force an individual to incriminate himself or herself" (607).

In determining whether the remarks were made to an "agent of the state," Iacobucci, J. stated that it is important to "focus on the effect of the relationship between the informer and the authorities on the particular exchange or contact with the accused" (608). Consider: is the exchange materially affected by the relationship between the informer and the authorities (608)? "Would the exchange have taken place, but for the intervention of the state or its agents?" Iacobucci, J. decided that the informer was acting as an agent of the state. The police set up the meeting and specifically instructed the friend to elicit information from the accused about the death of his grandmother. The conversation would not have taken place but for the intervention of the police.

If the person is an agent of the state, the next question to be considered by the Court is whether the statements were elicited by the informer. Iacobucci, J. defined "elicited" in terms of the question "Is there a causal link between the conduct of the state agent and the making of the statement by the accused?" (611). He listed two sets of factors that should be examined. The first set involves "the nature of the exchange between the accused and the state agent. Did the state agent actively elicit information such that the exchange could be characterized as akin to an interrogation?" Iacobucci stressed that the "focus should be on the form of the conversation." In this case, parts of the conversation resembled an interrogation. The informer also admitted that he went to see the accused for the purpose of attempting to elicit a statement. The fact that the police did not instruct him to do so was not relevant. The second set of factors that must be considered in determining if the statements were elicited include the "nature of the relationship between the state agent and the accused" (611). "Did the state agent exploit any special characteristics of the relationship to extract the statement?" Was there a relationship of trust? Did the state agent exploit or manipulate the accused (611)? In this case, the informer exploited a friendship in his attempt to undermine the accused's right to remain silent. He also undermined the advice given to the accused by his lawyer (614). The Court concluded that the statement should be excluded under section 24(2) of the *Charter*.

Out-of-Custody Admissions to Undercover Police Officers

As McLachlin, J. stated in *Broyles*, the right to silence under section 7 with respect to undercover officers and police trickery applies only to the eliciting of statements after detention (for a criticism of this position, see Penney 2004a, 323-29; Nowlin 2004). The use of what has come to be known as the "Mr. Big" or "Crime Boss" investigative technique, where undercover police officer pose as members of criminal organizations in order to obtain admissions from suspects, has come under increasing scrutiny and criticism (see Keenan and Brockman, forthcoming; Nowlin 2004 and 2006; Smith, Stinson and Patry 2009; and Moore, Copeland and Schuller 2009). After examining over 80 court decisions in which the technique was used, Keenan and Brockman conclude:

Box 10.3 The Meaning of "Eliciting"

An undercover police officer (Jones) was "arrested" with the accused Liew on drug charges. After being placed in a police cell, Liew was the first to speak:

> Liew: That Lee is hot.
> Jones: What?
> Liew: That Lee is hot.
> Jones: Fuck.
> Liew: Did you pass the money?
> Jones: Fuck. The cops got it.
> Liew: How much?
> Jones: $48 000.00
> Liew: Ah, fuck.
> Jones: What happened?
> Liew: The cops watching us.
> Jones: Yeah. They got my fingerprints on the dope.
> Liew: Lee and me too.

Did the undercover police officer "elicit" the underlined statements? What other facts might be relevant to your answer? See *R. v. Liew*, [1999] 3 S.C.R. 227.

There are many reasons to relegate the Mr. Big investigative technique to the archives. A world that encourages the police to cross the line into illegal activities can more readily result in noble cause corruption and wrongful convictions. . . [R]esearch has shown that false confessions continue to occur with regular and disconcerting frequency, and that interrogation-induced false confessions are becoming one of the more salient causes of erroneous convictions. . . Mr. Big provides targets with many incentives to lie about or exaggerate their involvement in crimes, whether they committed them or not (forthcoming).

Mulgrew (2005) suggests that "the risk of wrongful conviction is sky-high" in these undercover operations.

Other authors have suggest that "Mr. Big" tactics should not be allowed at all, or if allowed, should only be used with prior judicial authorization. Penney writes that "if we were truly concerned about protecting suspects from unwarranted invasion of privacy and betrayals of trust," police would be required to obtain judicial authorization for such operations (and also in-custody operations) (2004a, 328). When justifying the newly created section 25.1 of the *Code*, which allows police officers to break the law under limited circumstances with permission from their superiors, then Justice Minister McLellan explained that judges should not "be in the business of being called upon to sanction illegal conduct which they may later have to deal with" (Chwialkowska 2001, C4). Apparently the Mr. Big operation operates without authorization under section 25.1, because in most cases the officers claim that their make-believe offences do not have the requisite *mens rea* (Gorbet 2004, 51); however, Keenan and Brockman provide a number of examples where these so-called make-believe offences have crossed the line (forthcoming).

THE PRIVILEGE AGAINST TESTIMONIAL SELF-INCRIMINATION

Common Law

The common law position in Canada, before the introduction of what is now section 5 of the *Canada Evidence Act*, was the same as the law in the United States—a witness could refuse to answer a question on the ground that the answer might be self-incriminating. As is obvious from the many U.S. TV shows available in Canada, witnesses in that country can "take the Fifth." The Fifth Amendment to the Constitution of the United States provides that "no person shall be compelled in any criminal case to be a witness against himself." The U.S. courts have given this amendment a broad interpretation, and it now stands for the proposition that a witness cannot be compelled to give evidence where that evidence, or any evidence derived from it, can be used to incriminate him or her. A witness in the United States can refuse to answer a question on the basis that it might incriminate him or her.

Canada Evidence Act, Section 5(2)

The common law in Canada was changed in 1893, so that once a person becomes a witness, she or he cannot refuse to answer a question simply because it may incriminate her or him. Section 5(1) of the *Canada Evidence Act* states that "no witness shall be excused from answering any question upon the ground that the answer to such question may tend to incriminate him or may tend to establish his liability to a civil proceeding." The only protection for the witness is under section 5(2) of the *Canada Evidence Act* and, since 1982, under section 13 of the *Charter*. Section 5(2) of the *Canada Evidence Act* provides the witness with the option of objecting to a question on the ground that the answer may incriminate her or him; the witness then must answer the question, but the answer will not be admissible against the witness in any subsequent criminal proceeding or trial (except in a prosecution for perjury). The testimonial right against self-incrimination is now also dealt with under section 13 of the *Charter*. In *Kuldip*, Lamer, C.J.C. stated that section 5(2) of the *Canada Evidence Act* and section 13 of the *Charter* "offer virtually identical protection: a witness who testifies in any proceeding has the right not to have his or her testimony used to incriminate such witness at a later proceeding" (620).

Section 13 of the Charter

Section 13 of the *Charter* states that:

> A witness who testifies in any proceedings has the right not to have incriminating evidence so given used to incriminate that witness in any other proceedings, except in a prosecution for perjury or for the giving of contradictory evidence.

The section makes reference to protection against self-incrimination "in any other proceedings." The words "other proceedings" are necessary to prevent the inconsistency of an accused confessing to the crime at his or her own trial, but that evidence not being able to be considered against him or her.

The meaning of "any other proceeding" was considered in the case of *Dubois*. Dubois was charged with second-degree murder, and testified at trial that he had killed the deceased in circumstances amounting to justification. He appealed his conviction, and the Alberta Court of Appeal ordered a new trial. At his second trial, Dubois did not testify. The Crown read in 60 pages of transcripts from Dubois' first trial, including the admission that he had killed the deceased. On appeal from his conviction, the accused argued that section 13 of the *Charter* precluded the admission at the second trial of evidence the accused gave at the first trial on the same charge. Mr. Justice Lamer, for the majority of the Supreme Court, said that section 13 had to be read in conjunction with section 11(c) of the *Charter*, the right not to be compelled to give evidence against oneself, and with section 11(d), the right to be presumed innocent, which gives the accused "the initial benefit of a right of silence and the ultimate benefit...of any reasonable doubt" (357). The fact that the accused has a right to silence and does not have to respond until the Crown presents a "case to meet" underlies sections 11(c), 11(d) and 13 of the *Charter*. The purpose of section 13, when seen in this context, "is to protect individuals from being indirectly compelled to incriminate themselves, to ensure that the Crown will not be able to do indirectly that which section 11(c) prohibits" (358).

In concluding that a retrial for the same offence is included in "any other proceedings," Mr. Justice Lamer stated that to allow the Crown to use the evidence against the accused at his second trial would violate section 11(c) and 11(d) of the *Charter*. The accused, he concluded, would be "conscripted to help the Crown in discharging its burden of a case to meet, and is thereby denied his or her right to stand mute until a case has been made out" (365). More recently, the Supreme Court of Canada confirmed this decision (*Henry* para. 2).

In a somewhat rare event, the Supreme Court of Canada in *Henry* reversed two of its earlier decisions. In reaching their decision in *Henry*, the Court reviewed the purpose of section 13 of the *Charter*, in the context of sections 11(c) and (d)—"to protect individuals from being indirectly *compelled* to incriminate themselves, to ensure that the Crown will not be able to do indirectly that which s. 11(c) prohibits" (emphasis in original; *Henry* para. 2 quoting from *Dubois*). The protection is not there to allow the accused to try out a succession of different stories "in the hope of a better result because the second jury is kept in the dark about the inconsistencies" (para. 2). Accused who choose to testify at their trials are not being compelled to incriminate themselves. Therefore if accused choose to give evidence at their first trial, such evidence can be used against them at their second trial, if they chose to testify again. The Court confirmed that *Dubois* was rightly decided. The evidence from the first trial cannot be used against the accused at his or her second trial unless they elect to testify a second time. If the accused testifies again at his or her second trial, the accused's evidence at the first trial can be used to test the accused's credibility, or to incriminate the accused. Such cross-examination is no longer restricted to testing the accused's credibility, as it had been in the Court's earlier decisions. However, where a person is compelled to give evidence at someone else's trial, that evidence cannot be used against the witness at his or her own subsequent trial for any reason.

The Supreme Court of Canada's decision in *S. (R.J.)* addressed the impact and protection of section 13 of the *Charter* when a witness is compelled to answer questions. Two young offenders, RJS and JPM, were charged separately with the same breaking and entering, and theft. The Crown called JPM to testify at the trial of RJS, and counsel for JPM applied to have the subpoena quashed on the basis that JPM had a right to silence under section 7 of the *Charter*, and that forcing him to testify at RJS's trial would violate that right under section 7. The trial judge found that JPM had an absolute right to silence, and the subpoena was quashed. On appeal by the Crown, on the basis that the subpoena should not have been quashed, the Ontario Court of Appeal ordered a new trial. The issue was "Is a person separately charged with an offence compellable as a witness in the criminal trial of another person charged with that same offence, or would compellability in this context violate section 7 of the *Charter*?" (19).

Box 10.4 Possible Perjury

Can a witness who testifies that he or she, and not the accused before the court, committed the crime, be cross-examined as to his or her knowledge of section 13 (that is, knowledge that the evidence cannot be used against the witness if the witness is charged with the offence?) In *Jabarianha*, the Supreme Court of Canada commented: "the probative value of a witness's knowledge of s. 13 of the *Charter* will generally be overborne by its prejudicial effect. Given that witnesses like other persons are presumed to know the law, an interrogation on this question is usually irrelevant while having the potential to cast doubt on the credibility and honesty of a witness. It follows that Crown counsel should rarely be permitted to cross-examine on a witness's knowledge of s. 13" (para. 18). Such cross-examination may be allowed if the Crown provides "some evidence of a plot to lie or to obtain favours." However, the mere fact that the witness and the accused are friends would not be sufficient for the probative value to outweigh its prejudicial effect (para. 27).

What protection is there to discourage a witness from lying to the court or giving contradictory evidence? Section 132 of the *Criminal Code* makes it an indictable offence to commit perjury, and section 136 makes it an indictable offence to give contradictory evidence. The maximum penalty for both offences is 14 years.

All nine judges at the Supreme Court of Canada agreed that there was no absolute right to silence for a witness who is called to give evidence. To reach any different conclusion would mean that those who drafted the *Charter* forgot to add a provision similar to the Fifth Amendment of the United States' Constitution. The judges also agreed that section 5 of the *Canada Evidence Act* does not violate section 7 of the *Charter*. Thus, a witness can be required to answer questions, and the protection afforded the witness is set out in section 13. The judges differed (five to four) on the type of protection a witness had under section 7 of the *Charter*. The minority thought that the protection in section 13 (**evidence immunity**) was sufficient to protect a witness's rights under section 7. Chief Justice Lamer, for the majority, did not agree, finding instead that the principles of fundamental justice under section 7 of the *Charter* provide additional protection to a witness—limited **derivative-use immunity**.

Limited Derivative-Use Immunity

The majority in *S. (R.J.)* discussed the "derivative-use immunity" founded on section 7 of the *Charter*. Before the *Charter*, section 5(2) of the *Canada Evidence Act* offered witnesses only testimonial and simple-use immunity; derivative-use immunity (immunity from the use of clues or facts that existed before the testimony and that were discovered as a result of compelled testimony) did not exist (32–4). In rejecting a rule of absolute derivative-use immunity under the *Charter*, Iacobucci, J. stated:

> Derivative evidence which could not have been obtained, or the significance of which could not have been appreciated, but for the testimony of a witness, ought generally to be excluded under section 7 of the *Charter* in the interests of trial fairness. Such evidence, although not created by the accused and thus not self-incriminatory by definition, is self-incriminatory nonetheless because the evidence could not otherwise have become part of the Crown's case (81).

In determining whether the evidence "could have been obtained but for a witness's testimony," the inquiry is "into logical probabilities, not mere possibilities." Could the evidence have been located? "That is, would the evidence, on the facts, have otherwise come to light? Logic must be applied to the facts of each case, not to the mere fact of independent existence" (82). Derivative evidence should not be automatically excluded, but rather the exercise of the trial judge's discretion "will depend upon the probative effect of the evidence balanced against the prejudice caused to the accused by its admission" (quoted from *Seaboyer* 82–3). It may be that the courts will examine this question somewhat differently in the light of the Supreme Court of Canada's decisions in *Grant* (2009).

Can Witnesses be Exempt from Testifying?

Generally speaking, evidence immunity and derivative-use immunity will be sufficient to protect witnesses. There are times, however, when this protection is not sufficient. Then, the principles of fundamental justice require that courts "retain the discretion to exempt witnesses from being compelled to testify, in appropriate circumstances" (*S.(R.J.)* 13). Lamer, J. provided examples, such as "when the compelled testimony might reveal an accused's defence strategy, or bring to light crimes of which the state was previously unaware" (12).

In claiming this additional protection, the onus is on the witness to satisfy the judge that "in all the circumstances the prejudice to his or her interests overbears the necessity of obtaining the evidence" (13). Mr. Justice Sopinka (Lamer, C.J.C. concurring) added that the factors the court should consider in making the decision whether a person, who is otherwise charged with an offence that will take place in other proceedings, should be required to testify, are:

1. The relative importance of the evidence to the prosecution in respect of which the accused is compelled;
2. Whether the evidence can be obtained in some other manner;
3. Whether the trial or other disposition of the charge against the accused whose evidence is sought to be compelled could reasonably be held before he or she is called to testify;
4. the relationship between the proposed questions to the accused witness and the issues in his or her trial;
5. whether the evidence of the accused witness is likely to disclose defences or other matters which will assist the Crown notwithstanding the application of section 5(2) of the *Canada Evidence Act*;
6. any other prejudice to the accused witness, including the effect of publication of his or her evidence (134).

The courts have the discretion to exempt witnesses from testifying if compelling them to testify would be unfair. L'Heureux-Dubé and Gonthier, JJ. raised the example of a witness compelled to testify at an investigation or trial where the Crown is actually interested in the witness as the subject of investigation. If a witness fails in his or her application to be exempted from testifying, the witness is still protected by sections 13 and 7 of the *Charter*. If charged later with an offence, the witness (now accused) could ask the court for a remedy (e.g., a stay of proceedings) under section 24(1) of the *Charter*.

Anti-Terrorism Investigative Hearings

Following the events of September 11, 2001 in the United States, the federal government in Canada added section 83.28 to the *Criminal Code*, allowing for judicial investigative hearings. The legislation had a sunset clause which resulted in it ceasing to apply in March 2007. Since then, the government has tried to re-introduce the legislation. The legislation had been used once (in the Air India investigation), and the Supreme Court of Canada held that the legislation did not violate the *Charter* (*Application under s. 83.28 of the Criminal Code*). In *Vancouver Sun (Re)*, the Supreme Court of Canada stated that such hearings should, whenever possible, be held in open court. For a commentary on these decisions, see Stewart (2005).

Section 83.28 of the *Code* allowed a peace officer, with the consent of the Attorney General, to apply to a judge for an order requiring someone to be examined on oath or not, where "there are reasonable grounds to believe that (i) a terrorism offence has been committed, and (ii) information concerning the offence, or information that may reveal the whereabouts of a person suspected by the peace officer of having committed the offence, is likely to be obtained as a result of the order" (section 83.28(4)(a)). An order could also be made under specified circumstances where there were "reasonable grounds to believe that a terrorism offence will be committed" (section 83.28(4)(b)).

At the hearing, the person was required to answer questions by the Crown and to produce any information that was requested, unless it was protected by privilege or other law relating to non-disclosure. Section 83.28(10) provided use immunity and derivative evidence immunity to the person, but not transactional immunity. The person could still be charged with an offence admitted during the proceedings, but the Crown would not be allowed to use any evidence derived from the proceedings (Stewart 2005, 380).

Testifying in Non-Criminal Matters

Individuals may be compelled to testify in regulatory, non-criminal matters. In *Branch*, staff of the British Columbia Securities Commission served summonses on Branch and Levitt, directors of Terra Nova, requiring them to attend to answer questions about company expenditures that exceeded $1.3 million. The directors argued that section 128(1) of the *Securities Act*, the authority for making these orders, violated sections 7, 8, 9, and 15(1) of the *Charter*. They were unsuccessful in their argument in the lower courts, and the Supreme Court of Canada dismissed their appeal. After examining the purpose of the securities legislation ("to protect the public from unscrupulous trading practices which may result in investors being defrauded"), the majority found that the purpose of the inquiry related to regulating the securities industry, not to incriminating the directors. The investigation was civil and regulatory in nature, not criminal. However, once compelled to testify, they would then be entitled to evidence immunity under section 13, and derivative-use immunity under section 7 in any subsequent criminal proceedings.

Non-Testimonial Statements Compelled by Statute

Various statutes require people to make statements that may, in effect, incriminate them. Generally, the courts have found that compelling these statements is not a violation of an accused's rights, but that the use of them in criminal proceedings may violate the accused's right to a fair trial, and may be excluded in criminal proceedings under section 24(1) or the common law. In *White*, the accused had given a statement to the police following an accident. The statement, which was required under the *Motor Vehicle Act* of British Columbia, and which White accurately believed she was compelled to provide, did not violate White's rights under the *Charter*. However, using the statement against White in a criminal trial would violate her rights under section 7. As discussed in Chapter 7, the Supreme Court of Canada approved of the use of either section 24(1) of the *Charter* or the common law to exclude the statement. In a similar case, the Ontario Court of Justice excluded a Use of Force Report, filed by Wighton as required under Police Services Regulation, which was designed to ensure that police officers had sufficient training in firearms, when he was charged criminally (*Wighton*).

There are, however, circumstances in which a statement compelled by statute can be used against a person in a criminal trial. In *Wilder,* the British Columbia Court of Appeal found that statements that Wilder was required to make under *Income Tax Act* were admissible at his criminal trial for tax fraud under the *Criminal Code*. Section 241 of the *Income Tax Act* "expressly provide[d] for the use of taxpayer information in litigation including criminal proceedings" (*Wilder* para. 27). In *Fitzpatrick*, the Supreme Court of Canada found that since legislation required the keeping of records of one's commercial fishing results which were an integral part of regulating the fishing industry, the fishing logs were admissible at Fitzpatrick's criminal trial of over-fishing. In *White* (the motor vehicle case), the Supreme Court of Canada distinguished *Fitzpatrick* on the grounds that in *Fitzpatrick*, there was:

> (1) the lack of real coercion by the state in obtaining the statements; (2) the lack of an adversarial relationship between the accused and the state at the time the statements were obtained; (3) the absence of an increased risk of unreliable confessions as a result of the statutory compulsion; and (4) the absence of an increased risk of abuses of power by the state as a result of the statutory compulsion (as summarized in *White* para. 51).

SUMMARY

An admission is a statement made by a person against his or her interests. A formal admission is made in the course of legal proceedings, while an informal admission is not. Informal admissions are admissible against the accused, and the trier of fact decides how much weight to give to them.

A confession is an informal admission made to a person in authority. The test of who is a person in authority is examined subjectively from the perspective of the accused, but there must be a reasonable basis for this view. Before the Crown can tender a confession in evidence against an accused, it must prove in a *voir dire* that the statement was voluntary. In order to make this determination, the court will examine: 1) threats or promises, 2) oppression, 3) operating mind of the suspect, and 4) other police trickery. Under the *Charter,* the requirement of voluntariness, in terms of an operating mind, means that the person has sufficient cognitive capacity to understand what is being said and to understand that the evidence can be used against them. It is not, however, necessary to determine that the accused was capable of making a wise decision.

Before the *Charter,* derivative evidence obtained as a result of an involuntary confession was still admissible at trial, as were those parts of the confession that were confirmed by the finding of the evidence. Under the *Charter,* however, the admission of such evidence is considered under section 24(2) as outlined in the *Grant* case.

The right to counsel and the right to silence under the *Charter* have also significantly affected the law of confessions. The police have a duty to advise an accused on arrest or detention of the right to counsel under section 10(b). This duty implies that the accused must be able to understand the right before it can effectively be waived, and, accordingly, the police must wait for an intoxicated suspect to sober up sufficiently to understand the warning before questioning begins. Where an accused asserts her or his right to counsel, the police must desist from questioning until the accused has had reasonable access to counsel. The police must further advise an accused person of the availability of duty counsel and legal aid, although there is not necessarily an obligation on the provinces to provide free 24-hour duty counsel. Finally, where an accused has been unsuccessful in attempts to contact counsel, the police must also advise of their obligation not to question the suspect until there has been reasonable opportunity to contact counsel. Thus, while a suspect can waive her or his right to counsel and make a statement to the police, there is a heavy burden on the authorities to establish in such circumstances that the waiver was an informed decision by an accused capable of rationally making it. Additional obligations on the police include the duty to re-warn an accused if there is a change in the nature of the legal jeopardy, such as when the nature of the charges being investigated has changed.

The right to silence is based on section 7 of the *Charter.* The right is not limited to trial, but extends to the investigative stage of criminal proceedings. Any tricks by the police used to obtain a statement from a detained suspect will be seen as a breach of the right to silence. Thus, the use of undercover officers as cellmates to attempt to elicit inculpatory statements is no longer permissible, even though such a witness is subjectively not a person in authority. The focus of the decisions under section 7 has been whether a suspect has been able to exercise a real choice whether to speak with the authorities. Although a suspect is free to incriminate himself or herself, the state must be limited in its ability to use coercive power to force or elicit that outcome. Such restrictions do not apply to admissions made in undercover operations that take place outside of custody, such as Mr. Big investigations, where the undercover police officers are not seen as persons in authority.

Section 13 of the *Charter* entrenches the right against testimonial self-incrimination included in the *Canada Evidence Act.* This provision does not allow witnesses to refuse to testify or to refuse to answer questions (except in limited circumstances), but does grant them protection from having their evidence used against them in any other proceeding (except for perjury). This provision will protect a witness who testifies at someone else's trial and implicates themselves, and then is charged with an offence and testifies in their own defence. It will also protect accused who decide not to testify on a retrial from having their testimony at an earlier trial on the same charges used against them. Where accused testify at both their first trial and retrial, their testimony at the first trial can be used at their second trial to cross-examine them as to credibility, or to incriminate them. Section 7 offers additional protection, preventing the use of evidence derived from the earlier testimony in subsequent proceedings against the witness.

QUESTIONS TO CONSIDER

(1)   Discuss the problems that might arise with the use of evidence provided by jail-house informants.

(2)   Under what circumstances would a parent of the accused be considered a person in authority?

(3)   What is the confession rule?

(4)   Discuss the various rationales behind the confession rule. To what extent has the Supreme Court of Canada endorsed one or more of these rationales?

(5)   In a trial by judge and jury, who decides whether a confession is to be believed? Who decides whether the confession was voluntary? What factors are considered in determining whether a confession is voluntary? What factors might create an atmosphere of oppression, when it comes to police obtaining statements from suspects?

(6)   How has the Supreme Court of Canada described a suspect's rights under section 10(b) of the *Charter*?

(7)   What is the rationale for section 13 of the *Charter*? Under what circumstances will it protect a witness from self-incrimination?

(8)   What is the "case to meet" principle? How do Mr. Justice Lamer's views on this issue conflict with recent moves to have "reciprocal disclosure" (as discussed in Chapter 4)?

(9)   Can a witness refuse to answer a question under section 13 of the *Charter*? Explain.

(10)   Can undercover police officers be used to elicit statements from accused who are being detained?

(11)   What are some of the problems with the Mr. Big investigative technique? Should the police be allowed to use it? What limits would you place on its use?

(12)   Under what circumstances can evidence given by an accused at her first trial be used against her at her second trial for the same offence?

(13)   What is derivative use immunity, and what protection does it give to a person?

(14)   Under what circumstances can a witness (who is not the accused in the proceedings, and who is not exerting a privilege) be exempted from testifying? What factors will the court consider?

(15)   What evidence protections exist for people who are compelled to give information in anti-terrorism investigative hearings?

BIBLIOGRAPHY

Akhtar, Suhail. "Whatever Happened to The Right to Silence?" (2008) 62 *Criminal Reports* (6th) 73.

Boilard, Jean-Guy. *Guide to Criminal Evidence.* Cowansville, Quebec: Les Editions Yvon Blais Inc., 1999, with updates.
        Chapter 1, on confessions, discusses the *voir dire* (for out-of-court statements, previous testimony, and subsequent use of the *voir dire*) and admissibility of confessions (evidence in rebuttal, criteria for admissibility, and section 24(2) of the *Charter*).

Boyle, Christine, Marilyn T. MacCrimmon, and Dianne Martin. *The Law of Evidence: Fact Finding, Fairness, and Advocacy.* Toronto: Emond Montgomery Publications Limited, 1999.
        Chapter 11 contains a section of the confession rule and the exclusion of evidence under section 24(2).

Bryant, Alan W., Sidney N. Lederman and Michelle K. Fuerst. *Sopinka, Lederman & Bryant – The Law of Evidence in Canada* 3rd ed. Canada: LexisNexis Canada, 2009.
        Chapter 8 discussions Confessions.

Chwialkowska, Luiza. "Judge's OK not Needed for Officers to Break the Law." (22 December 2001) *National Post* C4.

Cory, Peter. *The Inquiry Regarding Thomas Sophonow* (2001) www.gov.mb.ca/justice/publications/sophonow/toc.html; accessed October 19, 2005.

Craig, John D.R. "The Alibi Exception to the Right to Remain Silent." (1997) 39 *Criminal Law Quarterly* 227.

Delisle, Ronald Joseph, Don Stuart, and David M. Tanovich. *Evidence: Principles and Problems,* 8th ed. Toronto: Thomson Canada Limited, 2008.

Duncan, Bruce Clarkson. "Some Unanswered Questions" (1986) 50 *Criminal Reports* (3d) 305

Freidland, M.L., and Kent Roach, *Criminal Law and Procedure: Cases and Materials*, 8th ed. Toronto: Emond Montgomery Publications Limited, 1997.
        Chapter 2, Part 1 covers the questioning of suspects, confessions, the right to counsel, and jail-house confessions.

Gorbet, Marc S. 2004. "Bill C-24's Police Immunity Provisions: Parliament's Unnecessary Legislative Response to Police Illegality in Undercover Operations." (2004) 9 *Canadian Criminal Law Review* 35.

Harvie, Robert A., and Hamar Foster. "Whittled Away: The Increasing Convergence of Canadian and United States Law Respecting Police Interrogations." (1997) 39 *Criminal Law Quarterly* 12.

Hill, S. Casey, David M. Tanovich and Louis P. Strezos. *McWilliams' Canadian Criminal Evidence.* Aurora, ON: Canada Law Book Limited (available online on Criminal Spectrum).
        See Chapter 8 (Confessions), Chapter 15 (The Charter which includes the right to silence and the principle against self-incrimination), and Chapter 22 (Formal and Informal Admissions).

Hutchinson, Brian. "RCMP turns to 'Mr. Big' to Nab Criminals: Shootings, Assaults Staged in Elaborate Stings" (18 December 2004) *National Post* RB1.

Ives, Dale E. 2007. "*R v. Singh*: A Meaningless Right to Silence with Dangerous Consequences." (2007) 51 *Criminal Reports* (6th) 250.

Jonas, George. *The Scales of Justice.* Montreal: CBC Enterprises, 1983.
    "Hypnosis," by Barbara Betcherman, is the third of seven cases featured in this volume on the Canadian Broadcasting Corporation's series *Scales of Justice.* It is a dramatization of the *Horvath* case. "The Fruit of the Poisonous Tree," by Michael Tait, is a dramatization of the *Wray* case.

Kaufman, Fred, C.M., Q.C., *The Commission on Proceedings Involving Guy Paul Morin,* volumes 1 and 2. Ontario: Queen's Printer, 1998.

Keenan, Kouri and Joan Brockman. *"Mr. Big" Recruiting For the Criminal Underworld: A Critical Examination of Undercover Police Investigations in Canada* (Halifax: Fernwood, forthcoming).

Law Reform Commission of Canada. *Advisory and Investigatory Commission.* Working Paper #13. Ottawa, 1979.

Law Reform Commission of Canada. *Questioning Suspects.* Working Paper #32. Ottawa, 1984a.
    This Working Paper reviews the law of confessions, including the voluntariness rule (scope, meaning of persons in authority, and the burden of proof) and the rationale for the rule (discussed in the context of *Wray* and *Rothman*). It then turns to recommendations for reform and codification of the law governing the questioning of suspects.

Law Reform Commission of Canada. *Questioning Suspects.* Report #23. Ottawa, 1984b.
    This document reflects the Commission's final recommendations on the codification of the law governing the questioning of suspects, following consultations on Working Paper #32.

Leo, Richard A. 2009. "False Confessions: Causes, Consequences and Implications." (2009) 37 *The Journal of the American Academy of Psychiatry and the Law* 332.

McLachlin, B.M., and Andrea Miller. "*Rothman*: Police Trickery: Is the Game Worth the Candle?" (1982) 16 *UBC Law Review* 115.

McWilliams, Peter K., *Canadian Criminal Evidence*, 3rd ed. Aurora, ON: Canada Law Book Inc., updated to 1999.
    Chapter 14 discusses admissions (formal and informal), and Chapter 15, "Statements to Persons in Authority," covers the *Charter*, voluntariness test, rationale for the confession rule, exceptions to the rule, the meaning of persons in authority, examples of threats and inducements, the taking of statements, rights upon arrest, the *voir dire,* and the use of confessions. Chapter 16 is on "Statements Made by Others and Accepted by the Accused," Chapter 30 on "Comment on Failure of Accused to Testify," and Chapter 35 contains a section on the privilege against self-incrimination.

Marin, Rene J. *Admissibility of Statements*, 9th ed. Aurora, ON: Canada Law Book, Inc., 2004.

Moore, Timothy E., Peter Copeland and Regina A. Schuller. "Deceit, Betrayal and the Search for Truth: Legal and Psychological Perspectives on the 'Mr. Big' Strategy." (2009) 55 *Criminal Law Quarterly* 348.

Mulgrew, Ian. "So-called 'Mr. Big' Confessions Bad Situation." (19 July 2005) *Vancouver Sun* B1.

Nowlin, Christopher. "Excluding the Post-Offence Undercover Operation from Evidence–'Warts and All.'" (2004) 8 *Canadian Criminal Law Review* 381.

Nowlin, Christopher. "Narrative Evidence: A Wolf in Sheep's Clothing, Part II." (2006) 51 *Criminal Law Quarterly* 271.

Penney, Steven. "What's Wrong with Self-Incrimination? The Wayward Path of Self Incrimination Law in the Post-Charter Era. Part I: Justifications for Rules Preventing Self-Incrimination." (2003) 48 *Criminal Law Quarterly* 249.

Penney, Steven. "What's Wrong with Self-Incrimination? The Wayward Path of Self Incrimination Law in the Post-Charter Era. Part II: Self-Incrimination in Police Investigations." (2004a) 48 *Criminal Law Quarterly* 280.

Penney, Steven. "What's Wrong with Self-Incrimination? The Wayward Path of Self Incrimination Law in the Post-Charter Era. Part III: Compelled Communications, the Admissibility of Defendants' Previous Testimony, and Inferences from Defendants' Silence. (2004b) 48 *Criminal Law Quarterly* 474.

Presser, Jill R., "The Voluntary Confessions Rule Restated: Some Implications of *R. v. Hodgson*" (1998) 18 *Criminal Reports* (5th) 192.

Public Safety and Emergency Preparedness. *Annual Report on the RCMP's use of the Law Enforcement Justification Provisions* (www.psepc-sppcc.gc.ca/abt/dpr/le/rcmp-en.asp; accessed February 12, 2006).

Roach, Kent. 2007. "Unreliable Evidence and Wrongful Convictions: The Case for Excluding Tainted Identification Evidence and Jailhouse and Coerced Confessions." (2007) 52 *Criminal Law Quarterly* 10.

Sherrin, Christopher. "Jailhouse Informants, Part 1: Problems with Their Use." (1997a) 40 *Criminal Law Quarterly* 106.

Sherrin, Christopher. "Jailhouse Informants in the Canadian Criminal Justice System (Part 2): Options for Reform." (1997b) 40 *Criminal Law Quarterly* 157.

Smith, Steven M., Veronica Stinson, and Marc W. Patry. "Using the Mr. Big Technique to Elicit Confessions: Successful Innovation or Dangerous Development in the Canadian Legal System?" (2009) 15(3) *Psychology, Public Policy, and Law* 168.

Stewart, Hamish. Investigative Hearings into Terrorist Offences: A Challenge to the Rule of Law." (2005) 50 *Criminal Law Quarterly* 376.

Stuart, Don "*Burlingham* and *Silveira*: New *Charter* Standards to Control Police Manipulation and Exclusion of Evidence" (1995) 38 *Criminal Reports* (4th) 386.

Stuart, Don, Ron Delisle, and Tim Quigley. *Learning Canadian Criminal Procedure*, 9th Ed. Toronto: Carswell, 2008.
    Chapter 4 deals with interrogations.

Uniform Law Conference of Canada. *Report of the Federal/Provincial Task Force on Uniform Rules of Evidence.* Toronto: Carswell, 1982.
    Chapter 3 discusses formal admissions, Chapter 12 informal admissions, and Chapter 13 confessions.

Walker, Samuel G. "The Subjective-Objective Dimension in *R. v. Singh*: Rethinking the Distinction between the Common Law Confessions Rule and the Charter Right to Silence." (2009) 55 *Criminal Law Quarterly* 405.

Webber, Grégoire Charles N. "Legal Lawlessness and the Rule of Law: A Critique of Section 25.1 of the *Criminal Code*." (2005) 31 *Queen's Law Journal* 121.

Wylie, Michael I. "Corporations and Non-Compellability Rights in Criminal Proceedings." (1999-1) 33 *Criminal Law Quarterly* 344.

CASES

Application under s. 83.28 of the Criminal Code (Re), [2004] 2 S.C.R. 248. Section 83.28 of the *Code* (investigative hearings) does not violate s. 7 of the Charter.

Black v. The Queen, [1989] 2 S.C.R. 138. When the charges against an accused change (such as from attempted murder to murder), the accused is entitled to be re-warned under section 10(b) and is entitled to reasonable opportunity to contact counsel in respect of the new charges. Reasonable opportunity may extend, depending on the circumstances, until the normal business hours the next morning.

British Columbia Securities Commission v. Branch, [1995] 2 S.C.R. 3.

R. v. Brooks, [2000] 1 S.C.R. 237.

Broyles v. The Queen, [1991] 3 S.C.R. 595. The use by the police of a friend of the accused's to elicit inculpatory statements by the accused after the assertion of the right to silence violates section 7 of the *Charter*. The test is whether the statements would have been made if the state had not intervened. Here, the friend was an agent of the state and actively elicited the evidence.

R. v. Brydges, [1990] 1 S.C.R. 190. "In circumstances where an accused expresses a concern that the right to counsel depends upon the ability to afford a lawyer, it is incumbent on the police to inform him of the existence and availability of Legal Aid and duty counsel" (345). In addition, "information about the existence and availability of duty counsel and Legal Aid plans should be part of the standard section 10(b) caution upon arrest and detention" (347).

R. v. Burlingham, [1995] 2 S.C.R. 206.

R. v. Calder, [1996] 1 S.C.R. 660. An involuntary confession cannot be used against an accused for any purpose.

Clarkson v. The Queen (1986), 25 C.C.C. (3d) 207 (S.C.C.). The waiver of section 10(b) rights by an intoxicated accused will not be effective unless the accused is at least aware of the consequences of the waiver.

R. v. Dubois, [1985] 2 S.C.R. 350. A retrial of the same offence is included in the phrase "other proceedings" under section 13 of the *Charter*. The use of evidence of the accused given at the first trial against the accused at the second trial (where the accused refuses to testify) would violate the accused's right to be presumed innocent and to remain silent (section 11(d)) and the right not to be compelled to give evidence against him or herself (section 11(c)).

R. v. Fitzpatrick (1994), 90 C.C.C. (3d) 161 (B.C.C.A.). Statute required the accused to keep records of fish caught (failure to keep records was an offence). The records indicated that the accused had retained fish in excess of his quota, and he was charged with retaining fish in excess of his quota. The court held (two to one) that the legislation did not violate the accused's rights under section 7 of the *Charter*, or if it did, the violation was justified under section 1.

R. v. G.(B.), [1999] 2 S.C.R. 475. The traditional rule that the Crown may not make any use of an inadmissible statement is not abrogated by section 672.21(3)(f) of the *Criminal Code*.

R. v. Grandinetti, [2005] 1 S.C.R. 27. Undercover police officers who claimed that they might be able to influence the course of an investigation through their corrupt police contacts were not persons in authority.

Hebert v. The Queen (1990), 57 C.C.C. (3d) 1 (S.C.C.). An active solicitation by an undercover agent of the state once an accused has indicated the intention of remaining silent violates section 7 of the *Charter*.

R. v. Henry, [2005] 3 S.C.R. 609. If an accused testifies at both his or her first and second trial, the evidence given at the first trial can be used against him or her at the second trial.

R. v. Hodgson, [1998] 2 S.C.R. 449. The Court discusses the law surrounding the definition of person in authority, and the circumstances under which a trial judge should initiate a *voir dire* on the issue.

Horvath v. The Queen (1979), 44 C.C.C. (2d) 385 (S.C.C.). Restates the principle that "voluntariness" of confessions is to be assessed in the ordinary sense of that word, including an aspect of awareness and of "free will." Voluntariness is not limited to the absence of hope of advantage and fear of prejudice.

R. v. I. (L.R.) and T. (E.), [1993] 4 S.C.R. 504. The court discusses the common law rule on derived confessions.

Ibrahim v. The King, [1914] A.C. 599 (P.C.). An accused's confession must be voluntary and free of fear of prejudice and hope of advantage to be admissible.

R. v. Jabarianha, [2001] 3 S.C.R. 430.

R. v. Kuldip, [1990] 3 S.C.R. 618.

R. v. Liew, [1999] 3 S.C.R. 227. The Court found that an undercover police officer "arrested" with the accused had not "elicited" statements from the accused. [citation to be filled in at first pass]

R. v. Mannion, [1986] 2 S.C.R. 272. A statement by the accused in the accused's first trial can not be used to incriminate the accused in the accused's second trial. The case was overruled by the Supreme Court of Canada in *Henry*.

R. v. McIntyre, [1994] 2 S.C.R. 480.

R. v. Oickle, [2000] 2 S.C.R. 3. A detailed look at the confession rule and false confessions.

Prosko v. The King (1922), 37 C.C.C. 199 (S.C.C.).

R. v. Prosper, [1994] 3 S.C.R. 236. Where a detainee indicates a desire to avail himself or herself of the right to counsel, such as through a system immediate, free duty counsel, the authorities must provide reasonable opportunity to do so and must refrain from attempting to obtain evidence from the detainee until there has been reasonable opportunity to contact counsel. If duty counsel does not exist, or is not available, the period that will be considered "reasonable" will be longer, including, for example, until the legal aid office is open, until the detainee can contact a private lawyer willing to help, or until the detainee has been brought before a Justice for a bail hearing.

R. v. S. (R.J.), [1995] 1 S.C.R. 451. The Court discusses use immunity under section 13 and derivative-use immunity under section 7 when a witness is compelled to answer questions at someone else's trial and then the witness becomes an accused in his or her own right.

R. v. St. Lawrence (1949), 93 C.C.C. 376 (S.C.C.). Where evidence is discovered as a consequence of an inadmissible confession, that part of the confession confirmed by that evidence is admissible.

R. v. Seaboyer; R. v. Gayme (1991), 66 C.C.C. (3d) 321 (S.C.C.). Section 277 of the *Code*, which prohibits the use of evidence regarding sexual reputation to challenge or support the credibility of the accused, does not violate the *Charter*. Section 276, which sets out the circumstances in which sexual reputation could be used, violates sections 7 and 11(d) of the *Charter* and is not justified under section 1.

R. v. Singh, [2007] 3 S.C.R. 405.

R. v. Spencer, [2007] 1 S.C.R. 500.

Streu v. The Queen (1989), 48 C.C.C. (3d) 321 (S.C.C.). The accused's statement that he knew the property was "hot" was evidence of the fact that the property was stolen, even though the accused himself did not have personal knowledge that the property was stolen.

R. v. Turcotte, [2005] 2 S.C.R. 519.

Vancouver Sun (Re), [2004] 2 S.C.R. 332. Although applications for investigative hearings are *ex parte* and held *in camera*, the investigative should be held in open court whenever possible. Such a hearing may be held *in camera* if the test established in *Dagenais/Mentuck* (see Chapter 5) is met.

R. v. Voss (1989), 50 C.C.C. (3d) 58 (Ont. C.A.). There should be a *voir dire* to determine whether a confession is voluntary and a second *voir dire* to determine whether the accused's rights under section 10(b) of the *Charter* were violated.

Ward v. The Queen (1979), 44 C.C.C. (2d) 498 (S.C.C.). The voluntariness of statements is not limited to the absence of hope of advantage or fear of prejudice; there must be an "operating mind."

R. v. Wells, 1998] 2 S.C.R. 517. The court found that this was one of those "rare circumstances" in which the trial judge should have held a *voir dire* to determine whether the parents of boys who were victims of sexual assault were persons in authority.

R. v. White, [1999], 2 S.C.R. 417. The use in a subsequent criminal trial of evidence gathered by compulsion under provincial statute in criminal proceedings violates an accused's rights under section 7 (right against self-incrimination) and can be excluded under section 24(1) of the *Charter*.

Whittle v. The Queen, [1994] 2 S.C.R. 914.

R. v. Wighton, [2003] O.J. No. 2611(Ont. C.J.).

R. v. Wilder, [2000] B.C.J. No. 62.

Queen v. Wray, [1970] 4 C.C.C. 1 (S.C.C.). Evidence found as a consequence of an inadmissible confession is generally still admissible, unless its admission would be gravely prejudicial to the accused, its admissibility is tenuous, and its probative value is trifling.

R. v. Yakeleya (1985), 20 C.C.C. (3d) 193 (Ont. C.A.). The accused's trial is not an "other proceeding" in relation to the accused's preliminary hearing and therefore the accused can be cross-examined at trial on evidence given at the preliminary hearing.

CHAPTER 11: *Types of Evidence*

CHAPTER OBJECTIVES

In studying this chapter, you should develop an understanding of the following topics and concepts:

- the relationship and distinction between onus or burden of proof and standard of proof
- the distinction between presumptions of fact and reverse onus provisions
- the requirement of relevance
- types of evidence, especially the distinction between direct and circumstantial evidence
- competence and compellability of witnesses, and special rules for spouses
- the nature of corroboration
- the nature and justification of rape-shield provisions
- provisions protecting child witnesses, persons with disabilities, and the safety of witnesses

INTRODUCTION TO EVIDENCE

The first three chapters in Part II dealt with techniques for gathering evidence, and with the impact of the *Charter* on the admissibility of that evidence. This chapter discusses the rules relating to the burden of proof, the nature of evidence, how it is introduced at trial, issues surrounding the testimony of witnesses, and the impact of the *Charter* on these rules. In 1982, Mr. Justice Estey of the Supreme Court of Canada made the following comment on the rules of evidence:

> We start with the reality that the law of evidence is burdened with a large number of cumbersome rules, with exclusions, and exceptions to the exclusions, and exceptions to the exceptions (*Graat* 377).

These last three chapters address some of these rules, exclusions, and exceptions.

The Burden (or Onus) and Standard of Proof

The **onus or burden of proof** identifies which party must prove something, and in a criminal trial generally lies on the Crown. This burden of proof is sometimes referred to as the **legal** or **ultimate burden,** and is embedded in section 11(d) of the *Charter* (the right to be presumed innocent). The **standard of proof** addresses the level of proof required; overall in a criminal trial it is "beyond a reasonable doubt." In *Lifchus*, the Supreme Court of Canada explained this standard: "A reasonable doubt is not an imaginary or frivolous doubt. It must not be based upon sympathy or prejudice. Rather, it is based on reason and common sense. It is logically derived from the evidence or absence of evidence" (para. 39). Following this decision, the Court dealt with a number of pre-*Lifchus*, cases in deciding that the deficiencies in the charges to the jury were not such that they would cause "serious concern about the validity of the jury's verdict, [or] lead to the conclusion that the accused did not have a fair trial" (*Pan* para. 128).

There is also a **secondary** or **evidential burden,** that at times requires a party to raise a particular fact or issue by evidence, or by implication from the evidence. For example, the accused must raise the issues of self-defence or provocation to rely on them. Once the accused provides an evidential foundation for the issue of provocation, the Crown must prove beyond a reasonable doubt that provocation did not exist (*Cinous* para. 39).

In enacting legislation, Parliament sometimes concludes that there are some facts, uniquely within the knowledge of the accused, that would be too difficult for the Crown to prove beyond a reasonable doubt. As a result, some legislation creates either a **presumption of fact** or a **reverse onus,** such that the accused has to negate or establish a particular fact. Legislation that requires an accused to disprove an element of the offence, or that allows an accused to be convicted despite the existence of a reasonable doubt, will violate section 11(d) of the *Charter*, guaranteeing "the right to be presumed innocent until proven guilty according to the law." The courts will declare such legislation to be of no force and effect, unless it passes the test in section 1 of the *Charter*. For example, section 258(1)(a) of the *Criminal Code* states that if it is proved that an accused occupied the driver's seat of a vehicle, "the accused shall be deemed to have had the care or control of the vehicle…unless the accused establishes that the accused did not occupy that seat or position for the purpose" of driving. The Supreme Court of Canada found that an earlier version of the section, that required the accused to prove on a balance of probabilities that he or she did not enter the vehicle with the intention of setting it in motion, violated the accused's rights under section 11(d) of the *Charter*, in that the accused could be convicted even though the trier of fact had a reasonable doubt as to the guilt of the accused. The section was, however, justified under section 1 of the *Charter* (*Whyte*). In *Laba*, the Supreme Court of Canada found that section 394(1)(b) of the *Criminal Code*, requiring a person who was selling or purchasing precious metals to "establish that he is the owner or agent of the owner or is acting under lawful authority," violated section 11(d) of the *Charter*, and was not justified under section 1. The legislation was amended, and section 394(4) now states that presumptions will be made "in the absence of evidence raising a reasonable doubt to the contrary."

Other legislation introduces presumptions of fact, which may also offend section 11(d) of the *Charter*. For example, in *Downey*, an accused challenged section 212(3) of the *Criminal Code,* which stipulates that "evidence that a person lives with or is habitually in the company of a prostitute or lives in a common bawdy-house…is, in the absence of evidence to the contrary, proof that the person lives on the avails of prostitution." The Supreme Court of Canada found that the section violated section 11(d) of the *Charter*, but that such a violation was demonstrably justified under section 1.

Presumptions impair the rights of an accused to a lesser extent than reverse onus provisions. In *Laba,* Mr. Justice Sopinka explained that a presumption of fact is an evidentiary burden, whereas a reverse onus is legal burden:

> Parliament…[may choose] merely to place an evidentiary burden rather than a full legal burden of proving ownership, agency or lawful authority upon the accused. Under such a provision [evidentiary burden] the accused would simply be required to adduce or point to evidence, which, if accepted, would be capable of raising a reasonable doubt as to whether he was the owner or agent of the owner or was acting under lawful authority. If he or she succeeded in raising such a doubt, the accused would be acquitted (418–9).

What Is Evidence?

Legal dictionaries define evidence as something that serves to prove or disprove an assertion. Evidence may be introduced to show that the accused committed the offence charged. If the evidence convinces the trier of fact that the accused is guilty beyond a reasonable doubt, a conviction will result. Evidence may also be introduced to disprove, or raise a reasonable doubt about, the accused's guilt; if this succeeds, there will be an acquittal. Thus, evidence must be relevant to some fact in issue.

A Question of Relevance

Relevance is seldom formally defined, but it refers to the extent that the evidence tends to prove, disprove, or explain some issue properly considered in a trial. Relevant evidence advances the search for truth in the matter or issue before the court. A fact is not relevant if it does not have **probative value**. In this context, probative means that the fact is capable of proving or has the tendency to prove something. The Law Reform Commission of Canada, in its proposed code of evidence (1975b), defined relevant evidence as "evidence that has any tendency in reason to prove a fact in issue in a proceeding" (section 4(2)). Although there must be a connection between two facts for one to be relevant to another, the courts are more likely to think of relevance in terms of common sense than as any kind of philosophical construct.

The question of **relevance** is not one that can be answered with "objective" scientific precision. Madame Justice L'Heureux-Dubé of the Supreme Court of Canada explained:

> Whatever the test, be it one of experience, common sense or logic, it is a decision particularly vulnerable to the application of private beliefs. Regardless of the definition used, the content of any relevancy decision will be filled by the particular judge's experience, common sense and/or logic. For the most part there will be general agreement as to that which is relevant and the determination will not be problematic. However, there are certain areas of inquiry where experience, common sense and logic are informed by stereotypes and myth (*Seaboyer* 356).

Judicial decisions should be examined for the stereotypes and myths that underlie their reasoning. Answers to questions of relevance can result in bias, discrimination, or prejudice against particular groups of people (See Boyle *et al.* 1999, Chapter 1).

The first general rule regarding evidence is that only relevant evidence is admissible. Further, all relevant evidence is admissible unless it is excluded by a rule. For example, conversations between a lawyer and a client may be relevant, but such conversations are excluded by the rule relating to solicitor–client privilege (discussed in Chapter 12). Evidence obtained in violation of a *Charter* right may also be excluded under section 24(2) of the *Charter*, even though it is relevant.

Why Have Rules of Evidence?

The rules of evidence are concerned with admitting facts that are relevant and excluding those that are not relevant. But they also serve a number of other purposes, and they can be broadly classified into those concerned with some policy or principle external to the criminal trial process (extrinsic rules), and those concerned with the process of the trial itself (intrinsic rules) (Dufraimont 2008, para. 10). Extrinsic rules, often concerned with social policies (such as spousal privilege), each have their own peculiar history and were designed to foster their own specific set of values. As social and political times change, so might these rules, either through developments in the common law or changes in legislation. Mr. Justice Iacobucci of the Supreme Court of Canada described this division of labour:

> Judges can and should adapt the common law to reflect the changing social, moral and economic fabric of the country. Judges should not be quick to perpetuate rules whose social foundation has long since disappeared. Nonetheless, there are significant restraints on the power of the judiciary to change the law. As McLachlin J. indicated in *Watkins,* in a constitutional democracy such as ours it is the legislature and not the courts which has the major responsibility for law reform; and for any change to the law which may have complex ramifications, however necessary or desirable such changes may be, they should be left to the legislature. The judiciary should confine itself to those incremental changes which are necessary to keep the common law in step with the dynamic and evolving fabric of our society (*Salituro* 301).

Paciocco refers to rules which sacrifice relevant evidence because of competing values as "rules of subordinated evidence" (2001, 438).

Intrinsic rules of evidence are designed to ensure that:

(1) an accused is tried on evidence that is reliable (the search for truth);
(2) an accused has a fair trial (a right now guaranteed under section 11(d) of the *Charter*);
(3) trials are conducted efficiently (and do not deal with **collateral** or unrelated issues); and
(4) there is predictability in the trial process so that the accused can prepare for trial.

Many of the rules of evidence about the trial process developed historically to prevent uneducated jurors from being swayed by evidence that judges thought might appear far too persuasive (for a detailed analysis of this rationale, see Dufraimont 2008). According to the Law Reform Commission of Canada, the jury rationale is one reason why the law of evidence is so complicated today (1977, 4). Dufraimont suggests that another influence on the rules of evidence is the adversarial system. Rules of evidence are required to both facilitate the adversarial testing of evidence and to curb adversarial excesses (2008 para. 72-76).

Paciocco refers to the rules that reduce the expense and time of a trial as "rules of practical exclusion" (2001, 438), and rules which exclude inaccurate information as "rules of non-evidence" (438-39). In writing about the balance between the rights of the individual and society in the law of evidence, he also suggests that, generally, there are rules about "evidence of guilt" (Crown evidence) and rules about "evidence of innocence" (defence evidence) (2001, 434). He is of the opinion that it is acceptable to compromise our pursuit of truth to prove guilt, and to exclude evidence that might otherwise lead to a conviction (although this should be minimized); however, it is not acceptable to compromise our pursuit of truth and exclude evidence that might point to the accused's innocence (435).

Where Are the Rules of Evidence?

The law of evidence is a combination of **common law** (that is, rules and principles developed by judges on a case-by-case basis), statutes, the *Charter*, and accompanying judicial interpretations. Section 8(2) of the *Criminal Code* states that the common law of England that was in force in a province immediately before April 1, 1955, continues in force in the province with regard to evidence, except as it has been modified by an Act of Parliament.

Modifications to the common law can be found in the *Criminal Code*, as well as the *Canada Evidence Act*, and other federal and provincial statutes. The *Canada Evidence Act* makes several major changes to the common law. One such change was the abolition of the right of a witness to refuse to answer questions on the basis that the evidence given might incriminate the witness (discussed in Chapter 10). Some sections of the *Canada Evidence Act* merely restate the common law of England.

Various laws of evidence are contained in the *Criminal Code*. Section 212(3), referred to above, is one of several sections that puts the onus on the accused to disprove some aspect of an offence. Other sections of the *Criminal Code* restrict the type of questions that can be asked of a complainant (see for example sections 276 and 277, dealing with questions regarding the past behaviour of victims of sexual assault—discussed in this chapter under "Witnesses"). Some sections of the *Code* spell out conclusions that can be reached in the absence of evidence to the contrary. For example, section 338(3) states that "evidence that cattle are marked with a brand or mark that is recorded or registered in accordance with any Act is, in the absence of any evidence to the contrary, proof that the cattle are owned by the registered owner of that brand or mark."

THE LAW REFORM COMMISSION OF CANADA

Can the rules of evidence, now scattered throughout the *Criminal Code*, the *Canada Evidence Act*, and the common law, be brought together under one code? Can we reduce all the rules of evidence to a *Code of Evidence*? In the United States, Professor Wigmore summarized the law of evidence in 11 volumes. In Canada, the Law Reform Commission of Canada set out to codify the law of evidence in Canada. Between 1972 and 1975 it published 12 study papers on evidence (for example, on the topics of the compellability of the accused, opinion and expert evidence, illegally obtained evidence, corroboration, professional privilege, burden of proof, and so on), and in 1975, it released a draft evidence code "to establish rules of evidence to help secure the just determination of proceedings, and to that end to assist in the ascertainment of the facts in issue, in the elimination of unjustifiable expense and delay, and in the protection of other important social interests" (section 1). The proposed code was a radical departure from traditional approaches to codifying common law, in that the Commission claimed that the Code was to be treated as a "complete code"—judges were not to rely on the common law if there were gaps in the law, but were rather to look to the purpose of the code to determine the law. Section 3 stated that "matters of evidence not provided for by this Code shall be determined in the light of reason and experience so as to secure the purpose of the Code."

One might argue that reason and experience include how things were done before the proposed code, that is, under the common law. However, section 2 prevented the new code from being strictly interpreted (construed) when it was contrary to the common law. This section was meant to bypass one of the **rules of statutory construction** that had been developed by the courts over the years: if legislation alters the common law, it is to be strictly construed. The effect of such a rule is to preserve the common law, or the way things were. Section 2 tried to circumvent this rule, in favour of construing the code "to secure its purpose."

The proposed Code met with resistance from both judges and lawyers. In 1977, the Uniform Law Conference appointed a task force to examine the Law Reform Commission's proposed code, and to offer alternative solutions to the various problems surrounding the law of evidence in Canada. In their 500-page *Report of the Federal/Provincial Task Force on Uniform Rules of Evidence,* the task force came up with a proposed *Uniform Evidence Act*. It was intended to be comprehensive, but not a complete codification of the law of evidence. The code did not specify its own purpose; nor did it specify that it should be interpreted in light of its undisclosed purpose. As a result, the common law would apply, except where it was inconsistent with the proposed code. The *Uniform Evidence Act* was never enacted. At present, we still do not have a comprehensive code of evidence in Canada.

EVIDENCE: GETTING IT INTO COURT

Evidence can be classified in several different ways, and the classifications that exist often have overlapping categories. For example, evidence can be examined by the way in which it is introduced in court proceedings.

Testimonial or *Viva Voce* Evidence

The most common type of evidence in the courtroom is ***viva voce* evidence**—oral evidence given by a witness in court under oath or affirmation (see discussion below under "Witnesses"). An example of this evidence is eye-witness testimony. A complainant's testimony of identifying an accused and relaying what happened can be highly persuasive–in some cases, perhaps, too persuasive. In *Hibbert*, the majority of the Supreme Court of Canada found that the trial judge's warning to the jury that "the identification of the accused for the first time in the courtroom after a failure to positively identify him from a photo line up is to be accorded little weight," was not sufficiently strong (para. 50). Given that such evidence is "deceptively credible," the jury in this case should have been warned that there is a "very weak link between the confidence level of a witness and the accuracy of that witness." The impact of the complainant seeing the police arrest the suspect and seeing him on T.V. could not be undone (para. 52). The dissenting judges went through the trial judge's instructions in greater detail and found them to be adequate. Eye-witness evidence was so compelling for one accused that he was convinced by his counsel to plead guilty despite the fact that he had not committed the offence (*Hanemaayer* para. 11; see also Brockman 2010 in press).

Real Evidence

Real evidence is evidence, other than the *viva voce* testimony of a witness, observed by the judge or judge and jury. For example, physical articles or things found in the possession of the accused or at the scene of a crime, such as a gun, blood-stained clothes, documents of the crime, or a book or a movie in an obscenity charge, can be referred to as real evidence. Section 652 of the *Criminal Code* allows the judge or judge and jury to take a view of any place, person, or thing; such observations would be the examination of real evidence.

Commissioned Evidence

Sections 709 through 714 govern the taking of evidence by a commissioner in certain circumstances. For example, if a witness is too ill to attend trial or is outside Canada, the testimony of that witness can be taken by a commissioner and read into evidence at the trial (see section 711).

EVIDENCE: WHAT IS IT USED FOR?

Evidence is also sometimes classified according to its use or purpose at trial. It can be demonstrative or illustrative, and it can be direct or circumstantial.

Demonstrative and Illustrative Evidence

Evidence can be illustrative or demonstrative; that is, it can be used to illustrate or to demonstrate something. Often, the two are treated as referring to the same thing. When a distinction is made, the term **demonstrative evidence** is used to refer to evidence that stands on its own. Electronic cameras in a bank, which are not monitored, would have to stand on their own—there is no one to say that what is on film is what actually took place. The film is the best evidence of what occurred. In *Nikolovski*, the Supreme Court of Canada upheld a conviction that was based solely on the videotape of a robbery at a convenience store, when the store clerk could not identify the accused.

Illustrative evidence is used by a witness to illustrate something. For example, the police may take photographs of the scene of the crime and then use the photographs, when giving evidence, to assist in illustrating or explaining what they saw at the scene of the crime. Gardner (1996) discusses various types of illustrative evidence: charts, models, and computer animated re-creations (or CARs). He suggests that to introduce such evidence, (a) it must be relevant, (b) the witness has to identify that it accurately represents what it supposedly portrays, (c) it should not be misleading or have prejudicial value substantially greater than its probative value, and (d) the evidence should not be cumulative, in that it should add some value to the other evidence being presented (431-32).

Direct Evidence

Direct evidence is evidence used to prove a fact that is in issue. For example, the witness might testify, "I saw X point a gun at the bank teller and heard X ask for all of the money." This would be direct evidence of a crime. As Berger (2005, 50) explains, the issue for the trier of fact is whether the witness is to be believed.

Circumstantial Evidence

Circumstantial evidence is evidence introduced to prove one fact, from which another ultimate fact in issue can be inferred. The witness might say, "I saw X running away from the bank with a brown paper bag immediately after I heard the alarm." The witness did not see X rob the bank, but the evidence of running from the bank with a brown paper bag might be circumstantial evidence implying that X robbed the bank. The fact (running from the bank) may be true, but the inference (X robbed the bank) may be entirely false (X may have been late for a lunch appointment). The question for the trier of fact is not only whether the witness should be believed, but whether one can draw an inference of guilt from the evidence (Berger 2005, 50). Post-offence conduct that can amount to circumstantial evidence includes: "flight from the scene of the crime or the jurisdiction in which the crime was committed; attempts to resist arrest; failure to appear at trial; and acts of concealment such as lying, assuming a false name, changing one's appearance, and hiding or disposing of evidence" (*Turcotte* para. 38). It does not include refusing to answer some or all questions put to a person by the police, even though the person has volunteered some information. Since a person has a right not to respond to police, failure to do so has no probative value and is irrelevant (*Turcotte* para. 55-56), although there are some exceptions (see Chapter 10).

Circumstantial evidence is often associated with the **rule in *Hodge's Case*.** Hodge was tried in England in 1838 for the murder of a woman who had been robbed. Hodge was seen near the scene of the crime, and was seen some time later burying something that turned out to be the same amount of money the woman had been carrying. There was no direct evidence that he killed her, and he was found not guilty. The rule in *Hodge's Case* requires that the jury be warned that before they convicted an accused on purely circumstantial evidence, they must be satisfied not only that the circumstances are consistent with the accused having committed the offence, but also that the circumstances are inconsistent with any rational conclusion other than that the accused committed the offence.

The Supreme Court of Canada, in *Cooper*, stated that it is not necessary to use the exact same words in warning the jury as was used in *Hodge's Case*. Cooper was charged with conferring a benefit (expense-paid trips from Ontario to Florida) on a government official while doing business with the government. Mr. Justice Ritchie, restoring the jury's conviction, said,

> It is enough if it is made plain to the members of the jury that before basing a verdict of guilty on circumstantial evidence they must be satisfied beyond a reasonable doubt that the guilt of the accused is the only reasonable inference to be drawn from the proven facts (33).

More recently, the Supreme Court of Canada in *Griffin* (para. 33) confirmed that there are no longer "special instructions" for circumstantial evidence; rather, there are different ways for the trial judge to instill in the jurors that "in order to convict, they must be satisfied beyond a reasonable doubt that the only rational inference that can be drawn from the circumstantial evidence is that the accused is guilty" (see Dufraimont 2009 for a discussion).

England no longer requires that the rule be put to the jury. The law reformers in Canada and dissenting judges at the Supreme Court of Canada are of the view that it is not necessary in every case to put the rule to the jury (see Berger 2005 for a detailed analysis of the development of the law in Canada).

One of the concerns with circumstantial evidence is that an accused might be convicted when in fact he or she is innocent. This probably happened in 1953, when Wilbert Coffin was accused in Quebec of killing three hunters from the United States. There was no direct evidence that he killed the three hunters, although there was some circumstantial evidence. He admitted to stealing from the hunters' truck, and some of their possessions were found in Coffin's house. It was established that Coffin had spent about $600 on a trip to Montreal, and that one of the deceased had $650 in his wallet when he left for the hunting trip. The accused was badly defended, according to Edward Greenspan, Q.C. The jury took only half an hour to convict him, and he was hanged in 1956. His last words were, "I am not guilty and may God have mercy on my soul." Before his hanging, witnesses had called the police about seeing two Americans in the area who had been looking for the hunting party. This confirmed Coffin's sighting of their vehicle, but was not enough to convince the Minister of Justice to commute the death sentence. Twenty years later, this hanging was used to argue in favour of the abolition of capital punishment (Jonas 1983, 39–62).

Wrongful convictions, of course, are not a problem exclusive to circumstantial evidence. A person could also be wrongfully convicted, for example, based on direct evidence–an error by an eyewitness to the offence. As Berger (2005, 74) points out, the Rule in *Hodge's Case* illustrates judicial "valuation of sensory perception [e.g., eye witness testimony] over inferential reasoning" [circumstantial evidence], although he questions whether this is necessarily an appropriate approach, given some of the recent studies on the inaccuracies of eye witness testimony.

WITNESSES

Viva voce evidence, through the mouth of witnesses, is the most common way of introducing evidence at trial. This section looks at a few of the rules regarding witnesses: the taking of the oath or affirmation, competence and compellability, the use of prior statements, and the requirement of corroboration to support the credibility of some witnesses. In addition, we examine some of the rules surrounding adults and children who become complainants and witnesses because of sexual assaults committed against them.

The Oath or Affirmation

At common law, a witness was not allowed to give evidence unless it was under **oath**. The *Canada Evidence Act* changed the common law to allow for the giving of evidence upon **solemn affirmation.** However, before February 15, 1995, a person had to object to taking an oath "on grounds of conscientious scruples" before they could affirm. Today, section 14 states that "A person may, instead of taking an oath, make the following solemn affirmation: I solemnly affirm that the evidence to be given by me shall be the truth, the whole truth and nothing but the truth." Such evidence has the same effect as evidence under oath (section 14(2)).

At early common law in England, a witness had to understand that God would punish him or her for lying under oath. Since today the spiritual consequences of not telling the truth under oath are unknown, the courts have stated that it is important for witnesses to understand "the moral obligation to tell the truth;" they do not need to believe in divine retribution (*Truscott* 368, confirmed in *K.G.B.* para. 87).

The oath does not have to be on the Bible, but can be in the tradition of the witness. The importance of the oath is to impress on the witness the importance of telling the truth. In the early 1900s, Chinese witnesses were usually given the "fire oath"—their testimony was written on a piece of paper and the witness would light the paper, and as it burned, he or she would swear to answer the court's questions truthfully. Wong Foon Sing, accused of the murder of Janet Smith in Vancouver in the 1920s, took the "chicken oath," which was considered an even stronger inducement to tell the truth (Starkins 1984, 93 annotated in Appendix C; for a description of the chicken oath, see Box 11.1).

In a more contemporary case in Ontario (*Kalavar*), the court stated that a person who objects to swearing on the Bible can take another form of religious oath. The accused there wanted to argue that his right to freedom of religion was violated by the presence of the Bible in the courtroom to the exclusion of other holy books. The court concluded that it was not necessary to consider the constitutional argument, because the trial judge should have allowed the accused to take whatever oath fit the accused's particular religious persuasions. It is not appropriate to require these witnesses to affirm as the only alternative to swearing an oath on the Bible.

Compellability

A **compellable** witness is one who can be forced or required to give evidence. Today, the common law on compellability remains—the accused cannot be compelled or forced to testify at her or his own trial. Although this seems to be the only possible approach in our adversarial system, it is one that has been debated by academics.

Box 11.1 The Chicken Oath

In *R. v. Ah Wooey*, the accused asked that a Cantonese man be given the "chicken oath," in addition to the fire oath, in a murder trial. The case report described the scene at the courthouse in New Westminster, British Columbia in 1902:

> The witness having signed his name twice, and a cock having been procured, the Court and the jury adjourned to a convenient place outside the building where the full ceremony of administering the oath was performed, as follows: By a block of wood, punk sticks, not less than three, and a pair of Chinese candles were stuck in the ground and lighted. The oath was then read out loud to the witness [Being a true witness, I shall enjoy happiness and my sons and grandsons will prosper forever. If I falsely accuse (prisoner) I shall die on the street, Heaven will punish me, earth will destroy me, I shall forever suffer adversity, and all my offspring be exterminated. In burning this oath I humbly submit myself to the will of Heaven which has brilliant eyes to see], after which he wrapped it in Joss-paper as used in religious ceremonies, then laid the cock on the block and chopped its head off, and then set fire to the oath from the candles and held it until it was consumed.

Competence

Historically, there were several categories of people who were not allowed to testify at a trial (i.e., they were not considered **competent** to testify). For example, at one time, the common law did not allow anyone who had an interest in a crime to testify. The accused, especially, had an interest, and was not allowed to testify; that is, the accused was not a competent witness. Section 3 of the *Canada Evidence Act* now makes an accused a competent witness, and thus one who is allowed to testify.

Effective January 2, 2006, section 16.1 of the *Canada Evidence Act* provides that a person "under the age of fourteen years is presumed to have the capacity to testify," and "shall not take an oath or make a solemn affirmation." The evidence of such a person "shall be received if they are able to understand and respond to questions," (section 16.1(3)) and if they promise to tell the truth (16.1(6)). Such evidence has "the same effect as if it were taken under oath" (section 16.1(8)). Anyone challenging the capacity of such a witness has the burden of satisfying the court there is an issue, and if there is an issue, the court will conduct an inquiry into the person's ability to understand and respond to questions. The proposed witness shall not be asked "any questions regarding their understanding of the nature of the promise to tell the truth for the purpose of determining whether their evidence shall be received by the court" (sections 16.1(4) to 16.1(7)).

As with persons under the age of 14, competency is generally presumed; however, if the competence of a witness is raised at trial, the judge will hold a *voir dire* under section 16 of the *Canada Evidence Act* to determine whether the person is a competent witness. Section 16(2) provides that a person whose mental capacity is challenged, but who "understands the nature of an oath or a solemn affirmation and is able to communicate the evidence shall testify under oath or solemn affirmation." If such a person "does not understand the nature of an oath or a solemn affirmation but is able to communicate the evidence," they may, "notwithstanding any provision of any Act requiring an oath or a solemn affirmation, testify on promising to tell the truth." If none of these criteria are met, the person is not allowed to testify (section 16(4)).

The person who raises the issue has the burden of showing that the witness is incompetent. According to the Supreme Court of Canada, "testimonial competence comprehends: (1) the capacity to observe (including interpretation); (2) the capacity to recollect; and (3) the capacity to communicate" (*Marquard* para. 12). There is no clear authority on the standard of proof for such an application, however both McWilliams (1999, 34-22) and Sopinka (1999, 691) suggest that it be on the balance of probabilities. Courts in both British Columbia and Ontario have approved of the procedure of conducting the *voir dire* in the absence of the jury, if both Crown and defence counsel agree (*Ferguson, Gough*). Conducting the inquiry in the presence of the jury might result in prejudice to the accused if the witness is found to be incompetent to testify. However, it could be argued that conducting this *voir dire* in the presence of the jurors would enable them to better assess the credibility of the witness, if the person is allowed to testify.

Spouses

Under the common law, a spouse was considered to have an interest in the outcome of a criminal trial involving his or her partner, and was not allowed to testify; a spouse was not a competent witness. There were certain exceptions to this rule. For example, at common law, a spouse was both competent and compellable for the Crown where the accused had threatened the life or safety of the spouse. Sections 4(2) and 4(4) allow the Crown to compel (or force) unwilling spouses to testify against their spouse for additional offences, in respect of which Parliament has decided that the interests of bringing the accused to justice through the court system outweigh the benefits of maintaining or preserving the sanctity of the marriage.

Section 4(1) of the *Canada Evidence Act* modified the common law, such that a spouse is now a competent witness, but only for the defence (except as discussed above). That is, the spouse who chooses to testify for the defence is competent to do so.

The meaning of "spouse" under section 4 of the *Canada Evidence Act* has been interpreted to mean those legally married, but to exclude those living in a common law relationship. It would include same-sex marriages. The legislation applies if two people marry after the offence, since it is the sanctity of the marital relationship that is being protected (*Hawkins*). Spousal incompetence does not continue after divorce; former spouses are competent and compellable witnesses at the instance of the Crown.

The Supreme Court of Canada in *Salituro* addressed the question of whether separated spouses are competent to testify for the Crown. The accused was convicted of having forged his wife's signature on a cheque payable to both of them. He said he had her permission; she said he was lying. On appeal, the accused argued that his wife was not a competent witness for the Crown. The Ontario Court of Appeal dismissed his appeal on the basis that the rule regarding spousal incompetency should not apply where the spouses are irreconcilably separated. Mr. Justice Iacobucci wrote the decision of the Supreme Court, expressing the view that the historical reasons for the rule were incompatible with the *Charter* and the freedom of all individuals. He wrote:

> The common law rule making a spouse an incompetent witness involves a conflict between the freedom of the individual to choose whether or not to testify and the interests of society in preserving the marriage bond. It is not necessary to consider the difficult question of how this conflict ought to be resolved, because in this appeal we are concerned only with spouses who are irreconcilably separated.
>
> To give paramountcy to the marriage bond over the value of individual choice in cases of irreconcilable separation may have been appropriate in Lord Coke's time, when a woman's legal personality was incorporated in that of her husband on marriage, but it is inappropriate in the age of the *Charter* (303).

Since the common law rule was "out of step with the values of the *Charter*," the court altered the common law. Stuesser (2007) points out some of the absurdities of the present law, and recommends that spouses be treated as any other witness—competent and compellable for all offences.

Corroboration

The general rule at common law was that the testimony of one competent witness was sufficient to convict the accused (Law Reform Commission of Canada 1975a, 5). Judges developed exceptions to the rule in situations where they were suspicious that the evidence might be fabricated. In such cases, they required **corroboration.** Corroboration is defined at common law as "confirmation from some other source that the suspect witness is telling the truth in some part of his story which goes to show that the accused committed the offence with which he is charged" (*Vetrovec* 16). Corroboration may be in the form of statements by the accused, conduct of the accused showing guilt, articles found in the accused's possession, and so on. The corroborative evidence has the effect of bolstering the credibility of the witness.

The offences that required corroboration historically (incest, seduction, sexual intercourse with a female employee, and so on) tell us much about the judges' beliefs regarding the credibility of witnesses. Additionally, for a number of offences (rape, attempted rape, sexual intercourse with a female under 14, and indecent assault on a female), the judge had to warn the jury that it "was not safe to find the accused guilty in the absence of…corroboration, but that they [were] entitled to find the accused guilty if they [were] satisfied beyond a reasonable doubt that [the complainant's] evidence [was] true" (Law Reform Commission of Canada 1975a, 11).

Amendments to the *Criminal Code* in 1975 and 1982 removed the requirement of corroboration for many of these offences. Section 274 now states that corroboration is not required to convict an accused for a number of specified sexual offences, and that "the judge shall not instruct the jury that it is unsafe to find the accused guilty in the absence of corroboration." Prior to 1988, no one could be convicted on the unsworn evidence of a "child of tender years" unless it was corroborated by some other material evidence. Even following an amendment in 1988, some trial judges still considered unsworn evidence by children to be fragile, and gave the common law warning about such evidence (Bala 1993, 369). In 1993, Parliament attempted to correct this by introducing section 659: "any requirement whereby it is mandatory for a court to give the jury a warning about convicting an accused on the evidence of a child is abrogated."

Corroboration is still required today for three types of offences under the *Criminal Code*. Section 47(3) states that "no person shall be convicted of high treason or treason on the evidence of only one witness, unless the evidence of that witness is corroborated in a material particular by evidence that implicates the accused." Similarly, section 133 requires corroboration before convicting an accused of perjury, and section 292(2) requires it before convicting a person of procuring or aiding in the procuring of a feigned marriage.

At common law, the courts also required a warning to the jury about the dangers of convicting an accused on the uncorroborated evidence of an accomplice. In reviewing this area of the law in *Vetrovec,* Mr. Justice Dickson described corroboration as "one of the most complicated and technical areas of the law of evidence," and advocated a "common sense" approach:

> Rather than attempting to pigeon-hole a witness into a category and then recite a ritualistic incantation, the trial judge might better direct his [*sic*] mind to the facts of the case, and thoroughly examine all the factors which might impair the worth of a particular witness. If, in his judgement, the credit of the witness is such that the jury should be cautioned, then he may instruct accordingly. If, on the other hand, he believes the witness to be trustworthy, then regardless of whether the witness is technically an "accomplice", no warning is necessary (11).

According to Dickson, J., confirmatory evidence that a witness is telling the truth does not necessarily have to implicate the accused. This approach was confirmed by the Supreme Court of Canada in *Kehler,* where the confirmatory evidence did not directly implicate the accused. In *Khela* and *Smith,* the Court emphasized flexibility in explaining the *Vetrovec* warning.

Shielding Rape Victims

At common law, complainants in a rape trial could be cross-examined on their prior sexual conduct to test their credibility, and to establish consent. The judges (all men, at the time the rules were developed) relied upon their own preconceived notions about women who were raped, backed by esteemed jurists such as Wigmore, who had made the following observation, still contained in the 1970 edition of his work:

> Modern psychiatrists have amply studied the behaviour of errant young girls and women coming before the courts in all sorts of cases. Their psychic complexes are multifarious, distorted partly by inherent defects, partly by diseased derangements or abnormal instincts, partly by bad social environment, partly by temporary physiological or emotional conditions. One form taken by these complexes is that of contriving false charges of sexual offenses by men (quoted by L'Heureux-Dubé, J. in *Seaboyer* 342).

Wigmore was of the view that women who alleged rape should be assessed by psychiatrists to determine whether they were fabricating their complaints (Kobly 1992, 990). The relevance of past sexual history was based on the assumption that women who had consented to sex in the past were likely to have consented to the alleged act of rape (The Uniform Law Conference Report 1982, 66).

As Madame Justice L'Heureux-Dubé notes, many of these myths continued to exist. An Ontario study in 1988 found that:

> stereotypes are held by a surprising number of individuals, for example: that men who assault are not like normal men, the "mad rapist" myth; that women often provoke or precipitate sexual assault; that women are assaulted by strangers; that women often agree to have sex but later complain of rape, and the related myth that men are often convicted on the false testimony of the complainant; that women are as likely to commit sexual assault as are men and that when women say no they do not necessarily mean no (*Seaboyer* 341).

The common law remained unchanged in Canada until 1976, when the *Criminal Code* was amended to restrict the accused from asking the complainant questions about past sexual conduct with other persons; however, judicial interpretation of this section resulted in complainants actually being worse off under the amendments than under the common law (see Boyle 1981, and discussion in *Seaboyer* 349–52). In 1982, Parliament again amended the *Code*. Section 277 reversed the common law, and now prohibits the use of sexual reputation for the "purpose of challenging or supporting the credibility of the complainant." The Supreme Court of Canada in *Seaboyer* found that this section did not violate sections 7 and 11(d) of the *Charter*. However, section 276, which provided that sexual activity of the complainant with anyone other than the accused was not admissible except in specified circumstances, violated sections 7 and 11(d) of the *Charter,* and was not demonstrably justified under section 1 (*Seaboyer* and *Gayme*).

Following the decision in *Seaboyer*, Parliament again amended the *Criminal Code* in 1992 (see McIntyre 1994). Section 276(1) now creates an absolute bar (Boyle and MacCrimmon 1999, 219, although this may be open to debate—Stein 1998) to certain types of evidence, in proceedings for the listed sexual offences:

> evidence that the complainant has engaged in sexual activity, whether with the accused or with any other person, is not admissible to support an inference that, by reason of the sexual nature of that activity, the complainant
> > (a) is more likely to have consented to the sexual activity that forms the subject
> > matter of the charge; or
> > (b) is less worthy of belief.

Evidence of sexual activity for any purpose other than under section 276(1) (e.g., honest but mistaken belief in consent), other than the sexual activity that forms the subject matter of the charge, cannot be introduced unless the judge determines that the evidence:

> > (a) is of specific instances of sexual activity;
> > (b) is relevant to an issue at trial; and
> > (c) has significant probative value that is not substantially outweighed by the danger of prejudice to the proper administration of justice (section 276(2).

Section 276(3) lists the factors that the judge shall take into account in determining whether the evidence under section 276(2) is admissible:

> > (a) the interests of justice, including the right of the accused to make a full answer
> > and defence;
> > (b) society's interest in encouraging the reporting of sexual assault offences;

(c) whether there is a reasonable prospect that the evidence will assist in arriving at a just determination in the case;

(d) the need to remove from the fact-finding process any discriminatory belief or bias;

(e) the risk that the evidence may unduly arouse sentiments of prejudice, sympathy or hostility in the jury;

(f) the potential prejudice to the complainant's personal dignity and right of privacy;

(g) the right of the complainant and of every individual to personal security and to the full protection and benefit of the law; and

(h) any other factor that the judge, provincial court judge or justice considers relevant.

The *Code* creates a two-step process. Section 276.1(2) requires the accused to make a written application to the court for a hearing to determine whether the evidence is relevant. The application must set out "the detailed particulars of the evidence that the accused seeks to adduce, and the relevance of that evidence to an issue at trial." If proper procedures have been followed and the judge finds that the evidence "is capable of being admissible," the judge shall hold an *in camera voir dire* under section 276.2 (the second stage) to determine its admissibility.

The requirement at the first stage that the accused provide detailed particulars and relevance of the evidence sought to be introduced does not violate the accused's rights under sections 7, 11(c) or 11(d) of the *Charter*, even though the accused may be required to submit to cross-examination on the affidavit at the second stage (*Darrach* para. 55). According to the Supreme Court of Canada in *Darrach*, such evidence cannot be used to incriminate the accused at trial because of section 13 of the *Charter*; however, it can be used to challenge the accused's credibility (para. 66). Following the Court's decision in *Henry* (see Chapter 10), it is likely that the evidence at the *voir dire* is not admissible at the accused's trial unless the accused decides to testify. If the accused testifies at trial, the evidence at the *voir dire* could be used to challenge the accused's credibility and incriminate the accused.

The stipulation in section 276.2(2), that the complainant is not a compellable witness at the *in camera* hearing, does not violate the *Charter*. To allow the complainant to be cross-examined on her sexual history before a finding that it was relevant would defeat the purpose of the law (*Darrach* para. 68). Following the hearing, the judge must rule whether the evidence (or any part of it) is admissible under section 276(2), and give reasons, including which of the factors set out section 276(3) affected the decision. This issue is a question of law (section 276.5), and therefore subject to appeal. Section 276.3 makes it an offence to publish the contents of the application or the proceedings (with limited exceptions), and section 276.4 requires the judge to "instruct the jury as to the uses that the jury may and may not make" of evidence admitted under section 276.2. For commentary on how the courts are using this section, see Benedet (2009) and Gotell (2006), and Stuart (2009).

Assistance/Protection of Child Witnesses, Witnesses with Disabilities & Other Witnesses

Child witnesses have historically been subjected to rules of evidence that strongly implied they were not to be believed. As stated above, the requirement for corroboration of a child's unsworn evidence was eliminated in 1988. Between 1988 and 2006, Parliament added and amended a number of sections of the *Criminal Code* to assist children, and others who may have difficulty "by reason of a mental or physical disability," in giving evidence. Some of these provisions were extended to witnesses more generally. Although professionals in the criminal justice system think that such aids are beneficial to child witnesses, the aids are often unused (see Bala, Lindsay, and McNamara 2001 for a discussion).

Box 11.2 What Happened to Seaboyer and Gayme?

Charges against both were dropped. In each case, the respective complainant decided she was unwilling to testify in a subsequent trial; see MacCrimmon (1992, 299–300).

Not all complainants give up. One of the complainants in *Tyhurst,* who, when asked if she would testify at a third trial if the Crown successfully appealed the acquittal, said "I will testify at a fourth and fifth trial if necessary. I will testify until that son of a bitch is in jail. I will testify until that man is in jail or is dead from old age." See Box 5.6 on *Tyhurst,* in Chapter 5.

Support Person for Witnesses–Section 486.1

Section 486.1 allows the trial judge, on application by the prosecutor, or by a witness who is under the age of eighteen years, or who has a mental or physical disability, to "order that a support person of the witness' choice be permitted to be present and to be close to the witness while the witness testifies" unless the judge "is of the opinion that the order would interfere with the proper administration of justice." A support person may be allowed for other witnesses if the trial judge "is of the opinion that the order is necessary to obtain a full and candid account from the witness" (section 486.1(2)). In making a decision under 486.1(2), the judge "shall take into account the age of the witness, whether the witness has a mental or physical disability, the nature of the offence, the nature of any relationship between the witness and the accused, and any other circumstances that the judge . . . considers relevant" (section 486.1(3)). The judge shall not allow a support person unless the judge "is of the opinion that doing so is necessary for the proper administration of justice" (section 486.1(4)).

Testimony Outside the Courtroom or Behind a Screen–Section 486.2

Section 486.2 allows the trial judge, on application by the prosecutor, or by a witness who is under the age of eighteen years, or who has a mental or physical disability, to "order that the witness testify outside the court room or behind a screen or other device that would allow the witness not to see the accused," unless the judge "is of the opinion that the order would interfere with the proper administration of justice." Such accommodations may also be made for other witnesses if the trial judge "is of the opinion that the order is necessary to obtain a full and candid account from the witness" (section 486.2(2)). In making a decision under 486.2(2), the judge "shall take into account the factors referred to in subsection 486.1(3)" (section 486.2(3)). The trial judge may also make such an order to protect the safety of witnesses for specified offences (related to criminal organizations, terrorism and certain offences under the *Security of Information Act* (section 486.2(4) and (5)). A prerequisite to testifying under these orders is that the accused by allowed to communicate with his or her counsel while watching the witness testify (section 486.2(7)).

The Supreme Court of Canada considered the use of screens in the case of *Levogiannis,* where the trial judge allowed a 12-year-old boy to testify behind a screen so that the he could not see the accused, who was charged with sexual interference. The accused and his lawyer could see the complainant. The issue was whether the section contravened the accused's rights under sections 7 and 11(d) of the *Charter*. In examining this question, Madame Justice L'Heureux-Dubé said that it was important to consider the rights of the witnesses (333). The use of a screen did not impair or restrict the accused's ability to cross-examine the complainant (339). In addition, the trial judge has "substantial latitude" in deciding whether to use the screen. The section, designed to enable young complainants "to be able to recount the evidence, fully and candidly, in a more appropriate setting, given the circumstances, while facilitating the elicitation of the truth," does not violate the *Charter*. Note: the section applies to more witnesses now than it did at the time of this decision.

Accused Not Allowed to Cross-Examine some Witnesses–Section 486.3

On application by the prosecutor, or by a witness under the age of eighteen years, "the accused shall not personally cross-examine the witness, unless the judge . . . is of the opinion that the proper administration of justice requires the accused to personally conduct the cross-examination" (section 486.3(1)). Section 486.3(4) allows for a similar order for criminal harassment offences. Restrictions on the accused cross-examining other witnesses may be applied if the trial judge "is of the opinion that, in order to obtain a full and candid account from the witness . . . the accused should not personally cross-examine the witness" (section 486.3(2)). In making a decision under 486.3(2), the judge "shall take into account the factors referred to in subsection 486.1(3)" (section 486.3(3)).

Videotaped Evidence of Victim or Witness under 18—Section 715.1

Section 715.1 allows for the admission of a videotape, in which the complainant or other witness, under the age of 18 at the time of the offence, describes the acts complained of, if the videotape is made "within a reasonable time after the alleged offence," and if the complainant or witness adopts the content of the videotape while testifying.

In *L.(D.O.)*, a nine-year-old girl gave videotaped evidence that her grandfather had sexually assaulted her over a three-year period, the most recent incident being five months before the videotape. The video was played when she gave her evidence at trial. The accused challenged section 715.1 on the basis that it violated his rights under sections 7 and 11(d) of the *Charter,* and that it breached the principles of fundamental justice and his right to a fair trial. More specifically, he argued that section 715.1 infringed a number of evidentiary rules—the rule against hearsay, the rule against prior consistent statements, and the right to cross-examine a witness at the same time that he or she gave evidence. The Supreme Court of Canada ruled against all of his arguments, holding that section 715.1 did not violate the *Charter*.

In reaching this decision, L'Heureux-Dubé, J. stated that it was important to examine these constitutional questions "in their broader political, social and historical context in order to attempt any kind of meaningful constitutional analysis" (438). She reviewed the numerous studies on the issue of child sexual abuse, and the imbalance of power between children and adults. She examined the goals and purposes of section 715.1, in terms of establishing truth and curbing the trauma suffered by children through being compelled to testify (439). The section has the safeguard of allowing the trial judge to edit the videotape or to refuse to show it. "Properly used, this discretion to exclude admissible evidence ensures the validity of section 715.1 and is conversant with fundamental principles of justice necessary to safeguard the right to a fair trial" (461). As well, defence counsel is allowed to cross-examine the child on the videotaped statement.

Madame Justice L'Heureux-Dubé suggested a number of factors that the trial judge should consider in exercising his or her discretion to admit videotaped statements:

> (a) the form of questions used by any other person appearing in the videotaped statement;
> (b) any interest of anyone participating in the making of the statement;
> (c) the quality of the video and audio reproduction;
> (d) the presence or absence of inadmissible evidence in the statement;

(e) the ability to eliminate inappropriate material by editing the tape;

(f) whether other out-of-court statements by the complainant have been entered;

(g) whether any visual information in the statement might tend to prejudice the accused (for example, unrelated injuries visible on the victim);

(h) whether the prosecution has been allowed to use any other method to facilitate the giving of evidence by the complainant;

(i) whether the trial is one by judge alone or by a jury, and

(j) the amount of time which has passed since the making of the tape and the present ability of the witness to effectively relate to the events described (463).

The section was considered again by the Supreme Court in *F.(C.C.)*. After adopting the statements on the videotape, the child witness in that case made some contradictory statements under cross-examination. The Ontario Court of Appeal held that the contradicted evidence was not admissible, because it was not "adopted" by the child. In interpreting the meaning of "adopt" in the section, the Supreme Court of Canada stated that it was important to keep in mind the purpose of the legislation: "enhancing the truth-seeking role of the courts by preserving an early account of the incident and of preventing further injury to vulnerable children as a result of their involvement in the criminal process" (para. 41). The fact that the child, under cross-examination, contradicts the evidence on the video does not amount to non-adoption, but is part of the evidence that the trier of fact is to consider. The court should hold a *voir dire* to determine the admissibility of the videotape, and consider the factors (a) to (j) listed above in deciding whether the admission of the videotape would impair the accused's right to a fair trial (see Bala 1998, and Moore and Green 2000, for a discussion). Should a child be unable to testify at trial, the videotape may be admitted under an exception to the hearsay rule (discussed in Chapter 12).

SUMMARY

In a criminal trial, the Crown generally has the onus or burden of proving the guilt of the accused; the standard of proof on the Crown is generally beyond a reasonable doubt. Certain provisions create reverse onus situation, where the onus shifts to an accused to disprove something. Reverse onus provisions are *prima facie* unconstitutional, but may be justified under section 1 of the *Charter*. Statute or common law also creates situations in which presumptions of fact arise, such that on the proof of certain things, other things will be presumed to be true, in the absence of evidence to the contrary. Such presumptions may or may not be constitutionally valid.

For evidence to be admissible at trial, it must be relevant to some fact in issue. Relevant evidence is admissible unless it is specifically excluded by some rule of law. Such rules are based on various rationales which may be intrinsic to the trial (to increase the reliability of the evidence, to ensure a fair trial, to preclude trials being side-tracked by collateral issues, to allow a sense of predictability), or may relate to extrinsic reasons involving policy reasons.

Evidence may be classified in a number of ways. Evidence may be classified by its form, as testimonial, real, or commissioned. It may be classified by its purpose, as demonstrative or illustrative. It may be classified by its effect, as direct or circumstantial.

Testimonial, or *viva voce*, evidence is given under oath (broadly interpreted) or on solemn affirmation. Witnesses may be assessed on both their competence and compellability to testify. Competence refers to their legal capacity to testify, while compellability refers to whether they can be required by one or both parties to testify. Under the *Canada Evidence Act*, a spouse is a competent witness for the defence, but is generally neither competent nor compellable at the instance of the Crown, with certain exceptions. This protection exists as a matter of policy, to protect the sanctity of marriage and marital communications. Accordingly, it has been extended only to those who are legally married at the time of the trial, and may not apply where the parties are still married but separated.

Corroboration is evidence that tends to confirm some aspect of a witness's testimony. Historically, it was either required or recommended in respect of a number of offences, especially sexual offences. Contemporary amendments to the *Criminal Code* have abolished many of the situations in which corroboration was required, and prohibits judges from providing the traditional warnings to juries about the danger of convicting in the absence of corroboration. The *Code* now requires corroboration for only three offences.

The traditional scope of cross-examination of alleged victims of sexual offences as to their sexual history or character was based predominantly on rules reflecting unsustainable historic male biases. Attempts by Parliament to limit the scope of this questioning have met with mixed success, and the situation is likely not yet settled. Current legislation allows such cross-examination, but only for restricted purposes in limited circumstances, where the defence meets the burden of satisfying the court of the factual basis for the questions, their relevance, and of the satisfactory balance between their probative value and the proper administration of justice.

Provisions to protect child witnesses, persons with mental or physical disabilities, and other witnesses include the use of screens, videotaped evidence, the partial or total exclusion of the public from the court room, the presence of a "support person" for the witness, and restrictions on the right of an accused to personally conduct cross-examination.

QUESTIONS TO CONSIDER

(1)   What would be the advantages and disadvantages of codifying the law of evidence? What role should the common law play in such a codification?

(2)   What does "onus of proof" refer to? What does "standard of proof" refer to? What section of the *Charter* has an impact on the onus and standard of proof?

(3)   Give an example of direct evidence. Give an example of the danger of convicting an accused on direct evidence.

(4)   Give an example of circumstantial evidence. What are the dangers of convicting an accused on circumstantial evidence?

(5)   What is the rule in *Hodge's Case*? How is it relevant (or not relevant) in the Canadian context today?

(6)   Distinguish between competence and compellability.

(7)   What is corroboration?

(8)   What would be the advantages and disadvantages of requiring the accused to give evidence at his or her own trial?

(9)   hy are spouses generally not compellable witnesses against one another? Why are spouses compelled to give evidence against one another for certain offences?

(10)   What arguments could be made in favour of making a spouse a compellable witness for the Crown for all offences? What counter-arguments could be made? Should a spouse be a competent witness for the Crown in all cases?

(11)   What provisions exist to protect witnesses and under what circumstances can some of these provisions be used?

BIBLIOGRAPHY

Allman, Anthony. "A Reply to *R. v. Seaboyer*: A Lost Cause?" (1992) 10 *Criminal Reports* (4th) 153.

Bala, Nicholas. "Recognizing that Child Witnesses are Children." (1998) 11 *Criminal Reports* (5th) 227.

Bala, Nicholas. "Criminal Code Amendments to Increase Protection to Children and Women: Bills C-126 and C-128." (1993) 21 *Criminal Reports* (4th) 365.

Bala, N., R.C.L. Lindsay and E. McNamara. "Testimonial Aids for Children: The Canadian Experience with Closed Circuit Television, Screens and Videotapes." (2001) 44 *Criminal Law Quarterly* 461.

Benedet, Janine. "Probity, Prejudice and the Continuing Misuse of Sexual History Evidence." (2009) 64 *Criminal Reports* 72.

Berger, Benjamin. "The Rule in Hodge's Case: Rumours of its Death are Greatly Exaggerated." (2005) 84(1) *Canadian Bar Review* 47.

Boilard, Jean-Guy, *Guide to Criminal Evidence*. Cowansville, PQ: Les Editions Yvon Blais Inc., 1999 with updates.
Boilard's Introduction discusses relevance, admissibility, and probative force of evidence, and Chapter 6, "New Rules of Evidence Regarding Sexual Offences," contains a section on victims' previous sexual activity and corroboration. Chapter 8 is on "Competence and Compellability of Spouses," Chapter 10 on "Corroboration" and Chapter 12 on "Reasonable Doubt and Guilty Intent."

Boyle, Christine. "Section 142 of the *Criminal Code*: A Trojan Horse?" (1981) 23 *Criminal Law Quarterly* 253.

Boyle, Christine. *Sexual Assault*. Toronto: Carswell, 1984.
Chapter 1 covers the historical background to the reform of rape and indecent assault law in Canada. Chapter 2 discusses constitutional issues. Chapters 3 to 5 elaborate on the amendments to the *Criminal Code* which created new offences. Chapter 6 is on pre-trial decisions, Chapter 7 on evidence at trial, and Chapter 8 on sentencing.

Boyle, Christine, and Marilyn MacCrimmon. "*R. v. Seaboyer*: A Lost Cause?" (1992) 7 *Criminal Reports* (4th) 225.

Boyle, Christine, and Marilyn MacCrimmon. "The Constitutionality of Bill C-49: Analyzing Sexual Assaults as if Equality Really Mattered." (1999) 41 *Criminal Law Quarterly* 198.

Boyle, Christine, Marilyn T. MacCrimmon, and Dianne Martin. *The Law of Evidence: Fact Finding, Fairness, and Advocacy*. Toronto: Emond Montgomery Publications Limited, 1999.
Chapter 4 talks about types of evidence, Chapter 5, the burden and standard of proof, Chapter 6, witnesses, and Chapter 13, sexual history.

Brockman, Joan. "An Offer You Can't Refuse:" Pleading Guilty When Innocent (2010) 56(1) *Criminal Law Quarterly* (in press).

Bryant, Alan W., Sidney N. Lederman and Michelle K. Fuerst. *Sopinka, Lederman & Bryant – The Law of Evidence in Canada* 3rd ed. Canada: LexisNexis Canada, 2009.
See Chapter 1 (Foundational Principles of Evidence and the Importance of Evidentiary Rulings in Context), Chapter 2 (Types of Evidence and Conditions for the Receipt of Evidence), Chapter 3 (Evidential Burden and Burden of Proof), Chapter 4 (Presumptions), Chapter 5 (Standards of Proof), Chapter 13 (Competence and Compellability of Witnesses), Chapter 16 (The Examination of Witnesses), Chapter 17 (Corroboration), and Chapter 18 (Documentary Evidence).

Delisle, Ron Don Stuart, and David M. Tanovich, *Evidence: Principles and Problems*, 8th ed. Toronto: Carswell, 2007.

Dufraimont, Lisa. "Evidence Law and the Jury: A Reassessment." (2008) 53 *McGill Law Journal* 199.

CHAPTER 11: *Types of Evidence*

Dufraimont, Lisa. "R. v. Griffin and the Legacy of Hodge's Case." (2009) 67 *Criminal Reports* 74.

Gardner, Wayne. "Explanations and Illustrations: Demonstrative Evidence in the Criminal Courtroom." (1996) 38 *Criminal Law Quarterly* 425.

Gotell, Lise."When Privacy is not Enough: Sexual Assault Complainants, Sexual History Evidence and the Disclosure of Personal Records." (2006) 43 *Alberta Law Review* 743.

Hill, S. Casey, David M. Tanovich and Louis P. Strezos. *McWilliams' Canadian Criminal Evidence.* Aurora, ON: Canada Law Book Limited (available online on Criminal Spectrum).
  See Chapter 2 (on sources of law), Chapter 3 (discussion of the qualified search for truth), Chapter 4 (Relevance), Chapter 18 (Presentation of Witnesses), Chapter 20 (Real Evidence), Chapter 21 (Documentary Evidence), Chapter 24 (Evidential Burden), Chapter 25 Persuasive Burden), Chapter 26 (Presumptions), Chapter 27 (Assessing Credibility), Chapter 28 (Circumstantial Evidence: Drawing Reasonable Inferences), Chapter 29 (Identification Evidence), Chapter 30 (Adverse Inferences), and Chapter 31 (Confirmation and Corroboration).

Idling, Lynn A. "Crossing the Line: The Case for Limiting Cross-Examination by an Accused in Sexual Assault Trials." (2004) 49 *Criminal Law Quarterly* 69.

Jonas, George. *The Scales of Justice: Seven Famous Criminal Cases Recreated.* Toronto: Canadian Broadcasting Corporation, 1983.

Kobly, Peggy. "Rape Shield Legislation: Relevance, Prejudice and Judicial Discretion." (1992) 30(3) *Alberta Law Review* 988.

Law Reform Commission of Canada. *Corroboration.* Ottawa, 1975a.

Law Reform Commission of Canada. *Report on Evidence.* Ottawa, 1975b.

Law Reform Commission of Canada. *Evidence.* Ottawa, 1977.

MacCrimmon, Marilyn T. "Developments in the Law of Evidence: The 1990–91 Term: Social Science, Law Reform and Equality." (1992) 3 *Supreme Court Law Review* (2d) 268.

McIntyre, Sheila. "Redefining Reformism: The Consultations That Shaped Bill C-49." In Julian V. Roberts and Renate M. Mohr. *Confronting Sexual Assault: A Decade of Legal and Social Change.* Toronto: University of Toronto Press, 1994, 293.

Majury, Diana. "*Seaboyer and Gayme.* A Study of InEquality." In Julian V. Roberts and Renate M. Mohr. *Confronting Sexual Assault: A Decade of Legal and Social Change.* Toronto: University of Toronto Press, 1994, 268.

McWilliams, Peter K., *Canadian Criminal Evidence.* Aurora, ON: Canada Law Book Limited, 1999 with updates.
  Chapter 2, "Development of the Law of Evidence," discusses the common law origins of the law of evidence, statutory reform, and constitutional law. Chapter 5 discusses circumstantial evidence, Chapter 6, documentary evidence, Chapter 7, real evidence, Chapter 25 burden of proof and presumptions, and Chapter 26, corroboration. Chapter 34 discusses sworn and unsworn testimony, child witnesses, competency and compellability, including the competency and compellability of spouses.

Moore, Timothy E. and Melvyn Green. "Truth and the Reliability of Children's Evidence: Problems with Section 715.1 of the *Criminal Code.*" (2000) 30 *Criminal Reports* (5th) 148.

Paciocco, David M. "Evidence About Guilt: Balancing the Rights of the Individual and Society in Matters of Truth and Proof." (2001) 80 *Canadian Bar Review* 433.

Rauf, Naeem M. "*Thompson*: Charter Challenge to Videotaped Statements of Children—Two Views." (1989) 68

Criminal Reports (3rd) 331.

Stein, Daniel A. "Admissibility of Sexual Conduct Evidence After *D.(A.S.)*." (1998) 13 *Criminal Reports* (5th) 312.

Stewart, Hamish. "Spousal Incompetency and the Charter." (1996) 34 *Osgoode Hall Law Journal* 411.

Stuart, Don. "Twin Myth Hypotheses in Rape Shield Laws are Too Rigid and Darrach is Unclear." (2009) 64 *Criminal Reports* 74.

Stuesser, Lee. "Abolish Spousal Incompetency." (2007) 47 *Criminal Reports* 49.

Uniform Law Conference of Canada. *Report of the Federal/Provincial Task Force on Uniform Rules of Evidence.* Toronto: Carswell, 1982.

CASES

R. v. Ah Wooey (1902), 8 C.C.C. 25 (B.C.S.C.).

R. v. Cinous, [2002] 2 S.C.R. 3.

R. v. Cooper (1977), 34 C.C.C. (2d) 18 (S.C.C.). Where a case is predominantly circumstantial, it is not an absolute requirement that the jury be instructed in accordance with the rule in *Hodge's Case*—it is within the trial judge's discretion. A strong charge on the concept of proof beyond a reasonable doubt is the important element.

R. v. Darrach, [2000] 2 S.C.R. 443.

R. v. Downey, [1992] 2 S.C.R. 10. The provision in section 212(3) that a person living with or being habitually in the company of prostitutes is presumed to be living off the avails of prostitution violates the presumption of innocence in the *Charter,* but is saved under section 1 as a reasonable limit on that right in a free and democratic society.

R. v. Ferguson (1996), 112 C.C.C. (3d) 342 (B.C.C.A.). Discussion of the questions asked of a child on a *voir dire* to determine competency. The *voir dire* can be held in the absence of the jury.

R. v. F. (C.C.), [1997] 3 S.C.R. 1183. The Court discusses the meaning of "adopts the contents of the videotape" in section 715.1 of the *Criminal Code.*

R. v. Gough, [1999] O.J. No. 1480 Ontario Court (General Division). It is acceptable and advisable to conduct a *voir dire* in the absence of the jury to determine the competence of a witness (here, both counsel agreed to this procedure).

R. v. Graat (1982), 2 C.C.C. (3d) 365 (S.C.C.). Non-expert witnesses are permitted to give opinions that amount to "compendious recitations of facts," such as to apparent age, speed, sobriety, and so on.

R. v. Griffin, 2009 SCC 28.

R. v. Hanemaayer, [2008] O.J. No. 3087.

R. v. Hawkins, [1996] 3 S.C.R. 1043. A witness who marries the accused after the accused's preliminary hearing is covered by the spousal incompetency rule.

R. v. Hibbert, [2002] 2 S.C.R. 445.

Hodge's Case (1838), 168 E.R. 1136 (Assize). In a case of purely circumstantial evidence, the jury must be warned that to convict, they must be satisfied that the circumstances were both consistent with guilt and inconsistent with any other rational explanation.

R. v. Kalavar (1991), 4 C.R. (4th) 114 (Ont. Gen. Div.). An accused testifying on his or her own behalf has a right to swear a religious oath other than on the Bible.

R. v. Kehler, [2004] 1 S.C.R. 328.

R. v. Khela, [2009] 1 S.C.R. 104.

R. v. B. (K.G.) [K.G.B.], [1993] 1 S.C.R. 740.

R. v. Laba (1994), 94 C.C.C. (3d) 385 (S.C.C.). Section 394(1)(b) of the *Criminal Code*, which makes it an offence to sell or purchase minerals or precious metals unless the seller or purchaser "establishes that he is the owner or agent of the owner or is acting under lawful authority" violates section 11(d) of the *Charter*.

R.v. L.(D.O.) [1993] 4 S.C.R. 419. The admission of a videotaped statement by a child victim, made several months after the alleged offence, pursuant to section 715.1 of the *Criminal Code,* does not contravene sections 7 or 11(d) of the *Charter;* there is no constitutionally protected right to contemporaneous cross-examination.

R. v. Lifchus, [1997] 3 S.C.R. 320. The Court elaborates on the meaning of "beyond a reasonable doubt."

R. v. Levogiannis (1993), 85 C.C.C. (3d) 327 (S.C.C.). Those provisions allowing for a child victim in a sexual assault case to testify from behind a screen to prevent the child from seeing the accused (although not the converse) do not violate the *Charter*.

R. v. Marquard, [1993] 4 S.C.R. 223. Experts should not testify respecting the credibility of witnesses, beyond the limited extent of expert evidence on human conduct and psychological or physiological factors relevant to credibility and beyond the ordinary scope of the human experience of jurors, in which case a warning is appropriate. Also discussion of meaning of competence.

R. v. Nikolovski, [1996] 3 S.C.R. 1197. An accused can be convicted on the evidence of a security videotape even though the clerk who was at the store at the time of robbery could not identify the accused.

R. v. Pan; R. v. Sawyer, [2001] 2 S.C.R. 344.

R. v. Salituro (1991), 68 C.C.C. (3d) 289 (S.C.C.). The common law exception to the principle of spousal testimonial incompetency extends to cases in which the spouses remain married but are separated without any reasonable possibility of reconciliation.

R. v. Seaboyer; R. v. Gayme (1991), 66 C.C.C. (3d) 321 (S.C.C.). Those aspects of the "rape shield" law preventing the defence from adducing evidence of the past sexual conduct of the complainant except in limited circumstances contravene the right to make full answer and defence. However, such evidence will only be permitted where it is first established that it is for a legitimate purpose.

R. v. Smith, [2009] 1 S.C.R. 146.

R. v. Turcotte, [2005] 2 S.C.R. 519.

Reference re Truscott, [1967] S.C.R. 309.

Vetrovec v. The Queen (1982), 67 C.C.C. (2d) 27 (S.C.C.). The rule requiring a warning to the jury of the danger of convicting on the uncorroborated evidence of an accomplice is abolished, although the warning remains appropriate in some instances.

R. v. Whyte (1988), 42 C.C.C. (3d) 97 (S.C.C.). The presumption in the *Code* that a person occupying the driver's seat of a motor vehicle was in care and control of the vehicle, for the purposes of drinking-driving offences, is contrary to the presumption of innocence in the *Charter* but constitutes a reasonable limit within the meaning of section 1.

CHAPTER 12: *Exclusionary Rules*

CHAPTER OBJECTIVES

In studying this chapter, you should develop an understanding of the following topics and concepts:

- privilege, including solicitor–client and spousal privilege
- Wigmore's criteria for considering new classes or claims of privilege
- public interest immunity
- police informer privilege and the public safety exception
- the relationship of privilege to privacy interests, particularly in relation to the personal records of complainants
- the nature of hearsay evidence, and the distinction between evidence that is not strictly hearsay, and that which is admissible as an exception to the hearsay rule
- the distinction between character evidence and evidence that goes to the issue of credibility
- the nature of similar fact evidence, and the conditions under which it is admissible

THE EXCLUSION OF EVIDENCE BY PRIVILEGE OR IMMUNITY

Privilege refers to the right of a person or the state, and the corresponding duty of a witness, to withhold from the court evidence that would be relevant and admissible in the absence of that privilege. The concept of privilege is designed to recognize and give effect to certain policy considerations, even though it may actually interfere with the search for truth. In effect, privilege is an exception to the general rule that all relevant information is admissible. In deciding whether to extend privilege, the courts address the balance between the social value of preserving confidentiality in a relationship and the interests of the administration of justice.

Solicitor–Client Privilege

Solicitor–client privilege is probably the most commonly known privilege. It developed in the sixteenth century as a privilege that belonged to the lawyer, on the basis that a lawyer (as a professional) would not reveal a client's secrets (Sopinka *et al.* 1999, 728). In the eighteenth century, the search for truth came to be more important than the lawyer's professional image, and so the rationale switched to protection of the client—a client had to be able to consult (in confidence) with a lawyer if he or she was to have effective legal service (Sopinka *et al.* 1999, 729). This was confirmed by Chief Justice Lamer in *Gruenke*, who wrote, "the *prima facie* protection for solicitor–client communications is based on the fact that the relationship and the communications between solicitor and client are essential to the effective operation of the legal system" (305).

Communications with a lawyer acting in a professional capacity, made in confidence, are privileged (with a few exceptions), and neither party can be compelled to disclose them. The requirement that the privileged conversation be with a lawyer acting in a professional capacity excludes any consultation in furtherance of a crime, as well as casual, public conversations, such as those at cocktail parties. Confidentiality attaches to all communications made within the framework of the solicitor–client relationship, including those with anyone acting as an agent for the lawyer, such as a secretary, a private investigator, or a psychiatrist retained by defence counsel (*Perron*).

If the client later consents to disclosure, or divulges the communication to others, the privilege is lost, and the lawyer can be compelled to testify regarding the information. The privilege is that of the client, not the lawyer, and it exists until the client waives it. Solicitor–client privilege may be waived expressly or by implication (for example, if the client makes allegations against his or her lawyer, as in *Read*). The privilege continues after death.

Solicitor–client privilege is protected in two ways: (1) the court will not compel the production of evidence, or listen to evidence covered by solicitor–client privilege, and (2) lawyers who breach solicitor–client privilege may be disciplined by the Law Society, the provincial self-regulating organization of lawyers with the mandate to regulate lawyers' conduct (see Brockman 2004).

In *Jones*, the Supreme Court of Canada dealt with the public safety exception to solicitor–client privilege. A Vancouver lawyer sent his client (Jones) to see a psychiatrist Dr. Smith (pseudonyms used by the court), to assist in the preparation of Jones's defence on charges of aggravated sexual assault on a Vancouver prostitute. Conversations between an agent's lawyer and his or her client are covered by solicitor-client privilege. Jones described to Dr. Smith his plan to kidnap, sexually assault, and kill a prostitute as a "trial run," to see if he could "live with" his behaviour. Dr. Smith discovered that the court would not hear about his concerns over the danger that Jones presented to the public, when the court sentenced Jones after his guilty plea to aggravated sexual assault. Dr. Smith made an application to the British Columbia Supreme Court to determine whether he could disclose the information. When the case reached the Supreme Court of Canada, Mr. Justice Cory, for the majority, found that the privilege should be set aside:

> the facts raise real concerns that an identifiable individual or group is in imminent danger of death or serious bodily harm. The facts must be carefully considered to determine whether the three factors of seriousness, clarity, and imminence indicate that the privilege cannot be maintained. Different weights will be given to each factor in any particular case. If after considering all appropriate factors it is determined that the threat to public safety outweighs the need to preserve solicitor–client privilege, then the privilege must be set aside. When it is, the disclosure should be limited so that it includes only the information necessary to protect public safety (*Smith v. Jones* para. 85).

Solicitor-client privilege may also be set aside under the "innocence-at-stake" exception. In *McClure*, the Supreme Court of Canada stated that in order to establish this exception, the accused must first show that "the information he seeks from the solicitor-client communication is not available from any other source; and he is otherwise unable to raise a reasonable doubt" (*Brown* para 4, summarizing a *McClure* application). At stage one, the accused must demonstrate an evidentiary basis to conclude that a solicitor-client communication exists that could raise a reasonable doubt as to the accused's guilt. If the accused establishes this evidentiary basis, the judge should move to stage two, and "examine the communication to determine whether, in fact, it is likely to raise a reasonable doubt as to the guilt of the accused" (*Brown* para. 4, summarizing a *McClure* application). If the judge rules that the evidence must be disclosed, the person who loses the privilege is protected by use immunity and derivative use immunity, but not transaction immunity (*Brown* para. 100; see discussion in Chapter 10, and Layton (2002)).

Solicitor–client privilege does not put a lawyer above the law. For example, under the *Income Tax Act,* the Minister of National Revenue can make demands for certain documents, and these demands can be served on lawyers. It is an offence not to comply with the demand. For lawyers to claim solicitor–client privilege in relation to the documents, they must follow a specific procedure

---

Box 12.1 Is Boredom Privileged?

A parliamentary lawyer who quit his job because there was nothing for him to do, and who sued the federal government for "wrongful hiring," was accused of breaching solicitor–client privilege in "exposing" his job as a waste of public money. The government apparently took the view that it was improper for the lawyer to tell the truth about his situation. The Law Society of Upper Canada dismissed the complaint ("'Bored' Lawyer Agrees to Settlement." (27 May 1995) *Vancouver Sun* A14).

---

under the *Income Tax Act*. Failure to follow this procedure can result in a conviction for failing to comply with the demand (*Raynier*).

To what extent are communications between the police and Crown counsel covered by solicitor–client privilege? The Supreme Court of Canada established that solicitor–client privilege exists where police officers seek legal advice from Crown counsel. In *Campbell*, the police sought a legal opinion on the legality of a reverse-sting operation. The fact that the operation was later found to be illegal does not by itself take the advice outside the privilege. However, in this case the police waived solicitor–client privilege by referring to the legal opinion, to bolster their argument that they were acting in good faith. The evidence also revealed that the advice fell into the "future crimes" exception to privilege. For a further discussion of solicitor-client privilege for government lawyers, see MacNair (2003).

Problems with solicitor–client privilege occur when lawyers become the targets of investigation, respecting offences committed by lawyers, or offences they commit with their clients. Issues regarding search warrants to search a lawyer's office were discussed in Chapter 8.

Spousal Communications

Spousal communications are privileged under certain circumstances. At common law, a spouse was not competent to testify against his or her spouse (that is, was not allowed to testify with some exceptions), and therefore privilege was usually not an issue. Today, certain spousal communications are privileged under section 4(3) of the *Canada Evidence Act*. Witnesses cannot be compelled to disclose any communication made to them by their spouses during their marriage. Section 189(6) provides that the privilege applies to intercepted spousal communications (*Lloyd and Lloyd*).

The privilege is that of the person hearing the communication, and it ends with the marriage (on death or divorce). The privilege may also end when the spouses are irreconcilably separated. Although the Supreme Court of Canada held that a spouse who was irreconcilably separated was a competent witness for the Crown (*Salituro,* discussed in Chapter 11), the questions of whether such a witness was a compellable witness for the Crown, or whether privilege would apply, were not in issue, and thus remain unresolved.

What is the effect of subsections 4(3) and 4(2) of the *Canada Evidence Act*? If a witness is compellable at the instance of the Crown, does the privilege under 4(3) apply? The Quebec Court of Appeal ruled that the privilege does not apply when a spouse is compellable by the Crown for an offence listed in section 4(2) (in this case incest), as this would defeat the purpose of section 4(2) (*St.*

Jean). This is implied by the Supreme Court of Canada in *Wildman,* where the Court stated that at Wildman's second trial his wife would be a competent and compellable witness, because of the newly enacted section of the *Canada Evidence Act* (now section 4(4)). However, Boilard suggests that the privilege continues to exist where a spouse is competent and compellable (1999, 8–3).

In *Zylstra,* the Ontario Court of Appeal decided that a wife who testified in her husband's defence could assert spousal privilege; however, such an assertion should be in front of the jury, and the jury ought to be told that

(a) The privilege in s. 4(3) is a statutory privilege which all legally married witnesses are entitled to assert in a trial; and
(b) The privilege is one that belongs to the witness, not the accused person, and, as such, the decision whether to assert or waive the privilege lies with the witness, not the accused (*Zylstra* para. 7).

Stuesser (2007) recommends that spousal privilege be abolished in Canada, as it has in both England and Queensland, Australia.

Extending Privilege By Wigmore's Criteria

Chief Justice Lamer in *Gruenke* discussed two competing views of privilege. First, *prima facie* or class privilege assumes that privilege exists for a certain relationship (or class of situations), unless "the party urging admission can show why the communications should *not* be privileged" (286). Second, case-by-case privilege involves the presumption that there is no privilege, although it is open for one of the parties to argue that there ought to be privilege in a particular case. Before *Gruenke,* there had been some debate as to whether an expansion of the common law should take place on a class basis (for example, deciding whether all religious communications or all doctor–patient communications are privileged), or on a case-by-case basis. Lamer objected to the class or "pigeon hole" approach to rules of evidence (291), and favoured a more principled approach. This principled or case-by-case approach requires the person asserting the privilege to show that the communication should be protected as coming within Wigmore's four criteria:

1. The communications must originate in a confidence that they will not be disclosed.
2. This element of confidentiality must be essential to the full and satisfactory maintenance of the relations between the parties.
3. The relation must be one which in the opinion of the community ought to be sedulously fostered.
4. The injury that would inure to the relation by the disclosure of the communications must be greater than the benefit thereby gained for the correct disposal of litigation (quoted in *Gruenke* 284).

Lamer did not completely reject the development of a class privilege, but took the view that there were no policy reasons to introduce a class privilege for religious communications (290).

Public Interest Immunity

Public-interest immunity is different than Crown solicitor–client privilege, and serves different interests (*Sander* 50). At common law, governments could claim "Crown privilege" by asserting that the disclosure of certain information would be contrary to the public interest (see Cooper 1990 for

Box 12.2 Students' Research Privilege

Following a newspaper article about his work, Simon Fraser University criminology graduate student Russell Ogden, and the newspaper reporter who wrote the story, were subpoenaed to testify at a coroner's inquest into an assisted suicide. As part of his research, Ogden had interviewed several people who had participated in assisted suicides. Ogden refused to testify, to protect the confidentiality of his sources. Following submissions by counsel, the Coroner applied Wigmore's four criteria, and found that Ogden should not be compelled to testify. The key to the decision was the evidence supporting the contention that confidentiality was necessary to the success of the research, and the importance of the research.

It should be noted that the Coroner ruled that no such privilege extended to the reporter, and that she would be in contempt if she refused to testify (Daisley 1994, 28). For a detailed discussion of the Russell Ogden case and its impact on research ethics at Simon Fraser University, see Palys and Lowman (2000). In 2003, Exeter University in England was ordered to pay Ogden $143,000 in damages for reversing its earlier decision to support his efforts to keep his sources confidential, two years after committing to do so (Todd 2003, B2).

a discussion). Mr. Justice LaForest has suggested that the "public interest in non-disclosure of a document is not…a Crown privilege. Rather it is more properly called a **public interest immunity,** one that, in the final analysis, is for the court to weigh" (*Carey v. Ontario* 510–1).

Some of the law governing this area is codified in sections 37–39 of the *Canada Evidence Act*. Section 37(1) provides that "Subject to sections 38 to 38.16, a Minister of the Crown in right of Canada or other official may object to the disclosure of information before a court…on the grounds of a specified public interest." If a trial is not before a court of superior jurisdiction, the application must be made to the Federal Court of Canada or, in some cases, to a Superior Court in the province, within 10 days of the objection (section 37(3) and (4)). Sections 38 to 38.16 deal with objections to disclosure relating to international relations or national defence and security (see Stewart 2003a for discussion of the 2001 amendments), and section 39 deals with objections relating to "a confidence of the Queen's Privy Council," which includes such things as a memorandum, discussion paper, and so on, for consideration by the Council (defined to include the Cabinet and committees of the Cabinet).

Amendments to section 37 in 2001 (part of the anti-terrorism legislation; see Rosenthal 2003) provided more direction to the courts when considering such applications. If the disclosure encroaches on a specified public interest, but "the public interest in disclosure outweighs in importance the specified public interest," the court may order disclosure of some or all of the information, or a summary of the information, subject to conditions (section 37(5)). Section 37.21, which required these hearings to be in private, was repealed in 2004.

Police Informers' Privilege

Police informer privilege is a subset of public interest immunity (*McClure* para 28). At common law, the courts could not compel the disclosure of the identity of persons who provided the police with information about a crime. Informer privilege still is "an ancient and hallowed protection which plays a vital role in law enforcement" (*Leipert* para. 9). The purpose of the privilege is to protect citizens who assist in law enforcement, and to encourage others to provide such assistance. This privilege, which belongs to the informant, is so important to the functioning of the criminal justice system that "once established, neither the police nor the court possesses discretion to abridge it" (para. 14). When the informant is known to the police, the police can check with the informant to ensure that the disclosure of information will not reveal their identity. When the informant is unknown to the police (e.g., Crime Stoppers Tip), it is impossible to ensure that the disclosure of any detail, no matter how innocuous, does not identify the informant, and therefore no information should be disclosed (*Leipert* para. 19).

The only exception to the informer privilege is the innocence-at-stake exception. To establish this exception, the accused "must show some basis to conclude that without the disclosure sought his or her innocence is at stake" (*Leipert* para. 33). If a basis is shown, the procedure is that:

> the court may then review the information to determine whether, in fact, the information is necessary to prove the accused's innocence. If the court concludes that disclosure is necessary, the court should only reveal as much information as is essential to allow proof of innocence. Before disclosing the information to the accused, the Crown should be given the option of staying the proceedings. If the Crown chooses to proceed, disclosure of the information essential to establish innocence may be provided to the accused (*Leipert* para. 33).

In making a decision as to whether informer privilege exists, the trial judge may hold an *in camera, ex parte* hearing–that is, a private hearing without the presence of the accused or defence counsel (*Basi*).

THE PRIVACY OF COMPLAINANTS

The privacy of complainants is currently under attack from at least two different angles. Defence counsel have been asking (1) for personal records of complainants, and (2) for courts to allow their experts to examine complainants. The offence involved when such requests are made is inevitably sexual assault, and it may be worthwhile thinking about why similar demands are not being made in respect of other offences.

Complainants' Records–the Mills Regime

After the introduction of legislation prohibiting the use of a complainant's sexual history to test credibility and consent in 1992 (see discussion in Chapter 11, "Shielding Rape Victims"), police, the Crown, and defence counsel started to ask complainants for their medical records, psychiatric records, diaries, and so on, to test their credibility and determine the issue of consent (Feldthusen 1996; Kelly 1997). Two major decisions on this topic, *O'Connor*, and *A.(L.L.) v. B.(A.)*, were considered a disaster for victims of sexual assault, and for gender equality in the criminal justice system. Following these decisions, Parliament added sections 278.1 to 278.91 of the *Criminal Code*. Section 278.1 defines personal record to mean any record that contains:

personal information for which there is a reasonable expectation of privacy and includes, without limiting the generality of the foregoing, medical, psychiatric, therapeutic, counselling, education, employment, child welfare, adoption and social services records, personal journals and diaries, and records containing personal information the production or disclosure of which is protected by any other Act of Parliament or a provincial legislature, but does not include records made by persons responsible for the investigation or prosecution of the offence.

The procedure for obtaining such records in possession of any person, including the Crown, for offences listed in section 278.2 (sexual assault, sexual interference, incest, prostitution offences, and so on), is set out in section 278.3. The accused must apply to the trial judge, naming the records sought, the person holding the records, and the reasons the accused believes the records are relevant. At this first stage, the trial judge will hold an *in camera* hearing under section 278.4, in which the person in possession of the record, or the complainant or witness, may be heard, to determine if the record should be produced to the judge for review. If the trial judge is satisfied that the accused has established the relevance of the record, and that the production is "necessary in the interests of justice" (section 278.5(1)), taking into account the factors listed in section 278.5(2), the judge may then order that it be produced for examination. At the second stage, the judge will review the record in the absence of the parties, but may hold an *in camera* hearing under section 278.6 to determine whether the record or parts of it should be produced to the accused. Section 278.3(4) specifically lists 11 assertions that are *not* sufficient on their own to show relevance–for example, it is not enough that the record exists, that the record may disclose a prior inconsistent statement of the witness, or that the record relates to the credibility of the witness.

If the judge orders the production of the record to the accused, section 278.7 allows for conditions to be imposed, "to protect the interests of justice and, to the greatest extent possible, the privacy and equality interests of the complainant or witness...." Section 278.8 requires that the judge provide written reasons for both the first-stage and second-stage decisions.

In *Mills*, the Supreme Court of Canada heard from 18 interveners (see the Introduction for consideration of interveners), and the majority (with Chief Justice Lamer dissenting, in part) determined that sections 278.1 to 278.9 did not violate sections 7 or 11(d) of the *Charter*. The Supreme Court of Canada in *McNeil* confirmed the constitutionality of the "Mills regime," stating that the relevance standard in *Mills* "is tailored to counter speculative myths, stereotypes and generalized assumptions about sexual assault victims and about the usefulness of private records in sexual assault proceedings"(para. 31). See Busby (2000), Coughlan (2000), and Gotell (2006) for commentary on the use of complainants' records.

Examination of the Complainant by Defence Experts

In the sexual assault case of *Olscamp,* the accused, applied to have an expert of his choosing examine the complainant (a seven-year-old child) and her mother, in the event that the trial judge decided to hear the testimony of two experts who worked with the child in play-therapy, and of her mother, who had spoken to the therapists. The Ontario Court (General Division) found that such an application could be made under section 24(1) of the *Charter*, but decided that denying the accused the right to obtain an independent assessment of the witnesses did not violate his rights under section 7 of the *Charter*. The trial process provided the accused with many ways of testing the expert testimony, if the trial judge decided it was admissible. To allow another expert into the process would not necessarily clarify the issues. In addition, the effect of such examinations on complainants, and on the administration of justice, must be considered in deciding such applications. The Supreme Court of Canada has not yet ruled on this issue.

THE EXCLUSION OF HEARSAY

The rule against hearsay, as with privileged communications, is an exception to the general rule that all relevant evidence is admissible. Hearsay is generally inadmissible, but there are exceptions to this rule (and exceptions to the exceptions).

Hearsay is an out-of-court statement, made by someone other than the witness, led to prove the truth of what the statement asserts. A witness can give evidence of what she or he saw, heard, smelt, or otherwise experienced through senses. For example, a witness could testify, "I saw the car swerve down the road, cross back and forth over the centre line, and hit the pedestrian. I saw the driver stagger from the car, and when I examined her eyes I noticed that they were bloodshot, and that her breath smelled of alcohol." This is not hearsay. The witness would not, however, be allowed to testify that, "I didn't see anything, but my father told me that he saw the car swerve down the road, cross back and forth over the centre line, and hit the pedestrian. He said he saw the driver stagger from the car, and when he examined her eyes, he noticed that they were bloodshot, and that her breath smelled of alcohol." The second statement is hearsay, and is inadmissible as evidence to prove the truth of its contents (the manner in which the person was driving and the state of her eyes and the smell of her breath). The rationale behind the exclusion of hearsay is that the statement by the father is not made under oath, the father is not available for cross-examination, the trier of fact cannot observe his demeanour or assess his credibility, its admission would be unfair to the accused, and so forth (see Hill *et al.* Chapter 7 for a discussion of the difficulties surrounding the definition of hearsay, and the rationale for the hearsay rule). In 2006, the Supreme Court of Canada confirmed that the defining characteristics of hearsay are: "(1) the fact that the statement is adduced to prove the truth of its contents and (2) the absence of a contemporaneous opportunity to cross-examine the declarant" (*Khelawon* para. 35).

The nature of the evidence changes if the witness is asked, "Why did you arrest the driver of the vehicle?" For example, the following response is not hearsay: "I arrived five minutes after the accident, and my father told me that the driver was swerving down the road and hit a pedestrian. The driver was running from the scene of the accident, so I helped my father arrest her." Although such evidence cannot be used to prove the truth of its other contents (that the driver hit the pedestrian), it can be used to explain why the witness assisted in the arrest of the driver. Analytically, such a statement falls outside the definition of hearsay (*Smith*).

This type of question is commonly asked of police officers, to establish reasonable grounds for arrest or for a search. If a police officer is asked by the Crown or by defence counsel why he or she made the arrest or conducted a search, the police officer can respond by saying that, "X told me that the accused had just robbed a bank." The evidence is not used to establish that the accused robbed the bank, but to prove that the police officer had reasonable grounds to make the arrest, or had a lawful basis to conduct the search.

The situation in *Collins,* discussed in Box 12.3, involved evidence that was not, strictly speaking, hearsay. There are also true exceptions to the rule that hearsay is not admissible. For example, admissions are a major exception to the hearsay rule (see Chapter 10). There are also numerous other statutory exceptions (such as medical and business records, the prior testimony of unavailable witnesses) and common law exceptions (dying declarations, spontaneous declarations), which are discussed in Bryant *et al.* 2009, and Hill *et al.* online. For example, spontaneous statements made by an accused when found in possession of an illegal substance are part of the *res gestae,* and are admissible through the testimony of the police officer who heard them. In *Risby*, the accused denied knowledge of what the substance was, but his statement at the time of arrest was

Box 12.3 *R. v. Collins*

An example of evidence that is not hearsay, although it may initially appear to be such, comes from the *Collins* case. One of the questions that remained unanswered at the trial was why the police officer took a flying leap at Ruby Collins and grabbed her by the throat. After the issue was raised by the defence in cross-examination, the police officer was re-examined by the Crown, to establish grounds for his behaviour:

> Mr. Wallace (appearing for the Crown):
> Q: Yes. Constable Woods, you said in answer to a question by Mr. Martin that the object, the sighting of the object in Ruby Collins' hand confirmed your suspicions?
> A: That's correct.
> Q: Where—when did you formulate those suspicions?
> A: They were prior to arriving at Gibsons. We were advised—
> Mr. Martin (appearing for the appellant): That's hearsay, your honour. Anything what [*sic*] he was advised other than that is hearsay and that is certainly outside the ambit of my cross-examination, your honour.

This was not hearsay, of course, because it was an explanation as to why the police officer did what he did. The statement was not tendered as proof of its contents. Later in the case, the Court of Appeal stated that the "objection raised was groundless," and the answer would not have infringed the hearsay rule (*Collins* para.4).

admissible through the police officer's testimony. The Supreme Court of Canada recently delineated the "state of mind" or "present intentions" exception to the hearsay rule (*Griffin* paras. 55-58).

The Principled Exception to the Hearsay Rule

The Supreme Court of Canada expanded the exceptions to the hearsay rule in *Khan*, where Madame Justice McLachlin, for the Court, noted that the hearsay rule had become "unduly inflexible in dealing with new situations and new needs in the law" (541). One of these "new needs" is for more flexibility in admitting as evidence children's statements about sexual abuse. In *Khan*, the trial judge had decided that a five-year-old child, who was three and one-half years old at the time of an alleged assault by Dr. Khan, was incompetent to testify, and that her statement to her mother regarding the assault was inadmissible because it was hearsay. The Supreme Court of Canada held that the child's statement to her mother should have been admitted. Such third-party statements will be admissible if "sufficient evidence is first led to establish the reliability of the out-of-court statement, and of the circumstances which establish the need to introduce the content of the child's statement through hearsay" (545). The Court found that in *Khan*, both criteria were met:

It was necessary, the child's viva voce evidence having been rejected. It was also reliable. The child had no motive to falsify her story, which emerged naturally and without prompting. Moreover, the fact that she could not be expected to have knowledge of such sexual acts imbues her statement with its own peculiar stamp of reliability. Finally, her statement was corroborated by real evidence [semen was found on her clothing] (548).

Madame Justice McLachlin ruled that allowing such evidence, on the basis of necessity and reliability, was a more principled approach, founded on policy (540). She favoured it over the "pigeon-hole," or categorical approach, which lacks flexibility.

While the element of necessity may arise when a child is not allowed to testify, it can also be satisfied when a child testifies, but the child's memory is not clear. This is in fact what happened in the *Khan* case, when the College of Physicians and Surgeons held a hearing into Khan's conduct. The child was then eight, and did not have a very good recollection of the events. The discipline committee heard evidence from both the mother as to what the child had told her, and from the eight-year-old child. On appeal, the Ontario Court of Appeal found that the evidence of the mother was necessary to get an accurate recall of the events (Rosenberg 1993, 75). The Ontario Court of Appeal wrote, "the fact that the child testifies will be relevant to, but not determinative of, the admissibility of the out-of-court statement" (24).

Chief Justice Lamer referred to this flexible definition of necessity in his decision in *B.(K.G.)*, stating that he was not prepared "to adhere to a strict interpretation that makes unavailability an indispensable condition of necessity" (296). It makes sense to provide the trier of fact with both types of evidence—that is, the earlier statement by the child, through the recollection of her mother, and the later testimony of the child at trial (Rosenberg 1993, 73)—when the criteria of necessity and reliability are established.

With regard to reliability, McLachlin, J. in *Khan* said that there were a number of factors to consider: "Timing, demeanour, the personality of the child, the intelligence and understanding of the child, the absence of any reason to expect fabrication in the statement,... The matters relevant to reliability will vary with the child and with the circumstances, and are best left to the trial judge" (545).

By way of example of this principled exception to the hearsay rule, the British Columbia Court of Appeal allowed the evidence given at the first trial of an accused by his son, who was six years old, to be introduced when the accused was retried, when the child was ten. The trauma experienced by the child when his mother was beaten to death in his presence, plus the passage of time, had affected his ability to recall the details of the murder. The trial judge, on the basis of necessity and reliability, allowed those portions of the child's evidence from the first trial to be used as evidence, when the child could not recall some of the details at the second trial (*Hanna*).

In *Hawkins*, the Supreme Court of Canada concluded that the necessity and reliability exception could be used to introduce evidence by the accused's girlfriend at his preliminary inquiry, because he had subsequently married her, and she was therefore incompetent and no longer available to the Crown at the time of his trial. Her evidence at Hawkins' preliminary inquiry had been under oath, and she had been cross examined. In *Couture*, the trial judge used this exception to the hearsay rule to admit a woman's earlier statements to the police that her husband had confessed to her, since she and the accused were subsequently married, making her incompetent to testify. The British Columbia Court of Appeal ruled that her earlier statements should not have been admitted under the principled exception to the hearsay rule because she was not competent to testify. The Supreme Court of Canada found that her statements were inadmissible because to admit them would undermine the spousal incompetency rule and its rationales. The question, according to the majority, "is whether, from an objective standpoint, the operation of the principled exception to the hearsay rule in the particular circumstances of the case would be disruptive of marital harmony or give rise to the natural repugnance resulting from one spouse testifying against the other" (para. 66). See Coughlan (2007), and Ives (2007), for commentary.

There are other limits to the principled exception to the hearsay rule. In *R.(D.)*, the majority of the Supreme Court of Canada decided that out-of-court statements by the five-year-old victim, made to foster parents and a doctor, were inadmissible, because the statements were not sufficiently reliable. At the time of the trial, the girl had no recollection of what she had described. The foster parents and doctor had asked the girl what had happened, and she described a sexual assault by her father. However, since there was also evidence that she may have been sexually assaulted by her brother on the same day, and that the children sometimes lied to cover up their own sexual behaviour, the evidence of what she said to the doctor was inadmissible, because it was equally consistent with the hypothesis that she was sexually assaulted by her brother and was trying to cover up for him. Madame Justice L'Hueruex-Dubé, in dissent, said the statements were admissible. The child had no recollection at trial, so they were necessary. The circumstances under which they were made "provided a guarantee of reliability: the statements were made very shortly after the alleged attack; the two statements, made on separate occasions and to different people, were consistent with one another; they were also consistent with the medical evidence, which clearly indicated that [she] had been assaulted" (para. 68). In response to Mr. Justice Major's decision for the majority, L'Heureux-Dubé observed that the children in covering up their sexual activities never accused adults, they just denied they had done anything. She concluded that the possibility of fabrication was not sufficient to exclude the statement (paras. 70–1).

The Supreme Court of Canada went one step further in *R. v. B.(K.G.)*, on the question of necessity. The Crown's case depended on the evidence of the accused's friends, who had given videotaped evidence to the police in the presence of a lawyer and a parent, to the effect that K.G.B. had made statements that he thought he was responsible for the death of the deceased. When the witnesses testified at trial, they said they had lied in the videotaped interview. The Crown, of course, could use their prior statements to impeach the witnesses credibility, but the Crown wanted to use the evidence for the truth of the earlier out of court, videotaped statements—statements to the effect that K.G.B. had implicated himself in the killing of the deceased. The Crown argued that the test ought to be one of necessity and reliability. Mr. Justice Lamer, for the majority, agreed. He took the view that the historical rule on the use of prior inconsistent statement (discussed later in this chapter) was too inflexible. He stated that the Crown must, in a *voir dire*, establish on the balance of probabilities that there was sufficient evidence of necessity and reliability. With regard to necessity, Lamer described the friends of the accused as holding the prior statement hostage. It was unavailable when the friends refused to repeat it at trial. According to Lamer, if:

> (i) the statement is made under oath or solemn affirmation following a warning as to the existence of sanctions and the significance of the oath or affirmation, (ii) the statement is videotaped in its entirety, and (iii) the opposing party, whether the Crown or the defence, has a full opportunity to cross-examine the witness respecting the statement, there will be sufficient circumstantial guarantees of reliability to allow the jury to make substantive use of the statement [meaning use it as evidence of its truth]. Alternatively, other circumstantial guarantees of reliability may suffice to render such statements substantively admissible, provided that the judge is satisfied that the circumstances provide adequate assurances of reliability in place of those which the hearsay rule traditionally requires (795-796).

If the statement is made to a person in authority, the trial judge, in the *voir dire*, must also be satisfied that "the statement was not the product of coercion of any form, whether it involves threats, promises, excessively leading questions by the investigator or other person in a position of authority, or other forms of investigatory misconduct," or anything that would bring the administration of justice into disrepute (802). Ordinarily, in what has come be known as a KGB *voir dire*, the Crown will have to establish on a balance of probabilities that the statement is necessary and reliable. Whether the statement is believed, of course, is still up to the trier of fact.

This case has interesting implications for section 715.1, dealing with videotaped statements of victims of specified offences who were under the age of 18 at the time of the offence (discussed in Chapter 11). Section 715.1 clearly stipulates that a condition for the admissibility of the videotape is that the complainant, while testifying, adopt the statement on the video. In cases where a complainant recants, Rosenberg suggests that the videotape might be admissible anyway, under the common law developed by the Supreme Court of Canada (1993, 83–4). In *F.(W.J.)*, the Supreme Court of Canada was prepared, where the child witness was unable to speak and therefore to adopt the videotaped statements, to allow the out-of-court statements under the necessity and reliability exception (para. 39).

Chief Justice Lamer, in *B.(K.G.)*, summarized the effect of *Khan* on the law of evidence, and on the rule excluding hearsay, by quoting from himself in *Smith*:

> [The decision in *Khan*] should be understood as the triumph of a principled analysis over a set of ossified judicially created categories [and] signalled an end to the old categorical approach to the admission of hearsay evidence. Hearsay evidence is now admissible on a principled basis, the governing principles being the reliability of the evidence, and its necessity (774).

He continued:

> The movement towards a flexible approach was motivated by the realization that, as a general rule, reliable evidence ought not to be excluded simply because it cannot be tested by cross-examination (777).

In *R. v. Khelawon*, the majority of the Supreme Court of Canada confirmed that the party wanting to introduce hearsay under the principled exception to the hearsay rule, must, on the balance of probabilities, establish necessity and reliability (para. 47). Other factors may come into play, allowing the trial judge to exclude hearsay that meets this criteria "where its probative value is outweighed by its prejudicial effect" (para. 49). See Crisp (2008) for a commentary on *Khelawon*.

CHARACTER EVIDENCE

Historically, evidence of an accused's **character** was used to help determine whether the accused had committed the alleged offence. The question was whether the accused was the type of person who would commit the type of offence alleged. The common law courts abandoned this use of character evidence because of its highly prejudicial effects on juries (Sopinka *et al.* 1999, 472). Today, as a general rule, the Crown cannot introduce evidence of an accused's bad character at trial. The accused is on trial only for the offence(s) alleged, not for previous conduct. It is viewed as unfair to admit such evidence, even though in our everyday lives we often use past behaviour as a predictor of the present or future behaviour of people we know (see Box 12.4). It is human nature (according to social psychologists) to be more likely to think someone has engaged in particular conduct if we know they have engaged in such behaviour in the past. It is our tendency to think in this way that causes the unfairness of such evidence at trial.

Box 12.4 Nettler's Rule

"He who does you dirt a first time will do so again, if you stick around long enough"—Dr. Gwynne Nettler's rule for predicting human behaviour (Nettler 1970, 137).

Evidence of unrelated past activity by the accused is not admissible, because the jury or the judge may place too much weight on it. An accused is presumed to be innocent until proven guilty, and the trier of fact is not supposed to engage in reasoning that, since the accused committed armed robbery in the past, the accused is predisposed to committing armed robbery, and therefore likely committed the armed robbery charged. The prejudice of an accused's past record would generally outweigh any value the evidence would have at a trial.

Section 666 of the *Criminal Code* provides an exception to the rule against character evidence. If the accused leads evidence (through cross-examination of Crown witnesses, by testifying at trial, or by calling witnesses) of his or her good character, the Crown can then lead evidence of any previous convictions of the accused. This section was added to the *Criminal Code* because, at common law, bad character could not be proved by describing specific acts, but only by general description (Sopinka *et al.* 1999, 501).

Section 360 also provides an exception to the general rule that the Crown cannot introduce evidence of the accused's bad character. If an accused charged with an offence under section 354 or

---

Box 12.5 Are the Social Psychologists Taking Over the Laws of Evidence?

In 1995, the United States Congress passed amendments to the Federal Rules of Evidence, allowing the prosecution to introduce evidence of convictions of previous sexual assaults and assaults on children in prosecutions of these offences, "for consideration on any matter to which it is relevant." One of the justifications for the law is that:

> A person with a history of rape or child molestation stands on a different footing [from other offenders]. His past conduct provides evidence that he has the combination of aggressive and sexual impulses that motivates the commission of such crimes, that he lacks effective inhibitions against acting on these impulses, and that the risks involved do not deter him. A charge of rape or child molestation has greater plausibility against a person with such a background (Stuesser 1997, 179).

What are the arguments against such a rule? What type of social science research could you use to support your position?

356(1)(b) has been previously convicted of theft or possession of stolen property in the last five years, evidence of that conviction can be introduced at trial as tending to show that the accused knew the subject matter of the offence was unlawfully obtained. Several lower courts have reached contradictory conclusions on whether the section violates section 7 of the *Charter*.

Character evidence is also admissible where the evidence is relevant to an issue in the case. However, it is not to be used to determine the guilt of the accused on the basis that the accused is the type of person who would commit the alleged crime, and its probative value must outweigh its prejudicial effect. For example, in *G.(S.G.)*, the Supreme Court of Canada found that evidence of a sexual relationship between the accused and one of the boys, whom it was alleged she persuaded to kill the victim, was admissible to show the control she exercised over the three adolescent boys implicated in the murder. In addition, the stolen property in her home was admissible to show motive, in that it was alleged she thought the victim was a "rat." However, the trial judge would have to decide whether the probative value outweighed its prejudicial effect. Such evidence, if admissible, could also be used to assess the credibility of the accused.

CREDIBILITY

The courts distinguish between character and credibility. Character, as discussed above, has to do with whether the accused is the type of person who would have committed the offence—it addresses disposition. **Credibility** is concerned with how much weight will be given to a witness's statement, and with the extent to which a witness is telling the truth (this last aspect of credibility is sometimes referred to as **credit**). Accused persons do not put their credibility in issue unless they testify. If they do testify, their credibility is in issue just as that of any other witness.

Section 12 of the *Canada Evidence Act* allows any witness to be cross-examined on their previous convictions. Such evidence is used to discredit a witness—to attack the witness's credibility. A criminal record, by itself, is not sufficient to discredit a witness, but the court will look at the extent of the record, the type of offence(s), how recent the last offence was, and so on. It is still left to the trier of fact to determine the effect of any record on the witness's credibility. Of course, if a witness denies a prior criminal conviction, it can be proved, and their false denial may discredit them. Conversely, if a witness admits a record, some courts may consider that as supporting their veracity, although convictions for perjury and giving contradictory evidence will likely still count against the witness.

In *Corbett*, the Supreme Court of Canada decided that section 12 of the *Canada Evidence Act* did not offend section 11(d) (the right to a fair trial) or section 7 (the principles of fundamental justice) of the *Charter*. As Chief Justice Dickson wrote, "the effect of the section is merely to permit the Crown to adduce evidence of prior convictions as they relate to credibility" (397). Dickson dismissed sociological studies "which purported to demonstrate that jurors *are* (emphasis in the original) incapable of distinguishing between evidence that goes to guilt and evidence that goes to credibility" (401). Rather, he decided that jurors could make the distinction, basing that conclusion on "the experience of trial judges…a strong faith in juries…and common sense" (401–2). In addition, it was "logically incoherent" to decide that jurors could not follow instructions on how the evidence could be used, given that the right to a jury trial was entrenched in section 11(f) of the *Charter* (401), and "if we are to continue our belief that a trial by a jury…offers the fairest determination of guilt or innocence, then we must credit the jury with the intelligence and conscience to consider evidence of prior conviction only to impeach the credibility of the defendant if it is so instructed" (403). Dickson's faith, common sense, and logic are, unfortunately, contrary to social science research on this issue. In 1982, the Uniform Law Report cited studies that clearly illustrate that we are incapable of "restricting the use of such evidence to the issue of credibility" (397). Without delving into the details of these studies, Dickson concluded that the Attorney-General of Canada, who analyzed the studies "with great sophistication," cast their scientific method into question, and that there were other studies that cast doubt on their conclusions. Dickson stated: "It is not possible to undertake a complete analysis of all these studies for the purpose of this judgment, but the conflicting results and the inherent limitations of such investigations should cause the court to be wary of relying upon the data" (401). One might ask whether any amount of research would have convinced him that section 12 violated the accused's rights under the *Charter*. For commentary on *Corbett* applications, see Plaxton (2009a), Rose (2003), and Sankoff (2006).

An accused is entitled to receive a judge's ruling on a *Corbett* application (to have his or her record excluded) after the close of the Crown's case, and before the defence begins its case. The trial judge may hold a *voir dire,* and hear evidence from the defence about the evidence it proposes to call, to make a decision about the admissibility of the accused's record should the accused decide to testify (*Underwood*).

Reasonable Doubt and Credibility

In 1991, Cory, J. for the Supreme Court of Canada in *W.(D.)*, explained the relationship between reasonable doubt and credibility, and suggested the following non-mandatory instructions for juries and judges:

> First, if you believe the evidence of the accused, obviously you must acquit.
> Second, if you do not believe the testimony of the accused but you are left in reasonable doubt by it, you must acquit.
> Third, even if you are not left in doubt by the evidence of the accused, you must ask yourself whether, on the basis of the evidence which you do accept, you are convinced beyond a reasonable doubt by that evidence of the guilt of the accused (para. 28).

The case resulted in much commentary (Healy 2007, Murphy 2008, and Plaxton 2008) and numerous appeals. More recently, the Supreme Court of Canada confirmed its earlier decision that there is "nothing sacrosanct about the formula set out in *W. (D.)*," and that "the decisive question [is] whether the accused's evidence, considered in the context of the evidence as a whole, raises a reasonable doubt as to his guilt" (*Dinardo* para. 23). Also see *R.E.M.* and *H.S.B.*

Prior Consistent Statements

Witness cannot repeat their earlier out-of-court statements or call other witnesses to repeat their statements at a trial. The evidence is considered self-serving, of little or no probative value, and is generally prohibited (see *Ellard* para. 31; also see *Dinardo* para. 40). There are exceptions to the rule that prohibits self-serving, prior consistent statements. For example, where the witness is attacked for having recently fabricated the evidence, the witness can always respond by saying, "I told the same story to X shortly after the incident." The issue of recent fabrication was raised, for example. in *Stirling*.

Prior consistent statements are considered hearsay; therefore, they are not admissible for the truth of their contents. They may, however, be introduced to explain the narrative of a crime (*Dinardo* para. 37; see Dufraimont 2008 for a discussion of this decision). For example, the words of a victim in a 911 call may be introduced to explain why the police arrived at the complainant's house. According to the courts, it is important for the trier of fact to understand the narrative or story of the crime. The 911 call is not used as evidence of the offence. It could, however, be so used as evidence of the offence if the witness was unable to testify, if the Crown established necessity and reliability. Prior written or recorded statements may also be used by a witness to refresh his or her memory. Once the memory is refreshed, the witness provides the evidence *viva voce.*

Prior Inconsistent Statements

At common law, prior statements of witnesses that are inconsistent with their in-court testimony could be used to question their credibility, but could not be used as evidence of the truth of their contents (the history and rationale for this rule is discussed in *B.(K.G.)* by Chief Justice Lamer). The common law in this area was changed by the Supreme Court of Canada in *B.(K.G.)*, and now allows previous inconsistent statements to be admitted for the truth of their contents under circumstances of necessity and reliability (see earlier discussion under "The Principled Exception to the Hearsay Rule").

According to the Supreme Court in *B.(K.G.)*, section 9(2) of the *Canada Evidence Act* provides a two-stage process for dealing with a witness who contradicts his or her previous statements that were reduced to writing or recorded. The party invoking section 9(2) must show in a *voir dire* that its requirements are met (the section is applicable to the party calling the witness). That party will then inform the court whether the statement is being presented to test the credibility of the witness or for the proof of its contents (substantive use). If it is for substantive use, the *voir dire* continues, and the trial judge will have to decide on a balance of probabilities if necessity and the *indicia* (indicators) of reliability are established. If the prior inconsistent statement is to a person in authority, the judge will also address the question of whether the statement was voluntary or whether there are other reasons to exclude the statement. The trial judge does not decide on the truth of the statement at this time, as this is up to the trier of fact should the prior inconsistent statement be ruled admissible as evidence of its substantive content (*B.(K.G.)* 799). More details on the procedure to be used under section 9(2) (sometimes referred to as a *Milgaard* application) can be found in *Milgaard* (221-2).

Sections 10 and 11 of the *Canada Evidence Act* govern the cross-examination of witnesses on previous written or recorded statements and on previous oral statements, respectively.

SIMILAR FACT EVIDENCE

Similar fact evidence is evidence not directly related to the charges before the court, but which tends to show that the accused has committed very similar acts at other times, from which other facts may be inferred. The rule regarding similar fact evidence developed from the rules surrounding character evidence, and is usually traced back to the *Makin* decision of the English House of Lords in 1894. A couple, in the business of adopting children for sums of money inadequate to support them, were charged with causing the death of one of the children after a body was found in the garden of one of the several houses that the couple had occupied. The question arose at their trial whether the fact that a number of other bodies were found in the gardens could be introduced at their trial for the one murder. The court suggested that generally such evidence was inadmissible; however it would be admissible if its probative value outweighed its prejudicial effect, as it did in this case.

The Supreme Court of Canada subsequently developed categories of exceptions—similar fact evidence was admissible if it showed intent, illustrated a plan or system, and so on. The Court appeared to be renaming propensity evidence (evidence that a person has a tendency or disposition to commit a crime) in order to render it admissible. However, the Court followed the English decision in that evidence which showed nothing other than propensity was generally inadmissible; however, "exceptions to this rule will arise when the probative value of the evidence outweighs its prejudicial effect" (*B.(C.R.)* 734-35).

The Supreme Court of Canada clarified the law on similar fact evidence in *Handy*. According to Binnie, J., for the Court, "propensity evidence by any other name is still propensity evidence" (para. 58). It is presumptively inadmissible. It is only admissible if the Crown can "satisfy the trial judge on a balance of probabilities that in the context of the particular case the probative value of the evidence in relation to a particular issue outweighs its potential prejudice and thereby justifies its reception" (para. 55). General propensity or disposition is never admissible. However, if such evidence shows a propensity to commit a particular crime in a distinctive manner, it may be admissible. As such, the evidence must relate to a specific propensity and relate to a specific issue in the trial. According to Binnie, "probative value exceeds prejudice, because the force of similar circumstances defies coincidence or other innocent explanation" (para. 47). Factors which might connect the similar fact evidence to the allegations against the accused include: "(1) proximity in time of the similar acts; (2) extent to which the other acts are similar in detail to the charged conduct; (3) number of occurrences of the similar acts; (4) circumstances surrounding or relating to the similar acts; (5) any distinctive feature(s) unifying the incidents; (6) intervening events; (7) any other factor which would tend to support or rebut the underlying unity of the similar acts" (para. 82; citations omitted). The court must also assess the prejudice, including the inflammatory nature, of the similar fact evidence, and "whether the Crown can prove its point with less prejudicial evidence" (para. 83). The court is required to take into account moral prejudice ("the potential stigma of 'bad personhood'") and reasoning prejudice ("including potential confusion and distraction of the jury from the actual charge against the respondent") (para. 100). See Hill *et al.* Chapter 7, Plaxton 2009b, and Stewart 2003b for a discussion of the impact of *Handy* on the law of similar fact evidence.

Judges must also be careful to examine other possible explanations. In *Handy*, there was the possibility of collusion between the witness providing similar fact evidence and the complainant, as there was a "whiff of profit." The similar fact evidence witness had told the complainant that she had received $16,500 from the Criminal Injuries Compensation Board, and that "all you had to do was say that you were abused." According to Binnie, "A few days later the complainant, armed with this information, meets the respondent and goes off with him to have sex in a motel room" (para. 111). In these circumstances, the trial judge had to address the possibility of collusion. Where there is an "air of reality" to allegations of collusion, the Crown must "satisfy the trial judge, on a balance of probabilities, that the evidence of similar facts is not tainted with collusion." If it is tainted, the Crown has to show that "the probative value of the proffered evidence outweigh its prejudicial effect" before it is admissible. If it is not tainted, it would be for the jury to weigh the evidence (para. 112).

One of the dangers of similar fact evidence is that it can result in moral prejudice and wrongful convictions. According to Binnie, "the forbidden chain of reasoning is to infer guilt from *general* disposition or propensity. The evidence, if believed, shows that an accused has discreditable tendencies. In the end, the verdict may be based on prejudice rather than proof, thereby undermining the presumption of innocence" in sections 7 and 11(d) of the *Charter* (para. 139). See Harris 2004 for a discussion of the issue of wrongful convictions in the context of similar fact evidence. The trend in other countries is to admit more similar fact evidence than they have in the past (Mahoney 2009, 22-23) .

SUMMARY

The law has traditionally recognized certain categories of privilege, involving the right (and a corresponding duty) not to divulge certain types of information. Privileged communications are protected as a matter of policy, even though they may be relevant to a fact in issue at trial. Solicitor–client privilege is the most firmly established privilege, providing that all confidential communications between clients and their lawyers (or their agents) in their professional capacity are protected. Neither party can be forced to divulge the contents of the communication, and in fact the lawyer or agent has a duty not to disclose unless the client waives privilege, either expressly or implicitly. Communications between spouses are also privileged during the course of their legal marriage, and neither party can generally be compelled to disclose statements they hear from their spouses.

Communications may be privileged in other situations, although the courts have moved away from a categorical, or "pigeon hole," to a case-by-case consideration. It is generally accepted that for a communication to be privileged, it must satisfy Wigmore's four criteria: the communications must have been made in confidence, confidentiality must be essential to the relationship between the parties to it, the relationship must be one that society wishes to foster, and the injury to the relationship of disclosure must outweigh the advantage of disclosure to the trial process.

The Crown has certain protections that are similar to privilege. Public interest immunity allows an objection to be made to the disclosure of certain information in court, on the ground of a specified public interest. These provisions ensure the proper functioning of government, and balance the public interest in non-disclosure against the accused's right to make full answer and defence. Police informer privilege, a subset of public interest immunity, provides that the authorities will not be compelled to disclose the identity of a confidential informant. If the informant is anonymous, no information about the informant can be disclosed except under the innocence-at-stake exception.

Although not strictly a matter of privilege, the courts are generally not willing to order the production of a complainant's personal records unless they are shown to be of likely relevance, and the court has then reviewed the documents to ensure that they are material to the defence.

Hearsay is an out-of-court statement made by someone other than the witness, tendered as evidence of the truth of the statement's content. The general rule is that hearsay is inadmissible. Some evidence that appears to be hearsay is nevertheless admissible because it does not, strictly speaking, come within the definition. Thus, an out-of-court statement made by someone else may be tendered for another purpose, such as to show state of mind, or to support a finding of reasonable and probable grounds to conduct a search. Aside from evidence that may appear to be hearsay but is not, there are also several true exceptions to the hearsay rule, such as admissions. Again, the creation of exceptions to the rule against hearsay is an area of the law in which the courts have moved from a categorical approach to a principled, case-by-case approach. The Supreme Court of Canada has decided that hearsay statements that are necessary and reliable can be introduced as evidence.

Character evidence is evidence that tends to show that the accused is the type of person who would (or would not) have committed the alleged offence. It is generally inadmissible, although the Crown can lead evidence of the accused's bad character, if the accused puts his or her character in issue by leading evidence of good character. Credibility evidence is evidence that addresses whether the accused (or another witness) should be believed, or how much weight should be given to their testimony. All witnesses put their credibility in issue by testifying. Some evidence, such as that of a criminal record, is capable of going to both character and to credibility. The admissibility of such evidence often depends on the purpose for which it is introduced, and on artificial philosophical considerations of the ability of the trier of fact to use the evidence only for the proper purpose.

Witnesses may not generally tender evidence of prior consistent or self-serving statements. An exception exists in which the witness is allowed to counter allegations of recent fabrication. Witnesses may, however, be cross-examined on prior inconsistent statements, as going to credibility. It is now possible in some circumstances to tender a prior inconsistent statement as evidence of its truth, in situations where there are sufficient *indicia* of reliability and necessity.

Similar fact evidence is an exception to the rule prohibiting character evidence, allowing in some circumstances evidence that the accused has committed similar acts on other occasions to those charged, from which facts in issue may be inferred.

QUESTIONS TO CONSIDER

(1) Are communications between a client and lawyer for the purpose of obtaining legal advice, done by way of video conference, covered by solicitor client privilege?

(2) What are two ways to protect solicitor-client privilege?

(3) Mr. X assaults Mrs. X, causing bodily harm. The Crown compels a reluctant Mrs. X to testify against Mr. X. Can Mrs. X refuse to repeat what Mr. X had said to her after the assault? Why or why not?

(4) What is the "pigeon hole" approach to developing the common law? What are the advantages and disadvantages of it?

(5) What are the advantages and disadvantages of developing the common law through a principled approach?

(6) Develop a fact-pattern question that would require that you apply Wigmore's four criteria for recognizing a privilege.

(7) What limits does the police informer privilege put on Crown disclosure?

(8) How might personal records of a complainant be admitted at the trial of an accused?

(9) What is hearsay?

(10) Under what circumstances can a witness testify, "my sister told me that the accused said he was driving while impaired"?

(11) What is the principled exception to the hearsay rule?

(12) Is the principled approach to hearsay relevant to the traditional exceptions to the hearsay rule?

(13) What is the difference between character and credibility?

(14) How were prior inconsistent statement used historically? Give an example of how they might be admissible for the truth of their contents today.

(15) Describe the circumstances under which similar fact evidence be used. Create a fact pattern question in which you can apply this framework, and then apply it.

BIBLIOGRAPHY

Boilard, Jean-Guy. *Guide to Criminal Evidence.* Cowansville, PQ: Les Editions Yvon Blais Inc., 1999 with updates.

Part B of the Introduction discusses solicitor–client privilege, police-informer privilege and case-by-case privilege. Chapter 2, Prior Statements, covers *res gestae,* confirmatory statements, *Milgaard* (guidelines under section 9 of the *Canada Evidence Act*), cross-examination by the opposite party (sections 10 and 11 of the *Canada Evidence Act*), and the use of evidence given at a preliminary hearing or previous trial. Chapter 3 is on hearsay and exceptions to the rule. Chapter 4 discusses cases that have dealt with similar fact evidence and the requirements of contemporaneousness, similarity in form, and link.

Boyle, Christine, Marilyn T. MacCrimmon, and Dianne Martin. *The Law of Evidence: Fact Finding, Fairness, and Advocacy.* Toronto: Emond Montgomery Publications Limited, 1999.

Chapter 7 discusses self-serving evidence, Chapter 8, hearsay, Chapter 10, privilege, and Chapter 12, character and similar fact evidence.

Brockman, Joan. "An Update on Self-Regulation in the Legal Profession (1989-2000): Funnel In and Funnel Out" (2004) 19(1) *Canadian Journal of Law and Society* 55.

Bryant, Alan W., Sidney N. Lederman and Michelle K. Fuerst. *Sopinka, Lederman & Bryant – The Law of Evidence in Canada* 3rd ed. Canada: LexisNexis Canada, 2009.

See Chapter 6 ("Hearsay"), Chapter 10 ("Character Evidence"), Chapter 11 ("Similar Fact Evidence"), Chapter 12 ("Self-Serving Evidence"), Chapter 14 ("Privilege"), and Chapter 15 ("Public Interest Immunity").

CHAPTER 12: *Exclusionary Rules*

Busby, Karen. "Third Party Records Since *R. v. O'Connor*." (2000) 27(3) *Manitoba Law Journal* 355.

Carter, Ian. "Chipping Away at *Stinchcombe*: The Expanding Privilege Exception to Disclosure." (2002) 50 *Criminal Reports* (5th) 332.

Cooper, T.G. *Crown Privilege*. Aurora, ON: Canada Law Book Inc., 1990.
    Chapter 1 discusses the rationale for Crown privilege, and includes the use of the term and its historical development. Chapters 2 and 3 discuss the relevance of common law, and procedural issues at common law. Chapters 4 through 6 discuss substantive and procedural issues in respect of sections 37 through 39 of the *Canada Evidence Act*, including observations and criticisms. Chapter 7 looks at the historical background, rationale, and present day law on the informer privilege.

Coughlan, Steve. "Complainant's Records After *Mills*: Same as It Ever was" (2000) 33 *Criminal Reports* (5th) 300.

Coughlan, Steve. "The Principled Exception and the Forgotten Criterion." (2007) 47 *Criminal Reports* (6th) 61.

Crisp, Glen. "*Khelawon*." (2008) 39 *Ottawa Law Review* 213.

Daisley, Brad. "Clear Evidence needed to invoke Wigmore rules; Student's Research Contacts are Privileged: Coroner." (9 December 1994) *Lawyer's Weekly* 28.

Delisle, Ronald Joseph, Don Stuart, and David M. Tanovich. *Evidence: Principles and Problems,* 8th ed. Toronto: Thomson Canada Limited, 2007.

Dufraimont, Lisa. "*R. c. Dinardo*: Troubling Issues Regarding Prior Consistent Statements." (2008) 57 *Criminal Reports* (6th) 76.

Feldthusen, Bruce. "Access to the Private Therapeutic Records of Sexual Assault Complainants." (1996) 75 *Canadian Bar Review* 537.

Gotell, Lise. "When Privacy is not Enough: Sexual Assault Complainants, Sexual History Evidence and the Disclosure of Personal Records." (2006) 43 *Alberta Law Review* 743.

Gorman, Wayne. "Hearsay in Sexual Offence Prosecutions." (1997) 39 *Criminal Law Quarterly* 493.

Harris, Nikos. "Limiting Instructions: Preventing Wrongful Convictions or Causing Juror Confusion?" (2004) 20 *Criminal Reports* (6th) 117.

Healy, Patrick. "Credibility and the Presumption of Innocence." (2007) 11 *Canadian Criminal Law Review* 217.

Hill, S. Casey, David M. Tanovich and Louis P. Strezos. *McWilliams' Canadian Criminal Evidence*. Aurora, ON: Canada Law Book Limited (available online on Criminal Spectrum).
    See Chapter 7 (The Hearsay Rule); Chapter 9 (Character Evidence); Chapter 10 (Similar Fact Evidence and other Discreditable Conduct); Chapter 11 (Prior Consistent Statements); Chapter 13 (The Law of Privilege); Chapter 14 (Public Interest Immunity);

Hubbard, Robert W., Peter J. DeFreitas, and Peter M. Brauti. "Informer and Police Investigatory Privilege at the Preliminary Inquiry." (1999) 41 *Criminal Law Quarterly* 68.

Hubbard, Robert, Susan Magotiaux and Suzanne Duncan. *The Law of Privilege in Canada*. Aurora, Ont.: Canada Law Book, 2006 (looseleaf).
    The book contains chapters on informer privilege, public interest immunity, national interest immunity, parliamentary privilege, spousal privilege, religious communication privilege, doctor-patient confidentiality, the privilege against self-incrimination, solicitor-client privilege, and litigation privilege.

Ives, Dale E. "*R v. Couture*–The Demise of Spousal Hearsay?" (2007) 47 *Criminal Reports* (6th) 70.

Kelly, Katharine D. "'You Must Be Crazy If You Think You Were Raped': Reflections on the Use of Complainants' Personal and Therapy Records in Sexual Assault Trials." (1997) 9 *Canadian Journal of Women and the Law* 178.

Layton, David. "U.(F.J.): Hearsay, Reliability and Prior Inconsistent Statements Made by Co-Accused, Part I." (1999) 41 *Criminal Law Quarterly* 345; and Part II. (1999) 41 *Criminal Law Quarterly* 501.

Layton, David. "*R. v. Brown*: Protecting Legal-Professional Privilege." (2002) 50 *Criminal Reports* (5th) 37.

Lawler, Lilia E. "Police Informer Privilege: A Study for the Law Reform Commission of Canada." (1985–6) 28 *Criminal Law Quarterly* 92.

MacNair, Deborah. "Solicitor-Client Privilege and the Crown: When is a Privilege a Privilege?" (2003) 82 *Canadian Bar Review* 213.

MacCrimmon, Marilyn T. "Developments in the Law of Evidence: The 1990–1 Term: Social Science, Law Reform and Equality" (1992) 3 *Supreme Court Law Review* (2d) 268.

Mahoney, Richard. "Acquittals as Similar Fact Evidence: Another View." (2003) 47 *Criminal Law Quarterly* 265.

Mahoney, Richard. "Similar Fact Evidence." (2009) 55 *Criminal Law Quarterly* 22.

McCrea, Patricia McNeill. "Judicial Law Making: The Development of the Principled Exception to the Hearsay Rule—Implications for Preliminary Hearing Recantations."(1998) 61(1) *Sask. Law Rev.* 199.

Murphy, Ronalda. "*S. (J.H.)*: A New and Improved *W. (D.)*." (2008) 57 *Criminal Reports* (6th) 89.

Nettler, Gwynne. *Explanations*. New York: McGraw-Hill, 1970.

Palys, Ted and John Lowman. "Ethical and Legal Strategies for Protecting Confidential Research Information." (2000) 15(1) *Canadian Journal of Law and Society* 39.

Plaxton, Michael. "Credibility, Belief and W(D): Direction: some thoughts in light of *Y(CL)*." (2008) 53 Criminal Reports (6th) 219.

Plaxton, Michael. "The Shaky Foundations of Corbett." (2009a) 13 *Canadian Criminal Law Review* 91.

Plaxton, Michael. "Limiting Instructions and Similar Facts." (2009b) 63 *Criminal Reports* 63 (6th) 12.

Porter, Stephen and Leanne ten Brinke. "Dangerous decisions: A theoretical framework for understanding how judges assess credibility in the courtroom." (2009) 14 *Legal and Criminological Psychology* 119.

Prithipaul, Ravi. "Observations on the Current Status of the Hearsay Rule." (1997) 39 *Criminal Law Quarterly* 84.

Rose, Vernon Gordon. *Social Cognition and Section 12 of the Canada Evidence Act : Can Jurors "Properly" use Criminal Record Evidence*. (2003). Dissertation, Department of Psychology, Simon Fraser University. Ottawa : National Library of Canada.

Rosenberg, Marc. *"B.(K.G.)*: Necessity and Reliability–The New Pigeon-Hole." (1993) 19 *Criminal Reports* (4th) 69.

Rosenthal, Peter. "Disclosure to the Defence after September 11: Sections 37 and 38 of the *Canada Evidence Act*." (2003) 47 *Criminal Law Quarterly* 186.

Sankoff, Peter. "*Corbett* Revisited: A Fairer Approach to the Admission of an Accused's Prior Criminal Record in Cross-Examination." (2006) 51 *Criminal Law Quarterly* 400.

Sopinka, John, Sidney N. Lederman, and Alan W. Bryant. *The Law of Evidence in Canada*, 2nd ed. Toronto and Vancouver: Butterworths, 1999.

      Chapter 6, "Hearsay", includes a discussion of the origins of the rule, exceptions to it, *res gestae*, and dying declarations. Chapter 10, "Character Evidence," discusses character of the accused as well of as other witnesses. Chapter 11, "Similar Fact Evidence," covers the rule, judicial developments, and English and Canadian case law. Chapter 14, "Privilege," discusses privilege between solicitor and client, husband and wife, spiritual advisors, journalists, doctor and patient, and privacy of complainants' records. Chapter 15, "Public Interest Immunity," talks about the basis and determination for Crown immunity as well as for the protection of informants' identity.

Stewart, Hamish. "Public Interest Immunity After Bill C-36." (2003a) 47 *Criminal Law Quarterly* 249.

Stewart, Hamish. "Rationalizing Similar Facts: A Comment on *R. v. Handy*." (2003b) 8 *Canadian Criminal Law Review* 113.

Stuesser, Lee. "Similar Fact Evidence in Sexual Offence Cases." (1997) 39 *Criminal Law Quarterly* 160.

Stuesser, Lee. "Admitting Acquittals as Similar Fact Evidence." (2001) 45 *Criminal Law Quarterly* 488.

Sugunasiri, Shalin M. and Ronalda Murphy. "*R v. F. (W.J.)*: Hearsay Evidence and the Necessity of 'Necessity'." (2000) 43 *Criminal Law Quarterly* 181.

Todd Douglas. "Academic Wins Ruling on Assisted-Suicide Research." (2003 November 1) *Vancouver Sun* B3.

Uniform Law Conference of Canada. *Report of the Federal/Provincial Task Force on Uniform Rules of Evidence.* Toronto: Carswell, 1982.

CASES

A. (L.L.) v. B.(A.), [1995] 4 S.C.R. 536.

R. v. B. (C.R.) (1990), 55 C.C.C. (3d) 1 (S.C.C.). Similar fact evidence was admitted because its probative value outweighed its prejudicial effect.

R. v. B.(K.G.), [1993] 1 S.C.R. 740. A witness's prior inconsistent statements may be admissible for the truth of its contents on a principled basis–reliability of the evidence and its necessity. If the statement is made to a person in authority, the judge should also determine if the statement was voluntarily and its admission as substantive evidence would not bring the administration of justice into disrepute.

R. v. Basi, 2009 SCC 52.

R. v. Brown, [2002] 2 S.C.R. 185. The Court held that the trial judge should have held a *voir dire* to determine if the test in *McClure* was met (innocence-at-stake exception to solicitor-client privilege). If the evidence is disclosed, the person whose privilege is breached is protected by use immunity and derivative use immunity, but not transaction immunity.

R. v. Campbell, [1999] 1 S.C.R. 565. Solicitor–client privilege exists where police officers seek legal advice from Crown counsel.

Carey v. Ontario (1986), 30 C.C.C. (3d) 498 (S.C.C.). Objection to the disclosure of cabinet documents is governed by the common law rules relating to public interest. Criteria to be considered by the courts in considering whether to order production.

R. v. Collins, [1983] B.C.J. No. 2307.

R. v. Corbett (1988), 41 C.C.C. (3d) 385 (S.C.C.). Although the court has a discretion to exclude such evidence in the proper case, an accused who testifies may normally be questioned on his or her criminal record.

R. v. Couture, [2007] 2 S.C.R. 517.

R. v. Dinardo, [2008] 1 S.C.R. 788.

R. v. Ellard, 2009 SCC 27.

R. v. F.(W.J.), [1999] 3 S.C.R. 569. Statements to parents and relatives were necessary and reliable in circumstances where the child-victim was unable to respond to questions or adopt a video statement she had made earlier.

R. v. G. (S.G.), [1997] 2 S.C.R. 716. The use of character evidence to assess the credibility of the accused.

R. v. Griffin, 2009 SCC 28.

Gruenke v. The Queen, [1991] 3 S.C.R. 263. There is no blanket privilege in respect of priest–penitent communications. Whether a communication is privileged in a particular instance is to be decided on a case-by-case basis, considering the four "Wigmore criteria."

R. v. Handy, [2002] 2 S.C.R. 908.

R. v. Hanna (1993), 80 C.C.C. (3d) 289 (B.C.C.A.). Where a child witness on a second trial is unable to recall certain events, that witness's evidence from an earlier trial may be adduced as an exception to the hearsay rule, being both necessary and sufficiently reliable.

R. v. Hawkins, [1996] 3 S.C.R. 1043.

R. v. H.S.B., [2008] 3 S.C.R. 32.

R. v. Khan, [1990] 2 S.C.R. 531. As an exception to the hearsay rule, out-of-court statements of young children may be admitted in evidence where they meet both the requirements of necessity and reliability.

R. v. Khelawon, [2006] 2 S.C.R. 787.

R. v. Leipert, [1997] 1 S.C.R. 281. The innocence-at-stake exception is the only exception to the informer privilege.

Lloyd and Lloyd v. The Queen (1981), 64 C.C.C. (2d) 169 (S.C.C.). Spousal communications intercepted pursuant to a court authorized wiretap are still privileged.

Makin v. Attorney-General for New South Wales, [1894] A.C. 57.

R. v. McClure, [2001] 1 S.C.R. 445. The Court provides the test for setting aside solicitor-client privilege on the basis of innocence at stake.

R. v. McNeil, [2009] 1 S.C.R. 66.

R. v. Milgaard (1971), 2 C.C.C. (2nd) 206; leave to appeal to the Supreme Court of Canada refused [1971] S.C.R. x.

R. v. Mills, [1999] 3 S.C.R. 668. The Court found that sections 278.1 to 278.9 (disclosure of personal records) do not infringe on sections 7 and 11(d) of the *Charter*.

R. v. O'Connor, [1995] 4 S.C.R. 411.

R. v. Olscamp (1994), 91 C.C.C. (3d) 180 (Ont. Ct. Gen. Div.)

R. v. Perron (1990), 54 C.C.C. (3d) 108 (Que. C.A.).

R. v. R. (D.), [1996] 2 S.C.R. 291. The Court held that out-of-court statements made by a five-year old were inadmissible because they were not reliable.

R. v. R.E.M., [2008] 3 S.C.R. 3.

R. v. Raynier (1992), 79 C.C.C. (3d) 176 (B.C.C.A.). A lawyer who refuses to comply with an *Income Tax Act* demand on the basis of privilege, but who fails to invoke the statutory procedure to protect privileged documents, may be convicted of failing to comply with a demand.

R.. v. Read (1993), 86 C.C.C. (3d) 574 (B.C.C.A.). Where an accused on appeal makes allegations by way of affidavit against his or her former counsel, the accused may be taken to have waived solicitor–client privilege, and the former counsel may accordingly respond to those allegations.

R. v. Risby (1976), 32 C.C.C. (2d) 242 (B.C.C.A.); affirmed S.C.C. 39 C.C.C. (2d) 567n.

Canada (Attorney-General) v. Sander (1994), 90 C.C.C.(3d) 41 (B.C.C.A.). Re: objection to disclosure of information on the grounds of "public interest" under the *Canada Evidence Act*.

R. v. Smith, [1992] 2 S.C.R. 455.

Smith v. Jones, [1999] 1 S.C.R. 455.

R. v. St. Jean (1976), 32 C.C.C. (2d) 438 (Que. C.A.). Where a spouse is a compellable witness at the instance of the Crown, spousal privilege does not apply.

R. v. Stirling, [2008] 1 S.C.R. 272.

R. v. Underwood, [1998] 1 S.C.R. 77.

R. v. W.(D.), [1991] 1 S.C.R. 742.

R. v. Wildman, [1984] 2 S.C.R. 311. Section 4(2) of the *Canada Evidence Act* concerning the admission of evidence is procedural and therefore the spouse of the accused is competent and compellable at the new trial that was ordered even though the section was enacted after the first trial.

R. v. Zylstra (1995), 99 C.C.C. (3d) 477 (Ont. C.A.). The trial judge ruled that the accused's spouse could assert spousal privilege if she testified. Such an assertion should be done in front of the jury, and the jury should be given special instructions.

CHAPTER 13: *Judicial Notice, Secondary Sources, and Opinion Evidence*

CHAPTER OBJECTIVES

In studying this chapter, you should develop an understanding of the following topics and concepts:

- the nature of judicial notice
- the distinction between judicial notice of law and judicial notice of facts
- the distinction between adjudicative and legislative fact, in the context of judicial notice
- Monahan and Walker's suggestions respecting the use of social science research
- judicial recourse to secondary sources
- the distinction between lay opinion evidence and expert opinion evidence, and the prerequisites for each to be admissible
- the extent to which expert opinion evidence may be based upon hearsay or secondary sources
- the rule against oath helping

INTRODUCTION

Generally, all facts in a criminal trial must be established or proved by admissible evidence. **Judicial notice** is an exception to that rule, and allows the court to take notice of certain things without formal proof. For example, it would not normally be necessary in a criminal trial to prove that Ottawa is the capital of Canada, or that there are 12 months in a calendar year; the courts will take judicial notice of such things. **Secondary sources**, including dictionaries, the writings of learned authors, or other such reference material, are also used without formal proof. **Opinion evidence** involves (typically) the evidence of experts to assist the court in respect of matters beyond the normal scope of knowledge of the average person. Although judicial notice, secondary sources, and expert opinion evidence are three different ways the courts might receive facts, the same type of facts may be received in different ways. Madame Justice McLachlin in *Williams* (discussed in Chapter 5 under "The Impact of Systemic Racism on the Selection of Jurors") commented:

> In the case at bar, the accused called witnesses [expert testimony] and tendered studies [secondary sources] to establish widespread prejudice in the community against aboriginal people. It may not be necessary to duplicate this investment in time and resources at the stage of establishing racial prejudice in the community in all subsequent cases. The law of evidence recognizes two ways in which facts can be established in the trial process. The first is by evidence. The second is by judicial notice....Widespread racial prejudice, as a characteristic of the community, may...sometimes be the subject of judicial notice. Moreover, once a finding of fact of widespread racial prejudice in the community is made on evidence, as here, judges in subsequent cases may be able to take judicial notice of the fact. "The fact that a certain fact or matter has been noted by a judge of the same court in a previous matter has precedential value and it is, therefore, useful for counsel and the court to examine the case law when attempting to determine whether any particular fact can be noted": see Sopinka, Lederman and Bryant, *supra*, at p. 977. It is also possible that events and documents of indisputable accuracy may permit judicial notice to be taken of widespread racism in the community (para. 54).

The use of judicial notice, expert opinion evidence, and secondary sources as means of introducing fact-related and law-related information as evidence in court become even more crucial in *Charter* litigation, where the courts are increasingly making decisions that are more obviously political choices, in correcting historical biases in the judicial interpretation and development of the common law (see discussion in Chapter 11 on child witnesses and shielding rape victims). Lawyers may also introduce social science research through the **facta** they file with the court. A factum may include reference to secondary sources, original research, or both.

The introduction of judicial education to inform judges of the results of social science research raises some interesting questions for judicial notice and adjudication. Paciocco suggests that this "seems to run headlong into the basic premise of the criminal accusatory system of justice in which the facts relied upon by the judge should be established by admissible evidence" (1997, 36). The content of such educational workshops is often the same as the evidence one would expect from an expert witness at trial. However, with an expert witness, evidence is presented and tested by cross-examination. According to Paciocco, the solution to this dilemma is to incorporate this learning through the use of "common sense," which has the advantage of "liberating the fact-finding process from the bloat of 'social science,' making expensive, time-consuming and prejudicial expert testimony needless in many cases" (43). What is somewhat disturbing about Paciocco's approach is that he suggests judges who rely on common sense are less likely to be overturned than those who refer to the social science literature to support their decision (55-6). He does, however, recommend that judges rely on expert witnesses when requiring knowledge that is clearly beyond common sense (63-4). An obvious problem with this approach is that sometimes social science demonstrates that common-sense assumptions are untrue. In addition, expert witnesses may be inaccurate (see Goudge Inquiry, Appendix A).

Tacit Assumptions—Bias in Adjudication and Interpretation of the Law

What is often ignored about judicial notice is that sometimes it "takes hold tacitly rather than explicitly" (Uniform Law Conference 1982, 45). The Uniform Law Conference Report refers to the benign assumption that "everyone in court will assume that rain falls." This example is deceptive in two respects. First, it assumes that the opinions of everyone in court actually "count" (see discussion in Introduction). Second, it assumes that judicial notice is taken only of the obvious. In the past (and some would say, in the present), the following "facts" often took (take) hold "tacitly rather than explicitly" in courts: children lie about sexual abuse, women lie about rape, men are entitled to abuse women, battered women "ask for it," and so on. These assumed facts make their way into decisions without ever being articulated. This process of judicial reasoning was recognized by Mr. Justice Holmes in 1881:

> the life of law has not been logic: it has been experience. The felt necessities of the time, the prevalent moral and political theories, intuitions of public policy, avowed or unconscious, even the prejudices which judges share with their fellow-men, have had a good deal more to do than the syllogism in determining the rules by which men should be governed (1).

The criminal justice system depends upon judges and jurors performing the perhaps impossible task of setting aside tacit assumptions that might operate to create an unfair trial for the accused. Other tacit assumptions assist the trier of fact (sometimes appropriately, sometimes not) in deciding the credibility of witnesses, and how much weight to give to their evidence.

More recently, the courts and academics have debated whether these tacit assumptions form part of judicial notice or judicial bias. In *S. (R.D.)*, Judge Sparks, in acquitting a black youth of assaulting a white police officer and resisting arrest, stated orally:

> The Crown says, well, why would the officer say that events occurred the way in which he has relayed them to the Court this morning. I am not saying that the Constable has misled the court, although police officers have been known to do that in the past. I am not saying that the officer overreacted, but certainly police officers do overreact, particularly when they are dealing with non-white groups. That to me indicates a state of mind right there that is questionable. I believe that probably the situation in this particular case is the case of a young police officer who overreacted. I do accept the evidence of [R.D.S.] that he was told to shut up or he would be under arrest. It seems to be in keeping with the prevalent attitude of the day (para. 4).

Was the judge 1) engaging in judicial notice, 2) relying on her fund of legal and social context knowledge, or 3) engaging in judicial bias, allowing her personal bias to enter her decision? Paciocco (1998, 324) suggests that since there was no evidence before the court that "1. Police officers have been known to mislead courts. 2. Police officers do (on occasion) overreact. 3. Overreaction is particularly likely to occur when police are dealing with non-white groups," she must have taken judicial notice of these facts. Paciocco believes that the "common sense" or "general knowledge" information that judges have is constantly being used to decide adjudicative facts, and falls within the realm of judicial notice. Most assumptions of fact, however, are not articulated in the manner articulated by Judge Sparks (Paciocco 1998, 327). The decision of Judge Sparks, which was upheld by the Supreme Court of Canada, has been the subject of numerous commentaries and much debate. Four judges in the Supreme Court of Canada decision recognized that judges constantly rely on their fund of knowledge to assess human behaviour, whereas five were more reserved (or, perhaps unrealistic), believing that judges should try to rise above personal knowledge (Paciocco 1998, 341). Paciocco concludes his commentary on judicial notice by suggesting that judges should be allowed to take judicial notice of that which is known by reasonable persons, in order to liberate judicial notice from "the tyranny of the untutored thinking that can be endemic in the 'notoriety' requirement" (344).

JUDICIAL NOTICE OF LAW

At common law, judges were required to take judicial notice of statutes. Sections 17 and 18 of the *Canada Evidence Act* codified this aspect of the common law. Section 781(2) of the *Criminal Code* requires judicial notice to be taken of "proclamations, orders, rules, regulations and by-laws," and their publication. Other sections dispense with proof of certain matters without referring to the concept of judicial notice. For example, section 33 of the *Canada Evidence Act* states that "no proof shall be required of the handwriting or official position of any person certifying...the truth of any copy of or extract from any proclamation, order, regulation, appointment, book or other document."

JUDICIAL NOTICE OF FACTS

Generally speaking, triers of fact are supposed to consider only the evidence introduced at the trial; they are not to consider facts that they may have gathered on their own, and (theoretically) they are not supposed to use their own expertise on the issues before the court. However, some facts are so well known or notorious that they do not need to be proved or illustrated at trial, and the trier of fact is allowed to take judicial notice of these facts. According to Madame Justice McLachlin in *Williams*, judicial notice applies to two kinds of facts: "(1) facts which are so notorious as not be the subject of dispute among reasonable persons [see Box 13.1]; and (2) facts that are capable of immediate and accurate demonstration by resorting to readily accessible sources of indisputable accuracy" (para. 54; also cited in *Find* para. 48). In *Krymowski*, where the accused was charged with wilfully promoting

---

Box 13.1 Toronto Who?

In what McWilliams (1999, chapter 24-8) describes as "a spirit of fairness," a court in British Columbia took judicial notice that Toronto is in Canada (*Cerniuk*).

---

hatred "against an identifiable group, to wit Roma," the Supreme Court of Canada found that the trial judge should have taken judicial notice through dictionary definitions, and relied on other evidence at trial, to conclude that the reference to "gypsy" referred to the ethnic group known as Roma (para. 24).

Rationale and Implications

The Report of the Uniform Law Conference of Canada suggested two rationales for the doctrine of judicial notice—"to expedite the hearing [and] to sustain the credibility of the judicial system"—that have different implications for the doctrine itself (1982, 42). Some jurists (for example, Thayer and Wigmore) saw judicial notice as a time-saver. The facts are presented as *prima facie* evidence, and the parties are allowed to lead evidence to rebut the presumed facts if they disagree with them. Others (for example, Morgan and Cross) saw judicial notice as a device to prevent a party from presenting evidence on something that was obvious, and sparing the court the embarrassment of reaching a conclusion that was contrary to common knowledge (42). With this latter rationale, that which is "demonstrably indisputable" is not open to challenge. The Report agreed with the latter approach, suggesting that "judicial notice should be given conclusive effect," subject only to appeal (43).

In *Zundel*, the Ontario Court of Appeal recognized this debate, and decided that "the generally accepted modern view...is that where the court takes judicial notice of a matter, the judicial notice is final" (150). McWilliams disagreed with this conclusion, and wrote, "I find that to require a fact which is judicially noticed to be irrebuttable is to sacrifice justice and a fair trial to the expediency of logic" (1999 chapter 24-4). To support his position, he cited a 1910 decision of the British Columbia Court of Appeal, *R. v. Schnell*:

> Before leaving the subject matter, it may be as well to add that taking judicial notice does not import that the matter is indisputable. It is a *prima facie* recognition of the fact or practice—the matter may still be open to refutation. Where the line between what may be noticed and what is not to be noticed is to be drawn, is not easily definable. There is no general principle. It must rest in the discretion of the trial judge, and if too loosely exercised, it may be corrected by the Court of Appeal (chapter 24-5).

Drummond (2000, 5-6) suggests that up until recently, the courts in Canada have favoured Morgan's "conclusive" approach (illustrated by the Ontario Court of Appeal in *Zundel*); however, Thayer's "*prima facie*" approach (favoured by McWilliams) seems to be re-appearing. Hill *et al.* (online) suggest that the Canadian courts favour Morgan's approach.

The Supreme Court of Canada recently waded into this debate in *Spence*. It stated that courts should start with the stricter Morgan criteria, and that "the closer the fact approaches the dispositive issue, the more the court ought to insist on compliance with the stricter Morgan criteria." If the Morgan criteria are satisfied, "the 'fact' will be judicially noticed, and that is the end of the matter" (Para. 61). "If the Morgan criteria are not satisfied, and the fact is "adjudicative" in nature [that is closer to the dispositive issue], the fact will not be judicially recognized, and that too is the end of the matter" (para. 62).

A question related to the rationale of judicial notice is whether judicial notice is mandatory or discretionary. The Uniform Law Conference took the view that judicial notice ought to be discretionary, except for judicial notice of law (1982, 44). The implications of this discretion can be seen in the case of *Zundel*, where the trial judge decided that to take judicial notice of the Holocaust would be "gravely prejudicial" to the accused, and would prevent him from making full answer and defence to the charges (149). The Court therefore required the Crown to prove that the Holocaust had occurred. The Ontario Court of Appeal stated that the trial judge was entitled to take judicial notice of the Holocaust, but had the discretion to require the Crown to prove it. This decision was criticized by Delisle as allowing Zundel to "communicate to the public that our judicial institutions are less than sure of the Holocaust's existence, when what is needed is recognition and affirmation from all quarters" (1987, 95). Sopinka *et al.* (1999, 1066) defended the decision, because if the court had taken judicial notice of the Holocaust, it might have influenced the jury's decision on whether the accused had knowledge that he was spreading false news—an essential element of the offence under section 181. It seems preferable then to allow trial judges to use their discretion in each case as to whether or not to take judicial notice. Paciocco (1997, 40) suggests that trial judges should not have the discretion to refuse to take judicial notice of indisputable facts, but that counsel should be allowed to attempt to show the court why it ought not to take judicial notice.

Adjudicative (or Social) Facts

Commentators and law reformers have suggested that it is important to distinguish between adjudicative (social) facts and legislative facts (social authority) in formulating rules surrounding judicial notice. Dealing with adjudicative or social facts (facts that might assist the trier of fact in deciding a specific issue before the court), the Law Reform Commission of Canada recommended that:

> 83(1) Judicial notice shall be taken of facts that are so generally known that they cannot be the subject of reasonable dispute.
> 83(2) Judicial notice may be taken of facts that are so generally known within the territorial jurisdiction of the trial court that they cannot be the subject of reasonable dispute and of facts capable of accurate and ready determination by resort to sources whose accuracy respecting such facts cannot reasonably be questioned (1973, 44–5).

It also recommended that judicial notice be mandatory if a party requested the judge to take judicial notice, and each party was given "reasonable opportunity to make representations regarding the matter and as to the propriety of taking judicial notice." If the judge consulted a source, or the "advice of persons learned in the matter," the information and source would form part of the trial record. Each party would be given "reasonable opportunity to make representations respecting the validity of that information" (section 85(3)).

The Uniform Law Conference, in their critique of the Law Reform Commission's recommendation, suggested that the distinction between when judicial notice is mandatory and when it is discretionary is too complex, and that it ought to be discretionary in all of the instances outlined by the Law Reform Commission above (1982, 44). It recommended that judicial notice could be taken of an adjudicative fact if it was "notorious, a matter of common knowledge," and "indisputable"—widely accepted by reasonable persons as true. In the alternative, "it may be a fact which can be verified by reference to readily accessible authoritative sources, including expert testimony" (45). Such "research" was justified because judicial notice was not dependent on actual knowledge of the judge or jury, but on knowledge that was imputed by the law (45).

Although none of these suggested reforms have been introduced as legislation, the courts do take judicial notice of commonly known facts, such as the fact that wine is intoxicating, that Toronto is in Canada, that a large screwdriver can be used as a housebreaking instrument, and so on. Such notorious facts will not cause the courts any problems with credibility.

Social Authority (Legislative Facts)

The Law Reform Commission of Canada (1973, 4) recommended that judges be allowed to "take judicial notice of scientific, economic and social facts in determining the law or in determining the constitutional validity of a statute." Their proposed code in 1977 was modified to state that judicial notice "may be taken of any fact in determining the law or the constitutional validity of a statute" (section 83(3)).

The nature of facts used to decide questions of law are far less likely to be indisputable or verifiable, because of their link to policy decisions and choices that depend on competing values. Professor Davis, who first introduced the distinction between legislative and adjudicative facts in 1942, outlined the following scales upon which legislative facts might fall:

(1) from narrow and specific facts to broad and general facts,
(2) from central and critical facts to background or peripheral facts,
(3) from readily accepted assumptions or facts to controversial assumptions of facts,
(4) from factual propositions that are almost entirely factual to somewhat factual propositions that are mixed with judgement, policy or political preference,
(5) from provable facts to facts that can be neither proved nor disproved and therefore must be found through legislating the facts, presuming them, imposing the burden on one party, or making an informed or uniformed guess, and
(6) from facts about immediate parties or facts that are known only or mainly by them to facts having no relation to the immediate parties (1980, 932).

In practice, judges often draw conclusions of law without even articulating, much less requiring proof of, their underlying assumptions. The Uniform Law Conference Report was unanimous in suggesting that there be no legislation governing legislative fact, "because it is judicial reasoning rather than evidence" (57). The Report referred to legislative fact as something that "relates not so much to facts that have to be proven as to the **mental context** in which the judges view the case. . . .Legislative facts. . .constitute the **fund of knowledge** the judge has and upon which he [*sic*] must draw to determine the case" (46–7). The Report did not acknowledge that the so-called "fund of knowledge" might contain contradictory assumptions or false "facts." In reaching its conclusion, the Conference Report adopted Professor Thayer's opinion that judicial notice was part of judicial reasoning, not part of the law of evidence. Thayer wrote, "In conducting a process of judicial reasoning, as of other reasoning, not a step can be taken without assuming something which has not been proved"(quoted in *Spence* para. 50). Some jurists are of the view that it is needlessly confusing to incorporate this type of reasoning into the concept of judicial notice.

With regard to finding legislative facts, the Uniform Law Conference Report recommended that judges be allowed to educate themselves about legislative facts, "by all available means" (46). Some would argue that judges should be required, however, to inform the parties when they take judicial notice of legislative facts or social authority. Whether or not judges give notice to the parties, the assumptions underlying the law are constantly under attack today by the parties, or by interveners who try to provide a social, economic, and political context for the law that differs from the assumptions the courts have made about law and human nature. The common law and legislative changes to the *Criminal Code* revolving around sexual assault and child witnesses (see Chapter 11) are clear examples of how previous assumptions underlying the law have been challenged.

Social Framework

Social framework evidence is social science research that provides the social or psychological context for what people might do in certain circumstances. The Law Reform Commission of Canada and the Uniform Law Report dealt with the two traditional categories of adjudicative (or social) facts and legislative facts (social authority). Monahan and Walker's social framework appears to fall within adjudicative facts, in which the Commission included knowledge of "typical modes of behaviour, and the causal relations between commonplace events" (1973, 8–9). However, social framework evidence is now recognized in Canadian courts as distinct from social authority and social facts. Up until 2005, the Supreme Court of Canada in its decisions had referred to social framework evidence by name in only non-criminal matters. In 2005, Mr. Justice Binnie for the Court in *Spence*, although citing the literature on social framework evidence, incorrectly referred to it as social facts (paras. 56-58).

Can a Court Rely on Secondary Sources to Assist in Taking Judicial Notice?

It is quite clear that judges can rely on secondary sources (books, maps, dictionaries, and so on) to inform themselves before taking judicial notice. This may also be done when taking judicial notice of historical facts, through the use of history texts and archival material. Such sources must, however, be indisputable and capable of immediate demonstration; if there is contradictory or complex information in these sources, judicial notice cannot be taken. It is also possible for the courts to hear testimony before judicial notice is taken. In *Zundel,* the Ontario Court of Appeal quoted Professor Morgan:

> There is no artificial limit upon the sources of information which [the party] may furnish the judge, and none upon those which the judge may consult on his own motion. The opponent likewise is not restricted by rules of evidence in offering, or inducing the judge to consult, reliable repositories of relevant data. If the judge believes it doubtful whether the matter falls within the domain of judicial notice, or if the sources available are inadequate, he leaves the subject within the domain of evidence, and all the ordinary rules applicable to the process of resolving an ordinary issue of fact are enforced (151).

The Uniform Law Conference Report stated that judges should not be restricted in respect of legislative facts (particularly constitutional facts) to only to taking judicial notice, but should be allowed "to educate and inform [themselves] as to legislative fact by all available means" (46). This is a reference to secondary sources.

SECONDARY SOURCES

Secondary sources, such as books, journals, and other documentary material, can be used by judges for further information. This can be a controversial process, as the parties will sometimes not know if a judge has considered secondary sources, and if so, which ones. If the secondary sources are uncontroversial, a reference to them might be subsumed under judicial notice—for example, a judge might refer to a dictionary or a map for information. However, once a judge enters the realm of social science in textbooks and articles, the rules become less clear. Appellate courts are often referred to articles and textbooks in the *facta* filed by the parties. These references to social science data or information are often called "*Brandeis briefs*" (after a judge in the United States), and enter the adversarial system without the "benefit" of the adversarial process. For example, Trotter describes the social science research on false confessions used in *Oickle*: "Academic articles and book excerpts were simply filed with the Court by one of the interveners (the Criminal Lawyers Association of Ontario), without complaint, objection or comment from any of the parties, or from the Court itself" (2004 para. 17).

THE USE OF *FACTA* TO INTRODUCE SOCIAL SCIENCE RESEARCH

Monahan and Walker (1988) recommend that all social science research, whether it be used as social authority, social fact, or social framework, be introduced through *facta* or briefs. They, however, suggest different ways of evaluating such research.

Monahan and Walker (1988) suggest that social authority be obtained, evaluated, and treated as if it were law. They suggest that social authority be presented to the courts in written briefs, rather than through witnesses, and that judges be allowed to research the issue themselves, just as they would the law. They wrote, "when parties argue that a prior legal decision should be followed as precedent for the present case, they do not call as a witness the judge who wrote the prior decision. Likewise the authors of social science research should not ordinarily be called to court to describe their work" (467). Monahan and Walker are convinced that written briefs are a better means of communicating the information to the judge than expert witnesses. In addition, it is cheaper, and can be easily updated without recalling the expert (467–8).

Monahan and Walker suggest that social science research be evaluated the same way cases are:

(a) Cases decided by courts higher in the judicial structure have more weight than lower court decisions; (b) well-reasoned cases have more weight than poorly reasoned cases; c) cases involving facts closely analogous to those in the present case have more weight than cases involving different facts; and d) cases already approved by other courts have more weight than cases that have not met with such approval (468).

In translating this to the evaluation of social science, they state that confidence should be placed on social science research to the extent that it

a) has survived the critical review of the scientific community, b) has used valid research methods, c) is generalizable to the legal question at issue, and c) is supported by a body of other research (468).

In terms of how one court should treat social authority used by another court, again the authors use the legal analogy. Trial court evaluation of legislative facts is clearly reviewable by appellate courts, as are trial court decisions on law (486). Lower courts should treat higher courts' use of social authority as they would law; that is, they should follow them, but also recognize that changes to law and to empirical conclusions supporting the law can always occur with new circumstances and new empirical evidence.

Monahan and Walker (1988) suggested that social framework evidence be presented in briefs or through the judge's own investigations, that the judge evaluate the methodology and instruct the jury on this evaluation, and that the jury be allowed to apply social framework evidence to the issue in the trial.

When evaluating social fact evidence, Monahan and Walker (1988) recommend that the methodology of social science (defined as the research design and statistical analysis) be treated in the same manner as law or social authority, but that the application of the methodology to concrete situations be treated as fact (469). Methodology could be presented in briefs, and judges would be free to read about it. The judge would then evaluate the methodology and instruct the jury on his or her conclusions about the methodology (470). The jury or trier of fact would then examine the application of the methodology. Monahan and Walker provide the example of a survey of attitudes, where the judge would decide whether a random community survey would provide evidence of community standards respecting, for example, obscenity. Then the researcher would testify how this methodology was applied in the study, and the results would be weighed by the trier of fact.

OPINION EVIDENCE

Historically, witnesses were allowed to give opinions, provided they could show the basis of their opinion. The courts gradually came to prohibit the opinion of witnesses in areas where the jury was equally capable of forming an opinion from the facts. Eventually, witnesses were entirely limited to stating the facts, from which triers of fact would infer their own conclusions (McWilliams 1999 chapter 9–2; Hill *et al.* 2004 section 12:20). This, of course, is easier said than done, as every statement of fact contains an assumption or perspective. If the witness says, "Yes, it was Sue who ran from the bank," the witness is, in a sense, giving an opinion based on a number of facts gathered through observations of the person who ran from the bank. The facts on which the witness relies for

---

Box 13.2: "Just the Facts."

Professor Delisle provides the following example of the opinion rule taken to the extreme from a text on opinion evidence from Illinois (1942).

Question:	What happened then?
Answer.	The lady in the car that got hit stumbled out of her car and fell in a faint.
Defence Counsel:	Move to strike the opinions of the witness. Let him state the facts.
The Court:	Strike them out. The jury will disregard that answer. [To witness:] You must state the facts and not your conclusions regarding them. You can't give the jury your opinion as to *which* car got hit, *whose* car it was, *how* the lady got out of the car or *why* she fell, if she did fall,—you must state the facts (Delisle 1999, 612–3).

Note that in Canada, lawyers do not "move to strike" testimony, but rather they object to the admission of testimony, and the court rules on its admissibility.

---

the opinion as to the identification can be challenged, but the witness would not be prohibited from giving an "opinion" on the identification of the person who ran from the bank.

In *Graat*, the Supreme Court of Canada dealt with the issue of whether a non-expert witness could express an opinion about whether an accused was so intoxicated that his ability to drive was affected. Generally, witnesses cannot give opinions on the ultimate issue that the trier of fact is suppose to decide. A witness could not give an opinion, for instance, that an accused was guilty. However the Court in *Graat* decided that the witness (a police officer) could give evidence that the accused was too drunk to drive a car properly. In fact, the Court stated, anyone could give that evidence. The police officer was not giving evidence as an expert. It is still the responsibility of the trier of fact to weigh such evidence and to decide the issue before the court. The Court quoted Howland, C.J. in the Court of Appeal as saying that "impairment" is a:

> compendious way of describing a condition based on observed facts. It does not require the evidence of a doctor or other expert, nor should it be limited to persons who themselves drive cars. It is a subject about which most people should be able to express an opinion from their ordinary day-to-day experience (370).

Other facts on which a witness might give so-called opinion evidence are such things as apparent age, height, the emotional state of someone, estimates of speed and distance, and so on. However, a child might have difficulty estimating the age or height of an adult, and a person who does not drive or spend much time in a car might have difficulty estimating the speed of a vehicle. This so-called opinion evidence is often followed by cross-examination on its accuracy.

Expert Opinion Evidence

A major exception to the rule against opinion evidence relates to experts. The opinion of experts with special skill or knowledge is admissible if it is necessary to assist the trier of fact on an issue that the trier of fact must decide. Such witnesses must be qualified to give expert opinion evidence, in the sense that they posses special skill or knowledge on the subject matter in issue. These qualifications are not limited to academic qualifications, but may simply be the result of experience. For example, an undercover police officer may develop street knowledge about the packaging and sale of a narcotic, and then may be called to give expert evidence on this aspect of the drug trade.

Unless both sides agree that a witness is qualified to give evidence on a particular issue, a *voir dire* will be held to decide whether the witness may give an opinion in the area outlined. The expert must have greater specific knowledge on a matter than the trier of fact, and she or he is called to give evidence that is necessary to assist the trier of fact in reaching a decision. In other words, experts are limited to circumstances in which ordinary people are unlikely to form a correct opinion without such assistance.

In *Lavallee*, the Supreme Court of Canada found that expert evidence on the psychological effect of battering on wives and common law partners was relevant and necessary, because of the myths and misconceptions surrounding the circumstances of such women, and because the average person does not understand the state of mind of battered women (413). Wilson, J. summarized the principles applicable in such cases:

> 1. Expert testimony is admissible to assist the fact-finder in drawing inferences in areas where the expert has relevant knowledge or experience beyond that of the lay person.
> 2. It is difficult for the lay person to comprehend the battered-wife syndrome. It is commonly thought that battered women are not really beaten as badly as they claim; otherwise they would have left the relationship. Alternatively, some believe that women enjoy being beaten, they have a masochistic strain in them. Each of these stereotypes may adversely affect consideration of a battered woman's

claim to have acted in self-defence in killing her mate.

3. Expert evidence can assist in dispelling these myths.

4. Expert testimony relating to the ability of an accused to perceive danger from her mate may go to the issue of whether she "reasonably apprehended" death or grievous bodily harm on a particular occasion.

5. Expert testimony pertaining to why an accused remained in the battering relationship may be relevant in assessing the nature and extent of the alleged abuse.

6. By providing an explanation as to why an accused did not flee when she perceived her life to be in danger, expert testimony may also assist the jury in assessing the reasonableness of her belief that killing her batterer was the only way to save her own life (125).

The *Lavallee* case sheds light on the interrelationship between expert witnesses, judicial notice, and secondary sources. There was very little expert evidence on the battered women syndrome at trial. When Lavallee appealed to the Supreme Court of Canada, neither the factum filed on behalf of Lavallee nor that of the Crown contained any references to the literature on battered women or the so-called Battered Woman Syndrome. This lack of expert evidence was in sharp contrast to Madame Justice Wilson's decision, which included a thorough review of the literature (secondary sources) about battered women and their responses to their batterers. Although criticized by some, Wilson's conforms to the suggestions of Monahan and Walker, that such evidence be treated like legal precedent, which would allow judges to do their own research on the issue. In *Malott*, L'Heureux-Dubé, J. took the opportunity to expand on the importance of *Lavallee* when trial judges inquire into a battered woman's state of mind:

> To fully accord with the spirit of *Lavallee*, where the reasonableness of a battered woman's belief is at issue in a criminal case, a judge and jury should be made to appreciate that a battered woman's experiences are both individualized, based on her own history and relationships, as well as shared with other women, within the context of a society and a legal system which has historically undervalued women's experiences. A judge and jury should be told that a battered woman's experiences are generally outside the common understanding of the average judge and juror, and that they should seek to understand the evidence being presented to them in order to overcome the myths and stereotypes which we all share. Finally, all of this should be presented in such a way as to focus on the reasonableness of the woman's actions, without relying on old or new stereotypes about battered women (para. 43).

As more and more social framework information comes into the public domain, it is possible for triers of fact to take judicial notice of it, or to use their own experience and common sense. Skurka and Renzella (1998, 270) ask, "Can it be seriously argued that a jury with a modicum of common sense requires a behavioural expert to assist it in understanding that a child's delayed disclosure of abuse may be affected by its fear of the perpetrator?" Madame Justice McLachlin, in *François*, found for the majority that a jury's verdict of guilty on charges of rape was not unreasonable, given the thorough cross-examination by defence counsel. With regard to the complainant's memory "flashback" of the events, McLachlin was prepared to allow the jury to use its common sense:

> Without pronouncing on the controversy that may surround the subject of revived memory amongst experts, it is sufficient to say for purposes of this appeal that the jury's acceptance of the complainant's evidence on what happened to her was not, on the basis of the record, unreasonable. The complainant was cross-examined on the possibility of concoction. She denied the suggestion of cross-examining counsel that her recovered "memory" was a product of the pressures she was experiencing. Thus explored, the matter was left to the good judgment of the jury. It was open to the jury, with the knowledge of human nature that it is presumed to possess, to determine on the basis of common sense and experience whether they believed the complainant's story of repressed and

recovered memory, and whether the recollection she experienced in 1990 was the truth. To do so cannot be characterized as unreasonable (840).

Mr. Justice Sopinka clarified the law relating to expert opinion evidence in *Mohan*. Mohan, a medical doctor, was charged with four counts of sexual assault against four female patients aged 13 to 16. He wanted to call a psychiatrist to testify that he (the accused) did not fit the profile of three personality groups in which most sex offenders were found. In ruling that the trial judge was correct in excluding this evidence, Sopinka, J. examined the admissibility of the expert opinion evidence on four criteria:

> (a) relevance;
> (b) necessity in assisting the trier of fact;
> (c) the absence of any exclusionary rule;
> (d) a properly qualified expert (411).

Relevance is a question of law to be decided by the judge. Sopinka, J. ruled that the evidence must be logically relevant (that is, related to the issue in question), and also legally relevant, in that the evidence must be worth its cost in terms of its impact on the trial. "Evidence that is otherwise logically relevant may be excluded on this basis, if its probative value is overborne by its prejudicial effect, if it involves an inordinate amount of time which is not commensurate with its value or if it is misleading in the sense that its effect on the trier of fact, particularly a jury, is out of proportion to its reliability" (411). Sopinka stressed the danger of expert evidence, that "Dressed up in scientific language which the jury does not easily understand and submitted through a witness of impressive antecedents, is apt to be accepted by the jury as being virtually infallible and as having more weight than it deserves" (411).

Expert evidence must also be necessary to assist the trier of fact. Sopinka J. stated that the test was not whether the evidence would be helpful—that is too low a standard—but whether the opinion is likely to provide information "which is likely to be outside the experience and knowledge of a judge or jury" (413). Therefore, the "subject matter must be such that ordinary people are unlikely to form a correct judgement about it, if unassisted by persons with special knowledge" (413).

Sopinka, J. added that expert evidence must also be screened, in terms of whether it is precluded by any of the exclusionary rules. For example, evidence that goes entirely to disposition (as in the *Morin* case) is inadmissible unless the accused has put his or her character in issue. Expert opinion of this type may, however, be admissible if "either the perpetrator of the crime or the accused has distinctive behavioural characteristics such that a comparison of one with the other will be of material assistance in determining innocence or guilt" (423).

Sopinka's fourth requirement for expert evidence is that there be a properly qualified expert, whether those qualifications are from study or experience. If the expert is advancing a novel scientific theory or technique, it must be "subjected to special scrutiny to determine whether it meets a basic threshold of reliability and whether it is essential in the sense that the trier of fact will be unable to come to a satisfactory conclusion without the assistance of the expert" (414). Copeland suggests that under these criteria, expert evidence on eyewitness testimony is reliable and necessary because of the dangers of wrongful conviction from eyewitness evidence. She describes the Ontario Court of Appeal's rejection of such evidence in *McIntosh* as "misguided in principle and contrary to the research evidence with respect to jurors' knowledge about the process of eyewitness identification" (2002, 198). The Goudge *Inquiry into Pediatric Forensic Pathology in Ontario* (see Appendix A; and Paciocco 2009a) provides devastating examples of the harm and wrongful convictions that can result from so-called expert testimony. Also see *Trotta*.

Box 13.3 The Chances of Being Wrong With DNA Evidence

Jurors in the 1999 trial of Larry Fisher, convicted of raping and killing Gail Miller in 1969 in Saskatoon, Saskatchewan, were told that the odds that the sperm cells taken from Ms. Miller's clothing belonged to someone other than Fisher were about one in 950 trillion (Roberts 1999). See Holmgren 2005a and 2005b in Chapter 5 for the effect of such numbers on jurors.

In *Terceira* (appeal dismissed by the Supreme Court of Canada), the Ontario Court of Appeal found that the trial judge was correct in admitting DNA evidence at trial. The trial judge need only assess the reliability of a scientific methodology—whether it "reflects a scientific theory or technique that has either gained acceptance in the scientific community, or if not accepted, is considered otherwise reliable in accordance with the methodology validating it" (27–8). Once the judge is convinced that the methodology is sufficiently reliable to be put to the jury (a threshold of reliability satisfactory to the judge), and meets the four criteria in *Mohan*, it is up to the jury to apply the particular science to the facts of the case before it. That is, the jury will decide its ultimate validity and reliability (16). With DNA evidence, it might also be advisable to instruct the jury "not to be too overwhelmed by the aura of science infallibility associated with scientific evidence," and that they should "use their common sense in their assessment of all of the evidence on the DNA issue and determine if it is reliable and valid as a piece of circumstantial evidence" (28).

The Supreme Court of Canada elaborated on the use of novel science in *J.L.J.* A man, accused of sexually assaulting two young boys (ages three and five) who were in his custody, wanted to call an expert to give the opinion that "in all probability a serious sexual deviant had inflicted anal intercourse on [the] children. . .and no such deviant personality traits were disclosed" in the expert's testing of the accused (para. 9). Part of the testing included a penile plethysmograph, which measured the degree of sexual arousal given various stimuli. It indicated that the accused had "no deviation in respect of boys in general or prepubescent boys" (para. 14). The trial judge excluded the evidence, and the accused was convicted. The Quebec Court of Appeal found that the evidence should have been admitted, and ordered a new trial. The Supreme Court of Canada restored the trial judge's conviction.

In concluding that the opinion evidence was inadmissible, Binnie, J. for the Court was concerned that trial judges be vigilant about keeping "junk science" out of the courtroom, and not allowing experts to usurp the role of the judge or jury. When it comes to novel science, Binnie agreed with the criteria used by the U.S. Supreme Court in *Daubert*:

(1) whether the theory or technique can be and has been tested . . .
(2) whether the theory or technique has been subjected to peer review and publication . . .
(3) the known or potential rate of error or the existence of standards; and,
(4) whether the theory or technique used has been generally accepted (para. 33; see Hill *et al.* 2004 section 12:30.20.30 for a discussion).

Although the penile plethysmograph had been used to assess therapeutic results, it had never been used "in a court of law to identify or exclude the accused as a potential perpetrator of an offence" (para. 35). The court must apply special scrutiny to the use of this evidence for this purpose, as it is very close to the ultimate issue in the case (para. 37). Binnie, J. concluded that the trial judge had properly applied the *Mohan* criteria in excluding

Box 13.4 Magical Experts

Somewhat surprisingly, expert witnesses are not viewed with universal awe. As part of a trend in the United States to limit the wide-ranging scope of psychological or psychiatric witnesses, an amendment was proposed to a piece of New Mexican legislation relating to psychologists:

> When a psychologist or psychiatrist testifies during a defendant's competency hearing, the psychologist or psychiatrist shall wear a cone-shaped hat that is not less than 2 feet tall. The surface of the hat shall be imprinted with stars and lightning bolts.
>
> Additionally, a psychologist or psychiatrist shall be required to don a white beard that is not less than 18 inches in length, and shall punctuate crucial elements of his testimony by stabbing the air with a wand. Whenever a psychologist or psychiatrist provides expert testimony regarding the defendant's competency, the bailiff shall contemporaneously dim the courtroom lights and administer two strikes to a Chinese gong.

The amendment apparently passed by a voice-vote in committee, and was to be considered as part of the proposed legislation during debate in the House *(The New Mexican,* March 6, 1995).

the evidence. Similarly, in *Trochym*, the majority of the Supreme Court of Canada decided that post-hypnosis evidence does not meet the requirements of *J.L.J.*" (paras. 24 and 53), even though the case was concerned with the pre-trial application of a scientific technique, not the use of an expert witness (para. 33).

Opinion Based on Hearsay

Every expert witness's opinion will have a component of hearsay, because experts rely on a body of knowledge that is not brought before the court to be proved. The issue with regard to hearsay is usually the extent to which the expert can rely on hearsay about a particular person (for example medical records composed by others, histories taken by others, statements by others about the person, and so on) without evidence being presented by these other witnesses.

Madame Justice Wilson discusses the hazards of admitting expert evidence that was based on hearsay, especially the danger that the jury will accept the hearsay as going to its truth. In *Lavallee*, she summarized the law regarding opinions based on hearsay, relying on four propositions from the *Abbey* case:

1. An expert opinion is admissible if relevant, even if it is based on second hand evidence.
2. This second hand evidence (hearsay) is admissible to show the information upon which the expert opinion is based, not as evidence going to the existence of the facts on which the opinion is based.
3. Where the psychiatric evidence is comprised of hearsay evidence, the problem is the weight to be attributed to the opinion.
4. Before any weight can be given to an expert's opinion, the facts upon which the opinion is based must be found to exist (127–8).

The Supreme Court of Canada's decision in *Abbey* seemed to imply that all of this so called foundation evidence had to be proved. Wilson's interpretation of *Abbey* in *Lavallee*, however, was that the solution to the hearsay problem was to charge (instruct) the jury appropriately, not to withdraw the evidence from them. She wrote:

> as long as there is some admissible evidence to establish the foundation for the expert's opinion, the trial judge cannot subsequently instruct the jury to completely ignore the testimony. The judge, must, of course, warn the jury that the more the expert relies on facts not proven in evidence the less weight the jury may attribute to the opinion" (130).

She added, "the trial judge is to caution the jury that the weight attributable to the expert testimony is directly related to the amount and quality of admissible evidence on which it relies" (131).

The Supreme Court of Canada in *Giesbrecht* approved of Wilson's summary of the law in *Lavallee*, and added, "hearsay evidence is admissible to explain the basis of the expert opinion, but the jury must be cautioned that such evidence is not admitted for the truth of the statements on which the opinion is based and the statements are not proof of the facts contained therein" (235). Giesbrecht's counsel elicited statements from the experts that the accused had told them that the "Town of Altona collectively was going to kill him" (235). These statements could be used by the experts to formulate their opinions, but not to prove the accused's state of mind (an issue for the jury to decide). The trial judge cannot tell the jury to ignore the expert's opinion if there is some admissible evidence to establish its foundation, but the judge must "warn the jury that the more the expert relies on facts not proved in evidence the less weight the jury may attribute to the opinion" (236).

Oath Helping

As a general rule, a witness is not allowed to comment on whether another witness is telling the truth. As Mr. Justice MacKinnon of the Ontario Court of Appeal observed, "Trial by psychiatrists is not an attractive prospect" (*French* para. 28). The rationale for this rule is found in the judgement of Madame Justice McLachlin in *Marquard*:

> Credibility is a matter within the competence of lay people. Ordinary people draw conclusions about whether someone is lying or telling the truth on a daily basis....The expert's opinion may be founded on factors which are not in the evidence upon which the judge and juror are duty-bound to render a true verdict. Finally, credibility is a notoriously difficult problem, and the expert's opinion may be all too readily accepted by a frustrated jury as a convenient basis upon which to resolve its difficulties (228).

The exception to this general rule is when there are aspects of the witness's evidence (for example a mental deficiency) that go beyond the competence of lay people. The Supreme Court of Canada has allowed experts to provide social framework evidence surrounding issues of child witnesses, such as recantation, denial, recall, and so on. Such evidence is provided as background material, not as opinion evidence on a particular witness. In *Marquard*, the Court approved of expert testimony on why children might give contradictory stories; however, it is still up to the trier of fact to assess the credibility of the witness in light of this expert testimony. For problems associated with this approach, see Norris and Edwardh (1996).

SUMMARY

Judicial notice involves the court accepting certain things without formal proof. Courts are required to take judicial notice of law. Certain formal matters are specified by statute not to require proof. The court may take notice of facts that are so well known or notorious that they are generally accepted. It is not settled whether a fact that is judicially noticed is merely presumptive or whether it is irrebutable. Courts may refer to secondary sources or hear evidence to inform themselves in preparation for taking judicial notice. It is more controversial as to the extent that a court may refer to secondary sources (social science or other academic research) for the purpose of ascertaining facts to which such research relates, especially as an alternative to hearing evidence on those facts.

Monahan and Walker suggest that social authority, or social science research that is used to develop the law, be treated as if it were law and be presented to the court through briefs, just as is legal authority. Social framework or social science research that provides social or psychological context for what people might do in certain circumstances should be presented either through briefs, or through the court's own investigations. Monahan and Walker further suggest that social or adjudicative fact should be treated as law or social authority, but the application of methodology be treated as fact and assessed by the trier of fact.

Opinion evidence is generally inadmissible. Two main exceptions exist: lay opinion evidence and expert opinion evidence. A lay witness may give opinions that really amount to a compendious recitation of facts. For example, a lay witness may give an opinion on a person's sobriety, as a shorthand way of reciting all the physical observations that might lead one to conclude that the person was drunk. Lay witnesses may similarly give opinions about things such as apparent age, emotional states, estimates of speed, and so on. Expert opinion evidence is admissible if it is relevant, necessary to assist the trier of fact, not prohibited by any other exclusionary rule, and if it is given by a properly qualified expert who has special skill or knowledge beyond that of the trier of fact. Expert opinion may be based on hearsay, although the opinion may only be given weight if the facts on which it was based are found to exist. Expert opinion may also be based on secondary sources.

Oath helping is a form of opinion evidence, going to a witness's opinion of the veracity of another witness. It is generally prohibited.

QUESTIONS TO CONSIDER

(1) What are the two rationales for having a judge take judicial notice of a fact?
(2) What are the implications for the two rationales for judicial notice?
(3) What did the Uniform Law Conference mean when it suggested that legislative facts are "more judicial reasoning rather than evidence"?
(4) What is often ignored about judicial notice is that sometimes it "takes hold tacitly rather than explicitly" (Uniform Law Conference 1982, 45). What did the ULC mean by this?
(5) How did Monahan and Walker suggest that social authority be evaluated?
(6) Give an example of how a non-expert witness might give evidence based on his or her opinion.
(7) What are the four criteria a judge will consider before allowing an expert to give opinion evidence?
(8) What are the dangers of expert evidence?
(9) What is a trial judge required to tell a jury about hearsay evidence relied on by an expert witness?
(10) Create a fact pattern question that a judge must decide whether to admit expert opinion evidence, and develop an answer to your question.

BIBLIOGRAPHY

Anderson. Glenn R. *Expert Evidence*. Toronto: LexisNexis Canada, 2005.
     Includes chapters on Setting the Context for Expertise (Chapter 2); The Evolving Law of Expert Evidence in Canada (Chapter 3); The American Revolution of Expert Evidence (Chapter 4); Assessing the Reliability of Expert Evidence (Chapter 5); Novelty, Bias and Other Evaluation Issues (Chapter 6); and issues surrounding reforms (Chapters 8 and 9).

Boilard, Jean-Guy. *Guide to Criminal Evidence*. Cowansville, PQ: Les Editions Yvon Blais Inc., 1999 with updates.
     Chapter 11, "Truth Experts," reviews cases on the prohibition against witnesses being asked to comment on whether another witness is telling the truth (oath helping).

Boyle, Christine, Marilyn T. MacCrimmon, and Dianne Martin. *The Law of Evidence: Fact Finding, Fairness, and Advocacy*. Toronto: Emond Montgomery Publications Limited, 1999.
     Chapter 1 discusses fact finding and judicial notice, and Chapter 9 covers opinion or expert evidence.

Bryant, Alan W., Marc Gold, H. Michael Stevenson, and David Northrup. "Public Attitudes Toward the Exclusion of Evidence: Section 24(2) of the *Canadian Charter of Rights and Freedoms*." (1990) 69 *Canadian Bar Review* 1.

Bryant, Alan W., Sidney N. Lederman and Michelle K. Fuerst. *Sopinka, Lederman & Bryant – The Law of Evidence in Canada* 3rd ed. Canada: LexisNexis Canada, 2009.
     See Chapter 12 (Opinion Evidence) and Chapter 19 (Rules Dispensing with or Facilitating Proof).

Copeland, Jill. "Helping Jurors Recognize the Frailties of Eyewitness Identification Evidence." (2002) 46 *Criminal Law Quarterly* 188.

Davis, Kenneth Culp. "Facts in Lawmaking." (1980) 80 *Columbia Law Review* 931.

Delisle, Ronald, *Evidence: Principles and Problems,* 3rd ed. (Toronto: Thomson Canada Limited, 1999).

Delisle, Ron Don Stuart, and David M. Tanovich, *Evidence: Principles and Problems*, 8th ed. Toronto: Carswell, 2007.

Delisle, Ronald. "Annotation—*R. v. Zundel*." (1987) 57 *Criminal Reports* (3rd) 93.

Drummond, Susan G. "Judicial Notice: The Very Texture of Legal Reasoning." (2000) 15(1) *Canadian Journal of Law and Society* 1.

Gold, A. "Expert Evidence—Admissibility: Comment on *R. v. Mohan*." (1994) 37 *Criminal Law Quarterly* 16.

Hill, S. Casey, David M. Tanovich and Louis P. Strezos. *McWilliams' Canadian Criminal Evidence*. Aurora, ON: Canada Law Book Limited (available online on Criminal Spectrum).
     See Chapter 12 (Opinion Evidence) and Chapter 23 (Judicial Notice).

Holmes, Oliver Wendell. *The Common Law*. Cambridge: Harvard University Press, 1881, reprinted 1963.

Law Reform Commission of Canada. *Judicial Notice*. Study Paper #6. Ottawa, 1973.

McWilliams, Peter K., *Canadian Criminal Evidence*. Aurora, ON: Canada Law Book Limited, 1999 with updates.

Monahan, John and Laurens Walker. "Social Science Research in Law." (1988) 43(6) *American Psychologist* 465 (see Chapter 2).

Norris, John, and Maryls Edwardh. "Myths, Hidden Facts and Common Sense: Expert Opinion and the Assessment of Credibility." (1996) 38 *Criminal Law Quarterly* 73.

Paciocco, D. M. "Taking a 'Goudge' Out of Bluster and Blarney: An 'Evidence-Based Approach' to Expert Testimony." (2009a) 13 *Canadian Criminal Law Review* 135.

Paciocco, D. M. "Unplugging Jukebox Testimony in an Adversarial System: Strategies for Changing the Tune on Partial Experts." (2009b) 34 *Queen's Law Journal* 565.

Paciocco, David M. "The Promise of *R.D.S.*: Integrating the Law of Judicial Notice and Apprehension of Bias." (1998) 3 *Canadian Criminal Law Review* 319.

Paciocco, David M. "Judicial Notice in Criminal Cases: Potential and Pitfalls." (1997) 40 *Criminal Law Quarterly* 35.

Paciocco, David M. "Evaluating Expert Opinion Evidence for the Purpose of Determining Admissibility: Lessons from the Law of Evidence." (1994) 27 *Criminal Reports* (4th) 302.

Pilkington, M. "Equipping Courts to Handle Constitutional Issues: The Adequacy of the Adversary System and Its Techniques of Proof." In *Special Lectures of the Law Society of Upper Canada: Applying the Law of Evidence—Tactics and Techniques in the Nineties*. Toronto: Carswell, 1991.

Roberts, David. (23 November 1999) *The Globe and Mail* online.

Skurka, Steven, and Elsa Renzella. "Misplaced Trust: The Courts' Reliance on the Behavioural Sciences." (1998) 3 *Canadian Criminal Law Review* 269.

Sharpe, Robert J. (ed.) *Charter Litigation*. Toronto: Butterworths, 1987.

Sopinka, John, Sidney N. Lederman, and Alan W. Bryant, *The Law of Evidence in Canada*. Toronto and Vancouver: Butterworths, 1999.

Trotter, Gary. T. "False Confessions and Wrongful Convictions."(2004) 35 *Ottawa Law Review* 179.

Uniform Law Conference of Canada. *Report of the Federal/Provincial Task Force on Uniform Rules of Evidence*. Toronto: Carswell, 1982.

CASES

R. v. Abbey, [1982] 2. S.C.R. 24.

R. v. Cerniuk (1947), 91 C.C.C. 56 (B.C.C.A.).

Daubert v. Merrell Dow Pharmaceuticals, Inc., 509 U.S. 579 (1993). The U.S. Supreme Court set out the criteria to be used to decide whether or not to admit expert opinion based novel scientific evidence.

R. v. Find, [2001] 1 S.C.R. 863. A judge is allowed to take judicial notice of the fact that sexual crimes are widely abhorred, but not the inference from this fact that there is widespread bias against those accused of sexual assault.

R. v. François, [1994] 2 S.C.R. 827. A jury's verdict of guilty on rape charges was not unreasonable given the thorough cross-examination of the complainant. The jury could use its own experience and common sense to assess the complainant's story of repressed and recovered memory.

R. v. French, [1977] O.J. No. 945 (Ont. C.A.).

Giesbrecht v. The Queen (1994), 91 C.C.C. (3d) 230 (S.C.C.). Statement elicited by the accused from experts can be used by the experts to formulate their opinions, but cannot be used by the accused to prove the accused's state of mind (an issue for the jury to decide). The trial judge appropriately instructed the jury in saying the "The weight of such testimony is dependent entirely upon the truth of the facts stated....I charge you that it is your duty, before accepting the conclusion of the experts on this question, to examine carefully all the facts related to the expert in the hypothetical question to determine whether such facts have been proven to be true" (234).

R. v. Graat (1983), 2 C.C.C. (3d) 365 (S.C.C.).

R. v. J.-L.J., [2000] 2 S.C.R. 600.

R. v. Krymowski, [2005] 1 S.C.R. 101.

R. v. Lavallee, [1990] 1 S.C.R. 852.

R. v. Malott, [1998] 1 S.C.R. 123. The Court elaborated on how trial courts should deal with social science research on battered women when they are charged with spousal homicide.

R. v. Marquard (1994), 85 C.C.C. (3d) 193 (S.C.C.). An expert cannot testify whether he or she believes a child is telling the truth about allegations of aggravated assault. However, it is permissible to explain factors that may be relevant to credibility if they go beyond the ordinary experience of the trier of fact—in this case, the expert explained why children might at first lie to hospital staff about their injuries. It is inappropriate for an expert to testify that certain behaviour is evidence of long-term abuse when it is not an issue in the trial. While this might be relevant to explain the child's reaction to the injury, it was very prejudicial and "its prejudicial effect clearly outweighed any probative value," especially in light of the trial judge's comments that "passivity was a 'hallmark' of an abused child" (226–7).

R. v. McIntosh (1997), 117 C.C.C. (3d) 385 (Ont. C.A.). Rejection of expert evidence on eyewitness identification.

R. v. Mohan (1994), 89 C.C.C. (3d) 402 (S.C.C.). Expert opinion evidence must be both logically and legally relevant, reliable, and necessary in the sense that it involves information outside the knowledge and experience of the trier of fact. It must not be excluded by any rule of evidence, and must be given by a witness with special skill or knowledge.

R. v. Morin (1988), 44 C.C.C. (3d) 193 (S.C.C.). Evidence elicited by the Crown in cross-examination of a defence psychiatrist was inadmissible because its only purpose was to show that the accused had the disposition to commit the crime.

R. v. Oickle, [2000] 2 S.C.R. 3.

R. v. S. (R.D.), [1997] 3 S.C.R. 484.

Schnell v. British Columbia Electric Railway Co. (1990), 14 W.L.R. 586 (B.C.C.A.).

R. v. Spence, [2005] 3 S.C.R. 458.

R. v. Terceira (1998), 123 C.C.C. (3d) 1 (Ont. C.A.), appeal dismissed by the Supreme Court of Canada, [1999] 3 S.C.R. 866. Discussion of how the courts should deal with the admission of DNA evidence as novel scientific evidence.

R. v. Trochym, [2007] 1 S.C.R. 239.

R. v. Trotta, [2007] 3 S.C.R. 453.

R. v. Williams, [1998] 1 S.C.R. 1128.

R. v. Zundel (1987), 31 C.C.C. (3d) 97 (Ont. C.A.), leave to appeal refused, 61 O.R. (2d) 588n. An expert may base his or her opinion upon hearsay, and the opinion may still be entitled to weight. It should be noted that general history may be proved by recourse to historical treatises. Similarly, an expert may base an opinion on material generally accepted and acknowledged as reliable by other experts in the field. Judicial notice may be taken of historical facts. The court may have recourse to secondary sources on its own motion in deciding whether to take notice of an historical fact.

R. v. Zundel (1990), 53 C.C.C. (3d) 161 (Ont. C.A.). The court may take judicial notice of an historical fact.

APPENDIX A: *Commissions of Inquiry and Studies into the Criminal Justice System in Canada*

Bellemare, Jacques and Rob Finlayson. *Report on the Prevention of Miscarriages of Justice*. Federal-Provincial Territorial Heads of Prosecutions Committee Working Group, 2004. www.justice.gc.ca/eng/dept-min/pub/pmj-pej/pmj-pej.pdf; accessed January 17, 2010.

British Columbia. Cariboo-Chilcotin Justice Inquiry. *Report on the Cariboo-Chilcotin Justice Inquiry*. Victoria, 1993.

Brodeur, Jean-Paul, Carol LaPrairie, and Roger McDonnell. *Justice for the Cree: Final Report*. James Bay, PQ: Grand Council of the Crees, 1992.

Brown, Mona G., Monique Bicknell-Danaher, Caryl Nelson-Fitzpatrick, and Jeraldine Bjornson. *Gender Equality in the Courts: Criminal Law*. Winnipeg: Manitoba Association of Women and the Law, 1991.

Canada. *Multiculturalism and Citizenship. Eliminating Racial Discrimination in Canada*. Ottawa: Supply and Services Canada, 1989.

Canadian Bar Association. Committee on Imprisonment and Release. *Locking Up Natives in Canada: A Report of the Committee of the Canadian Bar Association on Imprisonment and Release*. Ottawa, 1988.

Cawsey, R.A. (Chair). *Justice on Trial: Report of the Task Force on the Criminal Justice System and Its Impact on the Indian and Métis People of Alberta*. Edmonton: Attorney General and Solicitor General of Alberta, 1991.

Clark, S. *The Mi'kmaq and Criminal Justice in Nova Scotia*. Halifax: Nova Scotia: Royal Commission on the Donald Marshall, Jr., Prosecution, 1989.

Cory, Peter deCarteret. *The Inquiry Regarding Thomas Sophonow: The Investigation, Prosecution and Consideration of Entitlement to Compensation*. Winnipeg: Manitoba Justice, 2001. www.gov.mb.ca/justice/publications/index.html; accessed January 17, 2010.

Donlevy, Mary, Barbara Fisher, Paul Bratty, Alister Browne, Kathleen Keeting, William Maurcie, Doreen Sterling, Diane Watson, and Ellen Wiebe. *Crossing the Boundaries: The Report of the Committee on Physician Sexual Misconduct*. Vancouver: College of Physicians and Surgeons, 1992.

Enns, John E. *Review of Prosecutions Policy on Disclosure*. Manitoba Justice, 2004. www.gov.mb.ca/justice/publications/index.html accessed January 17, 2010.

Enns, John E. *A Review of Crown to Defence Disclosure Compliance in the James Driskell Murder Trial and Appeal*. Manitoba Justice, 2004. www.gov.mb.ca/justice/publications/index.html; accessed January 17, 2010.

Enns, John E. A *Review of Police to Crown Disclosure Compliance in the James Driskell Murder Trial and Appeal*. Manitoba Justice, 2004. www.gov.mb.ca/justice/publications/index.html; accessed January 17, 2010.

Federal/Provincial/Territorial Working Groups of Attorneys General Officials on Gender Equality in the Canadian Justice System. *Gender Equality in the Canadian Justice System: Summary Document and Proposals for Action*. Ottawa, Department of Justice, April, 1992. There are six background reports.

Federal-Provincial-Territorial Heads of Prosecutions Committee Working Group. *Report on the Prevention of Miscarriages of Justice,* Justice Canada, 2004. www.justice.gc.ca/eng/dept-min/pub/pmj-pej/pmj-pej.pdf; accessed January 17, 2010.
      The Report examines international and Canadian wrongful convictions, and the reasons for them: tunnel vision, mistaken eyewitness identification and testimony, false confessions, faulty forensic procedures, and in-custody informers. Wrongful convictions can be reduced by the correct use of DNA and forensic evidence, properly qualified experts, and education of the participants in the system.

APPENDIX A

Gittens, Margaret, David Cole, Moy Tam, Toni Williams, Ed Ratushny, and Sri-Guggan Sri-Skanda-Rajah. *Report of the Commission on Systemic Racism in the Ontario Criminal Justice System.* Ontario, 1996.

Glaude, G. N. Report of the Cornwall Inquiry. Cornwall, Ontario, 2009. www.cornwallinquiry.ca/en/index.html; accessed January 17, 2010.
>   The Commission inquired into the investigation of allegations of sexual assault of young persons in Cornwall.

Goudge, Honourable Stephen T. *Inquiry into Pediatric Forensic Pathology in Ontario.* Toronto: Ontario Ministry of the Attorney General, 2008. www.goudgeinquiry.ca; accessed January 17, 2010.

Hamilton, A.C., and C.M. Sinclair (Commissioners). *Report of the Aboriginal Justice Inquiry of Manitoba Volume 1: The Justice System and Aboriginal People; Volume 2: The Deaths of Helen Betty Osborne and John Joseph Harper.* Winnipeg: The Queen's Printer, 1991.

Hickman, T.A. (Chair). *The Royal Commission on the Donald Marshall, Jr. Prosecution.* Halifax: Province of Nova Scotia, 1989. Also see commentaries on the Commission: H. Archibald Kaiser, "The Aftermath of the Marshall Commission: A Preliminary Opinion."(1990) 13(1) *Dalhousie Law Journal* 374; H. Archibald Kaiser, "Legitimation and Relative Autonomy: The Donald Marshall, Jr. Case in Retrospect."(1990) 10 *Windsor Yearbook of Access to Justice* 171; J.A. Mannette, "A Trial in Which No One Goes to Jail." (1988) 20(3) *Canadian Ethnic Studies* 169; J.A. Mannette, "Not Being a Part of the Way Things Work." *Canadian Review of Sociology and Anthropology* 27(5) (1990) 508; Bob Wall, "Analyzing the Marshall Commission: Why it was Established and How it Functioned." In Joy Mannette (ed.), *Elusive Justice: Beyond the Marshall Inquiry.* Halifax: Fernwood Publishing, 1992, 13; Bruce H. Wildsmith, "Getting at Racism: The Marshall Inquiry."(1991) 55(1) *Saskatchewan Law Review* 106.

Hughes, E.N. (Ted) (Chair), Alison MacLennan, John McAlpine, Stephen F.D. Kelleher, Marguerite Jackson and Wendy Baker. *Gender Equality in the Justice System.* Vancouver: Law Society of British Columbia, 1992, Chapter 7.

Hughes, Samuel H.S. *The Royal Commission of Inquiry into the Response of the Newfoundland Criminal Justice System to Complaints.* Newfoundland, 1992.

Jaffer, Mobina (Chair), Gwenn Cutler, Gail Edinger, Alayne Hamilton, Sharon Hurd-Romanell, Pearl McKenzie, Carol Matusicky, David Mortimer, Sue Penfold, Tracey Potkins, Vera Radyo, Diane Turner, Linda Light, and Cindy Eaton. *Is Anyone Listening? Report of the British Columbia Task Force on Family Violence.* Victoria: Minster of Women's Equality, 1992.

Kaufman, Fred, C.M., Q.C. *The Commission on Proceedings Involving Guy Paul Morin,* volumes 1 and 2. Ontario: Queen's Printer, 1998.

Kaufman, Fred C.M., Q.C. *In the Matter of an Application by Steven Murray Truscott Pursuant to Section 690 (Now 696.1) of the Criminal Code.* Report to the Minister of Justice. Ottawa: Department of Justice, 2004. http://justice.gc.ca/eng/pi/ccr-rc/sec690-art690/index.html; accessed January 17, 2010.

Krindle, R. Appointment of Independent Counsel--Review. Manitoba Justice, 2007. www.gov.mb.ca/justice/publications/index.html; accessed January 17, 2010.

Lamer, Antonio, P.C., C.C.*Commission of Inquiry Pertaining to the Cases of Ronald Dalton, Gregory Parsons and Randy Drunken.* Manitoba Justice, 2006. http://www.nlcoi.gov.nl.ca/nlcoi/default.htm; accessed January 17, 2010.

LeSage, Patrick Q.C., *Report of the Commission of Inquiry Into Certain Aspects of the Trial and Conviction of James Driskell.* Manitoba Justice, 2007. http://www.driskellinquiry.ca; accessed January 17, 2010.

Lewis, Stephen. *Stephen Lewis Report on Race Relations in Ontario.* Toronto: Government of Ontario, 1992.

MacCallum, E. P. *Report of the Commission of Inquiry into the Wrongful Conviction of David Milgaard.* Regina, Saskatchewan Department of Justice, 2008. www.justice.gov.sk.ca/milgaard/; accessed January 17, 2010.

McPhedran, Marilou (Chairperson), Harvey Armstrong, Rachel Edney, Pat Marshall, Roz Roach, Briar Long, and Bonnie

APPENDIX A

Homeniuk. *Task Force on Sexual Abuse of Patients.* An Independent Task Force Report Commissioned by The College of Physicians and Surgeons of Ontario, November, 1991.

Oppal, Mr. Justice Wallace T., *Closing the Gap: Policing and Community.* Report of Commission of Inquiry on Policing in British Columbia, 1994.

Ratushny, Judge Lynn. *Self Defence Review: Final Report.* (Submitted to the Minster of Justice and to the Solicitor General of Canada, Ottawa, Department of Justice, July 11, 1997). The Report and the Government's response can be found at the Department of Justice Internet site <canada.justice.gc.ca>.

Richard, Peter K.. *The Westray Story: A Predictable Path to Disaster.* Report of the Westray Mine Public Inquiry. Halifax, 1997.

Robins, Sidney. L. In the Matter of Steven Truscott: Advisory Opinion on the Issue of Compensation. Justice Canada, 2008. /www.attorneygeneral.jus.gov.on.ca/english/about/pubs/truscott/robins_report.pdf; accessed January 17, 2010.

Royal Commission on Aboriginal Peoples. *Bridging the Cultural Divide: A Report on Aboriginal People and Criminal Justice in Canada.* Ottawa, Minister of Supply and Services, 1996.

Saull, Richard A. (Chair). *Forensic Evidence Review Committee #1* (Homicides). Manitoba Justice, 2004. www.gov.mb.ca/justice/publications/index.html; accessed January 17, 2010.

Saull, Richard A. (Chair). *Forensic Evidence Review Committee #2* (Sexual Assault, Robbery and other cases). Manitoba Justice, 2005. www.gov.mb.ca/justice/publications/index.html; accessed January 17, 2010.

APPENDIX B: *Reading a Case and Legal Research*

This Appendix provides some suggestions on how to read your *Criminal Code*, what to look for in a case, and how to conduct research on legal questions or issues.

USING THE CRIMINAL CODE

The *Criminal Code* is essential to understanding criminal procedure and evidence. Always read the section in the *Code* when it is referred to, either in this text or in a case, so that you understand what is said in the language of the *Code*. The complete way of referring to section 184(2)(c)(iii) orally would be "section 184, subsection 2, paragraph c, subparagraph iii," although it is often shortened to "section one eighty-four, two, C, three," or more properly, "subparagraph one eighty-four, two, C, three." When writing about the section, it is acceptable to refer to "section 184(2)(c)(iii)."

Sometimes the section number in a case will not correspond with the section in the *Criminal Code*. This is the result of the federal government renumbering federal statutes from time to time (referred to as a revision). The most recent revision took place in 1985, the one prior to that was 1970, and the revision prior to that was in 1953–4; you will rarely run into references to that earlier version of the *Criminal Code* today.

A revision is like a housekeeping task done in addition to amending the *Code* from time to time. Amendments may change the wording of a section, delete a section, or add a new one. When the *Code* is amended and sections are added to it, decimal points are used for new section numbers. For example, section 344.1 can be added between sections 344 and 345. Section 344(5.1) can be added between section 344(5) and 344(6). Remember that subsections 344(2) and 344(6) come before 344.1. When statutes are revised, the decimal points disappear and the sections are renumbered sequentially. There are also times when the federal government will renumber a series of sections in between revisions in order to clean up the clutter of amendments, although this can cause considerable confusion to some. The sentencing provisions, which came into force on September 3, 1996 (with further amendments in 1999) are a good example of this. There were so many new sections added that the government decided to renumber the sections related to sentencing.

One useful feature of the *Criminal Code* is the Table of Concordance, which appears at the front of the *Code*. A case that was heard before the 1985 revisions will likely refer to section numbers in the *Criminal Code*, R.S.C. 1970, Chap. C-34. The Table of Concordance will help you find the same section as it is renumbered in R.S.C. 1985. For example, if an earlier case refers to an application made under section 527 of the R.S.C. 1970 *Code*, the Table of Concordance indicates that section 527 is now section 599 under the R.S.C. 1985 *Code*. Look at the Table of Concordance to see if you can follow this example.

The Table of Concordance does not always work, as there may have been several different amendments between 1970 and 1985 that are not in the Table. For example, the Table is not that useful in sorting out the changes in the offences associated with drinking and driving, because of the large number of amendments between the revisions in 1970 and 1985, and the fact that similar sections were moved to different locations in the *Code*. In such instances it is easier to use the Index at the back of the *Code* to find the appropriate section number or to use the Table of Contents at the front of the *Code*.

Although the *Code* was revised in 1985, the revisions were not proclaimed in force until December 12, 1988. The 1985 revisions took into account all revisions up until the end of 1985, so any sections added to the *Code* after 1985 were numbered with decimals.

Appendix B

Citing the *Criminal Code*

If you look at the beginning of your *Code*, you will notice it is cited as R.S.C. 1985, Chap C-46. That is, Revised Statutes of Canada 1985, chapter C-46. You should cite it as *Criminal Code*, R.S.C. 1985, Chap C-46. Never cite it as the *Canadian Criminal Code*, as "Canadian" is not part of the title. If you need to distinguish it from other criminal codes, cite it as the Canadian *Criminal Code*. Never cite it as the C.C.C., which is the abbreviation for *Canadian Criminal Cases*, a law report (see below). When citing a section of the *Code*, always use section numbers; never use page numbers.

CITING A CASE

A *citation* for a case provides you all the information you need to find it in a library. However, not all cases are *reported*; in fact, most cases in Canadian courts are *unreported*. Trial court decisions are rarely reported, and many appeal court decisions remain unreported. Even if a case is unreported, you may still be able to find a summary of it in one of the many services in Canada that summarizes cases. If you find a case in a law report, you can refer to it as a reported case.

There are numerous *law report*s in Canada. The *Canadian Criminal Cases* (C.C.C.) and the *Criminal Reports* (C.R.) are the two most popular criminal law reports. Some cases are reported in more than one law report; however, you need to provide only one citation. Generally, it is expected that when you refer to a Supreme Court of Canada case you will cite the *Supreme Court Reports* (S.C.R.), because those are the official reports from that Court.

Many cases can be found on the website of various courts and by services that provide online access to cases. The Supreme Court of Canada's Web site <www.lexum.umontreal.ca/csc-scc/en/index.html> contains the full text of its decisions from 1983 to present. The British Columbia Court of Appeal, Supreme Court, and Provincial Court cases can be found at their web site <www.courts.gov.bc.ca>. The Alberta Courts Web site <www.albertacourts.ab.ca> contains Court of Appeal, Court of Queen's Bench, and Provincial Court decisions. The Ontario Courts Web site is at <www.ontariocourts.on.ca> and contains some decisions, although a more comprehensive data base is found at <www.canlii.org>. The Federal Department of Justice <canada.justice.gc.ca> provides useful updates on amendments to the *Criminal Code*. In addition, private companies which charge for their services offer databases with full text decisions, netletters, commentaries on those decisions, and so on. Students at Simon Fraser University now have access to parts of the following databases: BestCase, Criminal Source, Criminal Spectrum, LawSource, Quicklaw and LexisNexis. Check your library to see which data bases you have access to.

Those in the business of law use a uniform system of citation, for example, *R. v. Chow* (2005), 195 C.C.C. (3d) 246 (S.C.C.), or *R. v. Chow*, [2005] 1 S.C.R. 384. Pay attention to the proper way of citing a case. When the year of a case is given in (square) brackets (e.g., [2005]), it forms part of the volume number, and you will need to know the year to find the case in the case reports. When the year of a case is given in parenthesis (e.g., (2005)), it refers only to the year of the decision, and you do not need to know the year in order to find the case. The C.C.C. and C.R. use years in parenthesis in their current series of reports, whereas the S.C.R. use brackets. The Supreme Court of Canada has also started using what they refer to as Neutral Citations. For *Chow*, it is 2005 SCC 24. This allows us to find the case before it is reported in the Supreme Court Reports. Names of cases should be in italics. Note the location of the comma in the above citations, and always put the comma in the correct location. It comes **before** brackets, but **after** parenthesis, depending on whether the year forms part of the volume number (brackets) or not (parentheses). More information about legal citations can be found at: http://library.queensu.ca/law/lederman/legalcit.htm; accessed January 17, 2010.

Although it was the practice in the past to cite page numbers when paraphrasing or quoting from a decision, the practice has now moved to citing paragraph numbers when they are provided. This is extremely convenient because if someone cites a paragraph number from a Supreme Court of Canada decision you can find the location of the citation in any of many case reports and also on the Supreme Court of Canada's website. In the text, we have used the practice of using para. numbers where they are available, for example (para. 45). If we are using page numbers, the reference will appear as (45).

Publishing companies that sell law reports and some courts provide their own summary of a case, which is not part of the judgment. These summaries are referred to as *headnotes*, and some are more accurate than others. There was a time when it was common to find headnotes that were quite inaccurate, but that is mostly a thing of the past.

READING A CASE

There are many ways to read a case, and you could employ one or more of the theoretical frameworks discussed in the Introduction. For example, you could read a case to identify all of the facts a judge assumed in arriving at certain conclusions as to what the law is. You might then look at the social science literature to determine whether the facts that the judge assumed are supported in the academic literature. Would the judge have made a different decision if social authority (discussed in the Introduction) had been presented to the court?

You could also read the case as legal actors (lawyers and judges) might read it. This is sometimes involves *briefing* a case, and the result is a *case brief*, or succinct summary of the most important aspects of the case. There are different ways to brief a case (See Box B.1). Here is one of the standard methods: Start by recording the style of cause—it tells you who the parties are in the court case. In criminal cases, the Queen (identified as R. or Regina, pronounced as is the capital of Saskatchewan) is usually (although not always) one of the parties, and the accused or defendant is the other party. Thus, the style of cause might be *R. v. Chow*. It is also useful to record a complete citation of the case so that you can find it again if you need to. For example, *R. v. Chow*, [2005] 1 S.C.R. 384.

A case brief then summarizes the relevant or material facts in the case, ideally in a sentence or two. This is easier said than done, as you have to determine what was relevant to the judges who made the decision. You may find that the judge has ignored a number of facts that you consider relevant. A case brief might then identify the issues before the court. Again, how the judges characterize the issues may have a great deal to do with the outcome of a case.

Having sorted out the issues, your case brief would then record the decision of the court on the identified issues and the reasoning leading to the decision. Appellate decisions are not always *unanimous*. There may be a *majority* decision and a *minority* decision (or dissent). If there is only one judge speaking for all of the judges, it is usually easier to sort out the basis of a decision. Some judges may agree with only parts of the majority decision and will then write their own decisions on another aspect of the case.

The aspect of a decision that composes the law and the material facts on which the judge(s) relied is the *ratio decidendi* (*ratio* for short) of a case, whereas those parts of the judgment that were not necessary for the decision are referred to as *obiter dicta* (or *obiter*). If the traditional method of finding the *ratio* works, it should provide some certainty and predictability within the legal system. There is, however, little agreement on exactly how the ratio is found (see for example, Goodhart 1959 and Gooderson 1952). For a discussion of the significance of *obiter* see *R. v. Henry*, 2005 SCC 76.

Appendix B

A standard case brief might include: Style of Cause, Facts, Issues, Principle (Ratio), Reasons, and Decision. Note that a case brief is for your assistance, and you may wish to include other information or leave some out. The important thing about a brief is that it completely summarizes all of the important information about a legal judgment in as few words as possible, in a way that is of use to you in recalling the case and using it in argument or analysis of the law.

Box B.1 Sample Case Brief

R. v. Béland, [1987] 2 S.C.R. 398.

Background: Appeal by Crown, as of right, from decision of Que. C.A. overturning conviction and ordering new trial.

Issue: Is polygraph evidence admissible at trial, to support the credibility of accused who have testified *viva voce* in their own defence?

Ruling: (5–2) No.

Facts: One participant in alleged conspiracy to commit robbery informed the police before any robbery occurred, and gave evidence for Crown, at trial of the remaining participants. The accused testified, denying that evidence. Defence applied to have accused undergo polygraph examination, and to tender results. Trial Judge refused, holding that such evidence was inadmissible *per* **Phillion** (S.C.C.), which held such evidence to be hearsay. Quebec C.A. distinguished **Phillion** on its facts, as Phillion did not testify, and his credibility was not in issue.

Reasoning: (Per **McIntyre, J**. for majority)—**Phillion** followed.

—Admission of polygraph evidence, which is solely to bolster the credibility of the accused, would contravene (1) rule against oath helping, (2) rule against prior consistent (i.e., self-serving) statements, (3) rule confining character evidence led through third parties to evidence of general reputation, and (4) rule restricting expert opinion evidence to instances requiring special skill or knowledge (which does not include the issue of credibility).

—It would usurp the role of the trier of fact, and lead to "the time-consuming and confusing consideration of collateral issues," resulting in disruption, delays, and complications. Admission of polygraph evidence would serve no useful purpose not already served.

(Per **LaForest, J.**, concurring in the result)

—There is considerable concern the trier of fact would accord inordinate weight to such evidence, as it is "cloaked under the mystique of science."

—The courts should not waste time on collateral issues.

Appendix B

(Per **Wilson, J.** for the minority, dissenting)

—**Phillion** distinguished (he had made a confession, and did not testify, but sought to lead polygraph evidence relating to the confession without taking the stand to do so).

—Rule against oath helping is irrelevant, as polygraph evidence does not determine the issue in the way medieval oath-helping did. The authorities do not establish a rule against an **accused** leading evidence to bolster credibility re: **innocence**, only re: the Crown witnesses testifying to guilt; these situations can be distinguished based on the right to make full answer and defence.

—Rule against prior consistent statements is irrelevant, because here not leading evidence of repetition of story, but of similarity of physiological responses to those of a truthful person.

—Rule re: character evidence irrelevant, because polygraph evidence is not character evidence; it doesn't go to likelihood accused committed the offence, but to the collateral issue of credibility. In the alternative, the rule restricting character evidence through third parties (in cases where the accused has testified) to general reputation is artificial and should be abandoned.

—Rule re: expert evidence is irrelevant, because polygraph operator does not give opinion evidence on credibility, but interprets physiological data and gives opinion on conformation of data to that of person telling truth, and on nature and accuracy of device; this doesn't determine credibility, which the jury can assess.

—The evidence is thus relevant, and there is no reason to exclude it. Cross-examination, opposing experts, and judicial cautions can be used if necessary as safeguards.

You might also ask yourself the following questions as you read a case:

(1) How did the case arise? Was it an appeal from a verdict at trial? If so, was the accused found guilty or not guilty (i.e., whose appeal was it)? Or, is it an appeal from an application? If so, what was the nature of the application? To which court? What was the result?

(2) Did the Court of Appeal allow or dismiss the appeal? What issues were before the Court of Appeal? How were they decided and why?

(3) Did the Supreme Court of Canada allow or dismiss the appeal? What issues were before the Court of Appeal? How were they decided and why? Was there a dissenting judgment? What were the reasons for the dissent?

(4) What are the implications of the decision? How might the case be judicially considered in the future?

You may have your own method of reading a case; if so, feel free to use that.

The case method approach to teaching law has been criticized on many grounds (see Jewell 1984). Michael Mandel, in his book *The Charter of Rights and the Legalization of Politics in Canada*, adds further criticism to the traditional legal method. In reading cases, one can easily forget that there are many ways to read a case that do not involve the traditional legal method. Try to identify and use these alternative ways as you read the text and the related materials.

Appendix B

LEGAL RESEARCH

Several books and online sources discuss how to conduct legal research (see bibliography). It is often easiest to start legal research by consulting a textbook on the subject, or journal articles and case comments found through the subject index at the library, or an index to legal periodicals such as the *Index to Canadian Legal Periodical Literature* or the *Canadian Abridgement's Index to Canadian Legal Literature*. There are also a number of websites that assist in legal research. Many textbooks and legal journals are now available online. See Criminal Source which contains the *Index to Canadian Legal Literature*, and Criminal Spectrum which contains *Criminal Law Quarterly*; McWilliams' *Canadian Criminal Evidence*, and Salhany's *Canadian Criminal Procedure*). Ask about these sources at your local library.

If you start by reading a textbook, it may be easier for you to first *characterize* your legal issue, so that you can determine possible key words to use when consulting a legal periodical index. Do not underestimate the value of browsing through an index. You may find related key words you did not think about, and you may find that the indexing service is inconsistent in its use of words from one year to the next.

CASES JUDICIALLY CONSIDERED

Judges consider past cases when they make their decision. Case law is built up by lower courts following the decisions of higher courts. This is known as the principle of *stare decisis* ("standing by former decisions"). It basically means that similar cases should be decided in a similar manner, and that the lower courts are bound by the decisions of higher courts. For example, the British Columbia Supreme Court is bound by decisions of the British Columbia Court of Appeal, and the British Columbia Court of Appeal is bound by decisions of the Supreme Court of Canada. Decisions from British Columbia are only *persuasive*, not binding, in Ontario courts, and so on. That is, they may be used to support an argument, but the courts are not required to follow them.

It is also important to examine whether higher courts have considered and overturned lower court decisions or whether courts have distinguished earlier cases. For example, let us assume that you have a 1999 case on similar fact evidence from the British Columbia Court of Appeal, and you want to know whether it is still the law today. There are several possibilities (assuming that there have been no amendments to the legislation, if legislation is relevant to the issue which it is not in this case). The Supreme Court of Canada could have *reversed* or *overruled* the lower court's decision. The Supreme Court of Canada may have *affirmed* the decision, and then you know that it is still the law today (unless of course the law has been modified by legislation or the Supreme Court of Canada has changed its mind, *reversing* itself or *distinguishing* the case).

Judges may consider a case but not follow it. Thus, the Ontario Court of Appeal may have considered the British Columbia Court of Appeal's decision and not followed it. This is actually quite common. The Ontario and British Columbia Courts of Appeal often reach different conclusions on what the law is, and eventually the Supreme Court of Canada usually decides the question. If the case(s) do not go to the Supreme Court of Canada, the law will be different in the two provinces until a future case brings the issue before the Supreme Court of Canada or the law is amended by legislation.

The quickest and easiest way to determine if a case has been judicially considered is through a service such as Quicklaw which allows you to "Quickcite" or "Note-up" a decision. Each database uses slightly different terminology to search for judicial consideration of a specific case.

Appendix B

BIBLIOGRAPHY AND SUGGESTIONS FOR FURTHER READING

Banks, Margaret A. and Karen E.H. Foti. *Banks on Using a Law Library* 6th ed. Scarborough, ON: Carswell, 1994.

Best, Catherine P. *Best Guide to Canadian Legal Research* (online: http://legalresearch.org; accessed January 17, 2010).

Black's Law Dictionary, or any other law dictionary, for translation of legal terms.

Castel, Jacqueline R. and Omeela K. Latchman. *The Practical Guide to Canadian Legal Research*, 2nd ed. Scarborough, ON: Carswell, 1996.

Eisen, Lewis S. *The Canadian Lawyer's Internet Guide.* online ed., Quicklaw, CLIG (online).

Fitzgerald, Maureen F. *Legal Problem Solving: Reasoning, Research & Writing.* 4th ed. Toronto: Butterworths,2 007.

Gooderson, R.N. "Ratio Decidendi and Rules of Law." (1952) *Canadian Bar Review* 892.

Goodhart, A.L. "The Ratio Decidendi of a Case." (1959) *Modern Law Review* 117.

Jackson, M. Drew, and Timothy L. Taylor. *The Internet Handbook for Canadian Lawyers.* 3rd ed. Scarborough, ON: Carswell, 2000.

Sullivan, Ruth. "Statutory Interpretation in a Nutshell." (2003) 82 *Canadian Bar Review* 51.

Tjaden, Ted. *Doing Legal Research in Canada.* Bora Laskin Law Library, University of Toronto; online: www.llrx.com/features/ca.htm; accessed January 17, 2010.

Yogis, John A. and Innis M. Christie (founding authors), Michael J. Iosipescu and Michael E. Deturbide, 6th ed. *Legal Writing and Research Manual.* Markham, ON: LexisNexis Butterworths, 2004.

Waddams, S.M. *Introduction to the Study of Law.* 6th ed. Scarborough, ON: Carswell, 2004.

Williams, Glanville (edited by A.T.H. Smith). 12th ed. *Learning the Law.* London: Stevens & Sons, 2002.

APPENDIX C: *True Canadian Crime and Other Misconduct*

You may find these books interesting to read as you work your way through this text. Where we have cited case reports, you might be interested in reading the case and comparing it to its description in the true crime book.

Anderson, Barrie and Dawn Anderson. *Manufacturing Guilt: Wrongful Convictions in Canada.* 2nd ed. Halifax: Fernwood Publishing, 2009.
> The authors examine eight cases of wrongful conviction (Donald Marshall, David Milgaard, Wilbert Coffin, Guy Paul Morin, Thomas Sophonow, Stephen Truscott, James Driskell, and William Mullins-Johnson), some of the explanations as to why wrongful convictions occur, and possible means to reduce them.

Anderson, Frank W. *A Dance with Death: Canadian Women on the Gallows 1754-1954.* Saskatoon, SK, Canada: Fifth House Ltd., 1996.

Auger Michel. *The Biker Who Shot Me: Recollections of a Crime Reporter.* Toronto: McClelland & Stewart, 2003. Translated by Jean-Paul Murray.
> Michel Auger, a crime reporter, was shot in the back on September 13, 2000, while walking in a parking lot across the street from the offices of *Le Journal de Montréal.* He tells stories about his encounters with criminals throughout his career as a journalist, including this shooting.

Belliveau, John Edward. *The Coffin Murder Case.* Toronto: Kingswood House, 1956; updated 1979.

Bird, Heather. *Not Above the Law: The Tragic Story of Joann Wilson and Colin Thatcher.* Toronto: Key Porter Books, 1985.

Birnie, Lisa Hobbs. *Such A Good Boy: How a Pampered Son's Greed Led to Murder.* Toronto: McCllelland-Bantam, 1992.
> The story of how 18-year-old Darren Huenemann convinced two of his friends, Muir and Lord, to kill his mother and grandmother in his grandmother's home in Tsawwassen, British Columbia in 1990, apparently to inherit a $4 million estate.
> Note: in November, 1993, the British Columbia Court of Appeal rejected Muir's appeal from sentence ([1993] B.C.J. No. 1688) and Lord's appeal from conviction ([1993] B.C.J. No. 2387), and in February, 1995, the Supreme Court of Canada dismissed Lord's appeal ([1995] 1 S.C.R. 747). In December, 1993, the British Columbia Court of Appeal rejected Huenemann's appeal from conviction ([1993] B.C.J. No. 2576).

Birnie, Lisa Hobbs and Sue Rodriguez. *Uncommon Will: The Death and Life of Sue Rodriguez.* Toronto: Macmillan Canada, 1994.
> The life and suicide of Sue Rodriguez, who suffered from Amyotrophic Lateral Sclerosis (ALS or Lou Gehrig's disease).

Bolan, Kim. *Loss of Faith: How the Air-India Bombers Got Away With Murder.* Toronto: McClelland & Stewart, 2006.

Bourrie, Mark. *Flim Flam: Canada's Greatest Frauds, Scams, and Con Artists.* Toronto: Hounslow Press, 1998.

Boychuk, Rick. *Autopsie d'un meurtre : l'histoire de Jeannine Boissonneault Durand.* Montréal: Éditions de l'Homme, 1994. English version: *Honour Thy Mother: The Search For Jeannine Durand.* Toronto: Penguin Books, 1995.
> Jeannine Boissonneault Durand was killed in 1968 and not identified until 1990. Her husband was subsequently convicted of her murder.

Boyd, Neil. *High Society: Legal and Illegal Drugs in Canada.* Toronto: McCllelland-Bantam, 1993; first published 1991.
> The history and politics surrounding why some drugs are legal and others are illegal. The dividing line, he concludes, has more to do with politics and economics than any concern over our health.

Boyd, Neil. *The Last Dance: Murder in Canada.* Toronto: McCllelland-Bantam, 1992, first published 1988.
> The history of murder in Canada and provides some fascinating insights into the crime of murder in the family, murder of acquaintances, killing for money and sex, and killing by the emotionally and mentally handicapped. Boyd relies on 96 cases from 1867 to 1962, and interviews of 35 men and five women who were serving time for murder.

Burnside, Scott and Alan Cairns. *Deadly Innocence*. New York: Warner Books Inc., 1995.
One of a number of books written about the kidnapping, sexual assaults and murders committed by Paul Bernardo and Karla Homolka.
Doug French, Donna French, Dan Mahaffy and Deborah Mahaffy v. Her Majesty the Queen and Paul Kenneth Bernardo [1995] S.C.C.A. No. 250. Supreme Court of Canada dismissed an application for leave to appeal regarding the use of videotapes at Bernardo's trial and thereafter. There were numerous rulings on the admissibility of evidence including similar fact evidence (stalking of young women), spousal abuse, solicitor-client privilege, and items seized from the accused's house. Other rulings were made on publication bans, psychiatric assessments, and restricting public access to videotapes. Most of these are accessible through Quicklaw.

Cahill, Bette. *Butterbox Babies*. Toronto: McCllelland-Bantam, 1992.
Lila Young ran the Ideal Maternity Home in East Chester, Nova Scotia in the 1950s, a home for unwed mothers. Cahill raises and researches questions about what really happened to the babies born at the Home. Were the mothers lied to? Were some babies sold, while others were allowed to starve to death?

Cairns, Alan. *Nothing Sacred: The Many Lives and Betrayals of Albert Walker*. Toronto: Seal Books, 1998.
Cairns tells the story of Albert Walker, a one-time respected Canadian Business man, who defrauded Canadians of millions of dollars, and then moved to England with his daughter, passing her off as his wife. Walker befriended and killed Ronald Platt, a man from England, and then assumed his identity.
Some of the fraud Walker engaged in is described in his bankruptcy case: *Walker (Re)*, [1998] O.J. No. 2690.

Callwood, June. *The Sleep-Walker*. Toronto: McCllelland-Bantam, 1990.
In May, 1987, Ken Parks, in an apparent state of sleepwalking, drove from Pickering to Scarborough, Ontario (a distance of 10 kilometres) to the home of his in-laws. He then killed his mother-in-law and wounded his father-in-law. Callwood tells the story of the families involved, the murder, the courtroom battles.
R. v. Parks (1992), 75 C.C.C. (3d) 287 (S.C.C.) and 56 C.C.C. (3d) 449 (Ontario Court of Appeal).

Cameron, Stevie. *The Pickton File*. Knopf Canada, 2007.

Cameron, Stevie. *Blue Trust: The author, The Lawyer, His wife, and Her Money*. Toronto: Seal Books, 1998.
The story of Montreal tax lawyer Bruce Verchere and off-shore bank accounts.

Cameron, Stevie. *On the Take: Crime, Corruption and Greed in the Mulroney Years*. Toronto: Seal Books, 1995.

Cameron, Stevie and Harvey Cashore. *The Last Amigo: Karlheinz Schreiber and the Anatomy of a Scandal*. Toronto: Macfarlane Walter & Ross, 2001.

Campbell, Marjorie Freeman. *Bloody Matrimony: Evelyn Dick and the Torso Murder Case*. Toronto: Penguin Books, 1974.
In 1946, Evelyn Dick was charged with the murder of her husband John Dick, whose torso was discovered outside of Hamilton, Ontario. This is the story of her two trials.
R. v. Dick (1947), 87 C.C.C. 101 (Ont. C.A.); leave to appeal to S.C.C. refused 89 C.C.C. 252.

Clark, Doug. *Unkindest Cut: The Torso Murder of Selina Shen*. Toronto: McClelland & Stewart, 1992.
The story of Selina Shen, and of the three trials of Rui-Wen Pan, who was convicted of her murder in 1992. In 1999, the Ontario Court of Appeal dismissed his appeal from conviction (*R. v. Pan*, 1999, 134 C.C.C. (3d) 1, as did the Supreme Court of Canada ([2001] 2 S.C.R. 344).

Croft, Roger. *Swindle!: A Decade of Canadian Stock Frauds*. Toronto: Gage, 1975.

Davey, Frank. *Karla's Web: A Cultural Investigation of the Mahaffy-French Murders*. Toronto: Penguin Books, 1995.

Desbrats, Peter. *Somalia Coverup: A Commissioner's Journal*. Toronto: McClelland & Stewart, 1997.

Deverell, William. *Fatal Cruise: The Trial of Robert Frisbee*. Toronto: McClelland & Stewart, 1991.
Deverell's account of his defence of Robert Dion Frisbee, who was accused of killing Muriel Collins Barnett, his employer and a wealthy widow, while on a Norwegian cruise ship near Victoria.
See *R. v. Frisbee* (1989), 48 C.C.C. (3d) 386 (B.C.C.A.), leave to appeal refused 50 C.C.C. (3d) vi (S.C.C.).

APPENDIX C

Doe, Jane. *The Story of JaneDoe: A Book About Rape.* Toronto: Vintage Canada, 2004.
> The rape of Jane Doe in 1986 by the balcony rapist in Toronto ended with a successful law suit against the Toronto Police force.
> *Jane Doe v. Toronto (Metropolitan) Commissioners of Police*, [1989] O.J. No. 471; 58 D.L.R. (4th) 396; *Jane Doe v. Board of Commissioners of Police for the Municipality of Metropolitan Toronto et al.*, (1998), 39 O.R. (3d) 487; [1998] O.J. No. 2681.

De Vries, Maggie. *Missing Sarah: A Vancouver Woman Remembers her Vanished Sister.* Toronto: Penguin Canada, 2003.

Dubé, Richard. *The Haven: A True Story of Life in the Hole.* Toronto: Harper Collins, 2002.

Dubro, James. *Dragons of Crime: Asian Mobs in Canada.* Toronto: McClelland & Stewart, 1992.

Dubro, James. *Mob Mistress.* Toronto: McClelland-Bantam, 1989.

Dubro, James and Robin F. Rowland. *King of the Mob: Rocco Perri and the Women who Ran his Rackets.* Markham, Ont.: Viking, 1987.

Dubro, James and Robin F. Rowland. *Undercover: Cases of the RCMP's Most Secret Operative.* Toronto: McClelland & Stewart, 1992.

Eastham, Michael (with Ian McLeod). *The Seventh Shadow: The Wilderness Manhunt for a Brutal Mass Murderer.* Toronto: Warwick Publishing, 1999.
> A retired RCMP Sargent tells the story of the killing of the Johnson and Bentley families in 1982 in British Columbia's Wells Gray Park. David Shearing pled guilty to six counts of second degree murder and was sentenced to life imprisonment with no eligibility for parole for 25 years.

Edwards, Peter. *The Big Sting: The True Story of the Canadian who Betrayed Colombia's Drug Barons.* Toronto: Key Porter Books, 1991.

Edwards, Peter. *Night Justice: The True Story of the Black Donnellys.* Toronto: Key Porter Books, 2004.
> The 1880 murder of James Donnelly, his wife Johannah, and three family members as the result of vigilante justice.

Edwards, Peter. *One Dead Indian: The Premier, the Police, and the Ipperwash Crisis.* Toronto: McClelland & Stewart, 2003.
> A critical analysis of Anthony (Dudley) George's death. He was shot by an Ontario Provincial Police Officer on September 4, 1995, during a protest designed to reclaim a traditional burial ground in Ipperwash Provincial Park, near Sarnia, Ontario.

Edwards, Peter. *Waterfront Warlord: The Life and Violent Times of Hal C. Banks.* Toronto: Key Porter Books, 1987.

Edwards, Peter and Michel Auger. *The Encyclopedia of Canadian Organized Crime: From Captain Kidd to Mom Boucher.* Toronto: McClelland & Stewart, 2004.

Eichenwald, Kurt. The Informant: A True Story. New York: Broadway Books, 2000.

Ferry, Jon and Inwood, Damian. *The Olson Murders.* Langley, B.C.: Cameo Books, 1982.

Finkle, Derek. *No Claim to Mercy: The Controversial case for Murder Against Robert Baltovich.* Toronto: Penguin, 1998.

Francis, Diane. *Contrepreneurs.* Toronto: Macmillan of Canada, 1988.

Friedland, Martin L. *The Case of Valentine Shortis: A True Story of Crime and Politics in Canada.* Toronto: University of Toronto Press, 1986.

Gadsby, Joan E. *Addiction by Prescription: One Woman's Triumph and Fight for Change.* Toronto: Key Porter Books, 2000.

APPENDIX C

Goulding, Warren David. *Just Another Indian: A Serial Killer and Canada's Indifference.* Calgary: Fifth House Publishers, 2001.

The crimes and prosecutions of John Martin Crawford, a serial killer who preyed on Aboriginal women in Alberta and Saskatchewan, and questions about racism in our criminal justice system.

Griffiths, John. *Prescription: A Doctor without Remorse.* Hancock House Publishers, 1995.

Charalambous, a B.C. doctor, was convicted of first-degree murder and conspiracy to commit first-degree murder in the death of one of his patients (Sian Simmonds), who was about to testify against him at a hearing before the College of Physicians and Surgeons. Griffiths tells this story and also the story of Charalambous's wife, seduced by him when she was only 15 and one of his patients.

See *R. v. Charalambous*, [1997] S.C.C.A. No. 365, application for leave to appeal to S.C.C. dismissed without reasons (October 16, 1997); application to reconsider dismissed without reasons (Jan 8/98). Also see *Charalambous v. College of Physicians and Surgeons of British Columbia,* (1987) 27 Admin. L.R. 289; and [1987] B.C.J. No. 1212 and [1988] B.C.J. No. 3052.

Griffiths, John. *Resurrection: The Kidnapping of Abby Drover.* (Toronto: Insomniac Press, 1999).

The story of Abby Drover who, at the age of 12, was confined for six months in a 5x6 foot underground prison and subjected to rape, starvation and psychological torture by her neighbour Donald Hay.

Haines, Max. *Canadian crimes.* Toronto: Viking, 1998.

Haines, Max. *Doctors Who Kill.* Toronto: Penguin Books Canada Limited, 1993.

Haines, Max. *Multiple Murderers.* Toronto: Toronto Sun, 1994.

Haines, Max. *Multiple murderers II.* Toronto: Signet, 1996, 1995.

Hall, Neal. *The Deaths of Cindy James.* Toronto, Canada: M&S Paperbacks, 1991.

Hansen, Ann. *Direct Action: Memoirs of an Urban Guerrilla.* Toronto: Between the Lines, 2001.

Hansen was one of the "Squamish Five" who targeted a number of sites, including a B.C. hydro substation, in their fight against capitalism in the mid 1980s.

Harris, Frann. *Martensville: Truth or Justice.* Toronto: Dundurn Press, 1998.

The author's interpretation of what was happening in Martensville, Saskatchewan during the trial of those charged with sexual assault of children at a day care.

See *R v. Sterling* (1993), 84 C.C.C. (3d) 65 (Sask. C.A.) regarding disclosure issues; (1993), 108 Sask.R. 243 (Sask. Q.B.) regarding conflict of interest by lawyer; (1995), 141 Sask.R. 1(C.A.) regarding appeal of sentence; (1995), 102 C.C.C. (3d) 481 (C.A.) for appeal regarding how the child witnesses were interviewed.

Harris, Michael. *The Judas Kiss: The Undercover Life of Patrick Kelly.* Toronto: McClelland & Stewart, 1995.

Harris, Michael. *Justice Denied: The Law Versus Donald Marshall.* Toronto: Macmillian of Canada, 1986.

The story of how Donald Marshall, Jr. was tried, convicted and sentenced to life imprisonment for the 1971 murder of Sandy Seale in Sydney, Nova Scotia, and of the efforts made to free and compensate him for his conviction in respect of that murder that he did not commit.

R. v. Marshall (1972), 8 C.C.C. (2d) 329 (N.S.C.A.); *R. v. Marshall* (1982), 66 C.C.C. (2d) 499 (N.S.C.A.).

Harris, Michael. *The Prodigal Husband: The Tragedy of Helmuth and Hanna Buxbaum.* Toronto: McClelland & Stewart, 1994.

In February 1986, Helmuth Buxbaum (a millionaire nursing-home businessman from London, Ontario) was convicted of arranging the death of his wife of 24 years. Harris raises questions about whether the justice system failed in this case. Buxbaum refused to let his lawyer Eddie Greenspan enter a plea of not guilty by reason of insanity, and later appeals from his conviction were unsuccessful.

See *R. v. Buxbaum* (1989), 70 C.R. (3d) 20 (Ont. C.A.) dismissing Buxbaum's appeal; (1989) 76 C.R. (3d) xxix (S.C.C.); *Greenspan, Rosenburg (Re),* [1987] O.J. No. 562 regarding legal fees; *Buxbaum (Litigation guardian of) v. Buxbaum,* [1997] O.J. No. 5166 (Ont. C.A.), regarding damages for nervous shock, application for leave to appeal to the Supreme Court of Canada against award of damages dismissed [1998] S.C.C.A. No. 78.

APPENDIX C

Harris, Michael. *Unholy Orders: Tragedy at Mount Cashel*. Toronto: Penguin Books, 1991.
    Michael Harris writes about how the physical and sexual abuse of children at the Mount Cashel Orphanage run by the Christian Brothers in Newfoundland went uncorrected for years, despite complaints to authorities.
    This book covers a number of cases: See the various decisions by the Newfoundland Supreme Court Trial Division in *R. v. Kenny* (1992), 95 Nfld. & P.E.I.R. 131, (1992) 94 Nfld. & P.E.I.R. 181, 92 Nfld. & P.E.I.R. 318, and the Newfoundland Court of Appeal (1996), 108 C.C.C. (3d) 349. Application for leave to appeal to Supreme Court of Canada similar fact evidence, collusion and corroboration dismissed. Also see *R. v. Burke* (1991), 92 Nfld. & P.E.I.R. 289 (Trial Division); (1994), 88 C.C.C. (3d) 257 (C.A.), [1996] 1 S.C.R. 474. *R. v. English* (1991) 95 Nfld. & P.E.I.R. 147 (Trial Division); (1994) 31 C.R. (4th) 303 (C.A.).

Harris, Michael. *Lament for an Ocean: The Collapse of the Atlantic Cod Fishery: A True Crime Story*. Toronto: McClelland & Stewart, 1998.

Hebert, Jacques. *I accuse the assassins of Coffin*. Montreal, Canada: Les Editions du Jour, 1964.

Hebert, Jacques. *The Coffin Affair*. Toronto, Canada: General Paperbacks, 1982.
    A commentary on the hanging of Coffin, who may not have committed the murders.

Hemsworth, Wade. *Killing Time: The Senseless Murder of Joseph Fritch*. Toronto: Viking, 1994.

Henton, Darcy with David McCann. *Boys Don't Cry: The Struggle for Justice and Healing in Canada's Biggest Sex Abuse Scandal*. Toronto: McCelland & Stewart, 1995.
    The story of David McCann, who was sent to St. Joseph's Training School for Boys in Alfred, Ontario, in 1958, and how he survived sexual and physical abuse.

Holmes, W. Leslie with Bruce L. Northorp. *Where Shadows Linger: The Untold Story of the RCMP's Olson Murders Investigation*. Surrey, B.C.: Heritage House Pub., 2000.

Hustak, Alan. *They Were Hanged*. Toronto: J. Lorimer, 1987.

Hoshowsky, Robert J. *The Last to Die: Ronald Turpin, Arthur Lucas, and the End of Capital Punishment in Canada*. Toronto: Dundurn Press, 2007.

Hyde, Christopher. *Abuse of Trust, The Career of Dr. James Tyhurst*. Vancouver: Douglas and McIntyre, 1991.
    The story of the trial and conviction of Dr. James Tyhurst. He was found guilty of sexually assaulting four women while they were his patients. His alleged therapy for depression included the wearing of a slave-style tunic and whippings with a leather whip. Note: Tyhurst was acquitted on December 14, 1993, after a retrial on charges of indecent and sexual assaulting of two of the patients. Jury deliberations lasted six days.
    R. v. Tyhurst (1992), 79 C.C.C. (3d) 238 (B.C.C.A.) allowing an appeal; *R. v. Tyhurst* [1993] B.C.J. No. 2615 (B.C.S.C.) (Voir dire concerning similar fact evidence); [1996] B.C.J. No. 202 (B.C.C.A.); application by Crown for leave to appeal to Supreme Court of Canada dismissed [1996] S.C.C.A. No. 174.

Jadelinn. *Spirit Alive: A Woman's Healing From Cult Ritual Abuse*. Toronto: Women's Press, 1997.

Jones, George and Barbara Amiel. *By Persons Unknown: The Strange Death of Christine Demeter*. New York: Grove Press: distributed by Random House, 1977.

Jones, George (ed.). *The Scales of Justice: Seven Famous Criminal Cases Recreated*. Toronto: CBC Enterprises, 1983.
    CBC scripts for the cases: *Latta, Coffin, Horvath, Demeter, Dick, Wray, and Pappajohn*.

Kaplan, William. *Presumed Guilty: Brian Mulroney, the Airbus Affair and the Government of Canada*. Toronto: McClelland & Stewart, 1998.

Kaplan, William. *The Secret Trial: Brian Mulroney, Stevie Cameron and the Public Trust*. Montreal: McGill-Queen's University Press, 2004.
    Kaplan examines the Airbus scandal and the combative relationship between former prime minister Brian Mulroney and investigative journalist Stevie Cameron.

APPENDIX C

Kaplan, William. *Bad Judgment: The Case of Mr. Justice Leo A. Landreville*. Toronto: Osgoode Society for Canadian Legal History, 1996.

Karp, Carl and Cecil Rosner. *When Justice Fails: The David Milgaard Story*. Toronto: McClelland & Stewart, 1991, 262 pp.
> The authors tell the story of how 17-year-old David Milgaard was convicted for the 1969 murder of Gail Miller in Saskatoon, Saskatchewan. Milgaard spent 22 years in prison for an offence he did not commit.
> *R. v. Milgaard* (1971), 2 C.C.C. (3d) 206 (Sask. C.A.) leave to appeal to the Supreme Court of Canada denied, (1971), 4 C.C.C. (3d) 566; *Reference re Milgaard* (1992), 71 C.C.C. (3d) 260 (S.C.C.)
> In February, 1995, the Supreme Court of Canada dismissed an appeal by two prosecutors who were trying to prevent Milgaard from suing them and police for the 22 years he spent in prison. (1993), 112 Sask.R. 241 (Sask. Q.B.), [1994] S.J. No. 439 (Sask. C.A.), [1994] S.C.C.A. No. 458 (S.C.C.). In 1999, Milgaard received a $10 million compensation package (see Chapter 4 in this textbook).
> In May, 1995, Milgaard launched a law suit against Robert Mitchell for allegedly calling him a murderer (Canadian Press, "Milgaard Sues Justice Minister," *Vancouver Sun,* May 6, 1995, A9.) See *Milgaard v. Mitchell* [1997] 3 W.W.R. 82 (Sask. Q.B.).

Kimber, Stephen. *Not Guilty: The Surprising Trial of Gerald Regan*. Don Mills, Ontario: Stoddart Publishing, 1999.

Knuckle, Robert. *The Flying Bandit: Bringing Down Canada's Most Notorious Armed Robber*. Burnstown, Ont: General Store Pub. House, 1996.

Kostelniuk, James. *Wolves Among Sheep: The True Story of Murder in the Jehovah's Witness Community*. Harper-Collins, 2000.

Krawczyk, Betty. *Lock Me Up or Let Me Go*. Vancouver, BC, Canada: Press Gang, 2002.
> A 73-year-old grandmother is jailed for a year for an environmental political protest.

LeBourdais, Isabel. *The Trial of Steven Truscott*. Toronto, Canada: McClelland and Stewart, 1966.
> The author raises questions about Truscott's trial and sentence to be hung.

Lowe, Mick. *Conspiracy of Brothers*. Toronto: MacMillan of Canada Ltd., 1988.
> This is a book about the murder convictions of members of Satan's Choice Motorcycle Club for the death of Bill Matiyak. Lowe examines the evidence that was presented and that was suppressed and raises questions about the convictions.

Lowe, Mick. *Premature Bonanza: Standoff at Voisey's Bay*. Toronto: Between The Lines, 1998.

Macdonald, Ian and Betty O'Keefe. *The Mulligan Affair: Top Cop on the Take*. Surrey, BC: Heritage House Publishing Company Ltd., 1997.
> The story of a corrupt police chief in Vancouver in the 1950s and the inquiry into the Vancouver Police Department.

MacIntyre, Linden and Theresa Burke. *Who Killed Ty Conn?* Toronto, Canada: Penguin Books Canada, Limited, 2001.
> Conn was the subject of a CBC *The Fifth Estate* documentary on the effects of child abuse. Following his escape from Kingston Penitentiary in 1999, he shot himself rather than to face a return to prison.

Makin, Kirk. *Redrum the Innocent*. Toronto: Penguin Books, 1993; republished 1998, 810 pp.; 630 pp.
> Makin tells the story of how Guy Paul Morin was acquitted at his first trial in 1986, and then convicted at his second trial in 1992, of the murder of nine-year-old Christine Jessop in Queensville, Ontario.
> Note: Morin was acquitted by the Ontario Court of Appeal in January, 1995, after DNA testing excluded him as the perpetrator of the offence. The Crown requested the acquittal. *R. v. Morin* (1995), 37 C.R. (4th) 395 (Ont. C.A.) Also see the Kaufman Commission in Appendix A.
> Also see *R. v. Morin* (1987), 36 C.C.C. (3d) 50 (Ont. C.A.); (1988), 44 C.C.C. (3d) 193 (S.C.C.); (1993), 78 C.C.C. (3d) 559 (Ont. C.A.), and *Re Ontario (Commission on Proceedings Involving Guy Paul Morin)* (1997) 154 D.L.R. (4th) 146 (Ont. Div. Ct.), [1998] O.J. No. 337 (Ont. C.A.).

Manweiler, Arnold (as told by James Burke). *If it Weren't for Sex. . . I'd Have to Get a Job: Confessions of a Private Investigator*. Toronto: McClelland & Stewart, 1984.

Marshall, W.L., and Sylvia Barrett. *Criminal Neglect: Why Sex Offenders Go Free*. Toronto: McCllelland-Bantam, 1992; first published 1990, 209 pp.
    The authors talk about a number of cases in which the system has failed victims of sexual assault. They also provide a profile of sex offenders and their motives, the link to pornography, the case for treatment, and the impact of sexual assault on the victims.

Martin, Brian. *Never Enough: The Remarkable Frauds of Julius Melnitzer*. Toronto, Canada: Stoddart, 1993.
    Case of a bank fraud in Canada.

Martin, Brian. *Buxbaum: A Murderous Affair*. Canada: General Paperbacks, 1986.
    A Canadian businessman, obsessed with drugs and prostitutes, pays a hitman to kill his wife.

Martineau, Pierre. *I Was a Killer for the Hells Angels: The Story of Serge Quesnal*. McClelland & Stewart, 2003.

Mathers, Chris. *Crime School: Money Laundering*. Toronto: Key Porter Books, 2004.

McIntyre, Mike. *Nowhere to Run*. Winnipeg, Canada: Great Plains Publications, 2003.
    The killing of an Aboriginal RCMP officer.

McIntyre, Mike. *To the Grave: Inside a Spectacular RCMP Sting*. Winnipeg, Canada: Great Plains Publications, 2006.
    A "Mr. Big" investigation provides evidence against Michael Bridges in a murder investigation. Also see: *R. v. Bridges*, [2005] M.J. No. 232 (Q.B.); [2005] M.J. No. 536 (Q.B.); [2006] M.J. No. 428 (C.A.).

McKirdy, Margaret. *The Colour of Gold*. Prince George, BC, Canada: Caitlin Press, 1997.

McNish, Jacquie. The big Score. Toronto: Doubleday Canada, 1998.

McNicoll, Susan. *British Columbia Murders: Mysteries, Crimes, and Scandals* . Canmore, AB, Canada: Altitude Publishing Canada Ltd., 2003.

McQuaig, Linda. *All You Can Eat: Greed, Lust, and the New Capitalism*. Toronto: Viking, 2001.

McQuaig, Linda. *Behind Closed Doors: How the Rich Won Control of Canada's Tax System—and Ended up Richer*. Markham, Ont.: Viking, 1987.

McQuaig, Linda. *It's the Crude, Dude: War, Big Oil and the Fight for the Planet* . Toronto: Doubleday Canada, 2004.

McQuaig, Linda. *The Quick and the Dead : Brian Mulroney, Big Business and the Seduction of Canada*. Toronto: Viking, 1991.

McQuaig, Linda. *Shooting the Hippo: Death by Deficit and other Canadian Myths*. Toronto: Viking, 1995.

McQuaig, Linda. *The Wealthy Banker's Wife: The Assault on Equality in Canada*. Toronto: Penguin Books Canada, 1993.

Milgaard, Joyce (with Peter Edwards). *A Mother's Story: The Fight to Free my Son David*. Toronto: Doubleday Canada, 1999.

Miseck, Lorie. *A Promise of Salt*. Regina: Coteau Books, 2002.
    A moving tribute to Miseck's sister Sheila Maureen Salter who was killed in Edmonton in 1995. Her story is in sharp contrast to the court case which describes her horrific death: R. v. Brighteyes, [1997] A.J. No. 325; [1997] 6 W.W.R. 313; 50 Alta. L.R. (3d) 77 (Alta. Q.B.).

Monet, Jean. *The Cassock and the Crown: Canada's most Controversial Murder Trial*. Montreal; Buffalo: McGill-Queen's University Press, 1996.
>A Roman Catholic priest is accused of killing his brother in Montreal in the 1920s. The story of the Delmore trial provides a picture of Quebec culture in the 1920s and the power of the Catholic church.

Moran, Bridget. *Judgement at Stoney Creek*. Vancouver: Tillacum Library, 1990.
>Coreen Thomas, a pregnant Carrier native from Stoney Creek reservation near Vanderhoof, British Columbia, was killed by a car driven by Richard Redekop. Moran tells how the investigation illustrates a system of justice in which aboriginals are treated differently from whites.

Mulgrew, Ian. *Final payoff: The True Price of Convicting Clifford Robert Olson*. Toronto: Seal Book, 1991.

Mulgrew, Ian. *Bud Inc.: Inside Canada's Marijuana Industry*. Toronto: Random House Canada, 2005.

Mulgrew, Ian. *Unholy Terror: The Sikhs and International Terrorism*. Toronto, Canada: Key Porter Books, 1988.
>Sikh terrorism and the bombing of Air India flight.

Mulgrew, Ian. *Who Killed Cindy James?*. Toronto, Canada: Seal Books, 1991.
>Examines the death of a Vancouver nurse.

Murdoch, Robertson. *A Touch of Murder Now and Then*. Prince George, B.C.: Caitlin Press, 1999.

Murphy, Mark G. *Police Undercover: The True Story of the Biker, The Mafia and The Mountie* (Toronto: Avalon Publishing, Inc. 1999).

Nikiforuk, Andrew. Saboteurs: Wiebo Ludwig's War Against Big Oil. Toronto??? Macfarlane Walter & Ross, 2001.

O'Brien, Dereck. *Suffer the Little Children: An Autobiography of a Foster Child*. St. John's, NF: Breakwater, 1991.
>Dereck O'Brien tells the story of the physical and emotional abuse he suffered as a foster child in Newfoundland. In 1989, he testified at the Royal Commission of Inquiry into the response of the Newfoundland justice system. The publisher points out that his book was "first banned by the Department of Justice in Newfoundland" and then went on to become a Canadian bestseller.

O'Malley, Martin. *Gross Misconduct: The Life of Spinner Spencer*. Toronto, Canada: Penguin Books, 1989.

Penfold, P. Susan, *Sexual Abuse by Health Professionals: A Personal Search for Meaning and Healing* (Toronto: University of Toronto Press, 1998).

Priest, Lisa. *Conspiracy of Silence*. Toronto: McClelland & Stewart, 1989, 222 pp.
>Helen Betty Osborne, a Cree Indian, was brutally murdered in 1971 in The Pas, Manitoba. Dwayne Johnston was convicted of the murder in 1987, although rumours as to who had committed the murder had circulated in the town for years. Another man was acquitted, a third granted immunity, and a fourth was not charged with the offence.
>See A.C. Hamilton, and C. M. Sinclair (Commissioners). *Report of the Aboriginal Justice Inquiry of Manitoba Volume 1: The Justice System and Aboriginal People; Volume 2: The Deaths of Helen Betty Osborne and John Joseph Harper*. Winnipeg, Manitoba: The Queen's Printer, 1991.
>*R. v. Johnston* (1988), 44 C.C.C. (3d) 15 (Man. C.A.), [1989] 3 W.W.R. lxxi.

Priest, Lisa. *Operating in the Dark: Accountability in our Health Care System: A Special Report*. Toronto: Atkinson Charitable Foundation, 1999.

Priest, Lisa. *Women Who Kill: Stories of Canadian Female Murderers*. Toronto: McClelland & Stewart, 1992, 288 pp.
>Priest interviewed lawyers, psychologists and psychiatrists, and 11 women across Canada who have killed—some accidentally others deliberately.

Pron, Nick. *Lethal Marriage: The Unspeakable Crimes of Paul Bernardo and Karla Homolka*. Toronto: McCllelland-Bantam, 1995, 544 pp. Also see Burnside and Cairns and Williams.
>Pron talks about the publication ban surrounding the case of Karla Homolka and the trial of Bernardo and Homolka in the killing of Leslie Mahaffy and Kristen French. See Burnside for cases.

Pron, Nick and Kevin Donovan. *Crime Story: The Hunt for the "Body Parts" Killer*. Toronto: McCllelland-Bantam, 1992, 350 pp. Also see Clark.
      Rui-Wen Pan was convicted of the murder of his former girlfriend Selina Shen in Ontario, after two trials resulted in hung juries. Pron and Donovan tell the story of the three trials.

Reynolds, John. Free Rider: How a Bay Street Kid Stole and Spent \$20 Million. Toronto: McArthur & Co., 2001.

Robinson, Jeffrey. *The Sink: Crime, Terror, and Dirty Money in the Offshore World*. Toronto: McClelland & Stewart, 2004.

Ross, Gary Stephen. *Stung: The Incredible Obsession of Brian Molony*. Toronto: McClelland & Stewart, 2002.

Sampson, Connie. *Buried in the Silence*. Edmonton: NeWest Press, 1995.
      Sampson tells the story of Leo LaChance, a Cree from a reserve near Prince Albert, Saskatchewan, the guilty plea and sentencing of white supremacist Carney Nerland, and the inquiry that followed by E.N. (Ted) Hughes.
      Nerland v. Saskatchewan (Lachance/Nerland Commission of Inquiry) (1993), 109 Sask. R. 38 (Sask. C.A.).

Sanger, Daniel. *Hell's Witness*. Toronto: Viking Canada, 2005.
      The life and questionable suicide of Dany Kane, a Hells Angels informant, who died during the RCMP's efforts to end a battle to control the illegal drug business in Quebec.

Schiller, Bill. *A Hand in the Water: The Many Lies of Albert Walker*. Harper Collins, 1998. Also see Cairns.
      Schiller interviewed Albert Walker four times while he awaited trial on killing a business partner. Walker, a man who embezzled millions from Canadians, fled to England where he befriended Ronald Platt, killed him, and assumed his identity. Schiller sheds some light on the paternity of the two children that Walker's daughter bore during the time they lived as husband and wife, but does not draw any conclusions.

Selleck, Lee; Thompson, Francis. *Dying for Gold: The True Story of the Giant Mine Murders*. Toronto, Canada: HarperCollins Canada, 1997.

Sher, Julian. *White Hoods: Canada's Ku Klux Klan*. Vancouver: New Star Books, 1983.

Sher, Julian. Until you are Dead: Steven Truscott's Long Ride into History. Toronto: Alfred A. Knopf Canada, 2001; Vintage Canada Edition, 2002 with updates.
      Steven Truscott was sentenced to hang in 1959 at the age of 14 for a murder he did not commit. His case was recently investigated by Fred Kaufman (see Appendix A).

Sher, Julian and William Marsden. *The Road to Hell: How the Biker Gangs are Conquering Canada*. Toronto: A.A. Knopf Canada, 2003.

Siggins, Maggie. *A Canadian Tragedy: JoAnn and Colin Thatcher: A Story of Love and Hate*. Toronto: Macmillan of Canada, 1985.
      Siggins tells the story of JoAnn and Colin Thatcher, from their childhoods through to the murder of JoAnn and the subsequent conviction of Colin.
      R. v. Thatcher (1984), 7 C.C.C. (3d) 446 (Sask.C.A.).
      R. v. Thatcher (1986), 24 C.C.C. (3d) 449 (Sask. C.A.).
      R. v. Thatcher (1987), 32 C.C.C. (3d) 481 (S.C.C.).

Siggins, Maggie. *A Canadian Tragedy: JoAnn and Colin Thatcher: A Story of Love and Hate*. Toronto: Macmillan of Canada, 1985.
      Siggins tells the story of JoAnn and Colin Thatcher, from their childhoods through to the murder of JoAnn and the subsequent conviction of Colin.
      R. v. Thatcher (1984), 7 C.C.C. (3d) 446 (Sask.C.A.); *R. v. Thatcher* (1986), 24 C.C.C. (3d) 449 (Sask. C.A.); *R. v. Thatcher* (1987), 32 C.C.C. (3d) 481 (S.C.C.).

Siggins, Maggie. *JoAnn & Colin Thatcher: A Story of Love and Hate: A Canadian Tragedy*. Toronto: McClelland & Stewart, 2001.

Siggins, Maggie. *Bitter Embrace: White Society's Assault on the Woodland Cree*. Toronto: McClelland & Stewart, 2005.

Siggins, Maggie. *Brian & the Boys: A Story of Gang Rape*. Toronto: Lorimer, 1984.

Siggins, Maggie. *Revenge of the Land: A Century of Greed, Tragedy, and Murder on a Saskatchewan Farm*. Toronto: McClelland & Stewart, 1991.

Smith, Barbara. *Fatal Intentions: True Canadian Crime Stories*. Toronto: Hounslow Press, 1994.

Speck, Dara Culhane. An Error in Judgement: Politics of Medical Care in an Indian White Community . Vancouver, Canada: Talonbooks, Limited, 1989.
> An examination of the death of an 11-year-old Native girl in xx, British Columbia.

Starkins, Edward. *Who Killed Janet Smith?* Toronto: Macmillan of Canada, 1984, 339 pp.
> Jane Smith, a 22-year-old nursemaid, in an exclusive Vancouver suburb, was killed in 1924. This is the story of the investigation of her murder. It talks of prejudice and privilege, as it existed in the 1920s.

Tadman, Peter. *Shell Lake Massacre*. Hanna, Alberta, Canada: Gorman & Gorman Ltd., 1992.

Truscott, Steven (as told by Bill Trent). *The Steven Truscott Story*. Richmond Hill, ON: Pocket Book Edition, 1971.

Truscott, Steven (as told by Bill Trent). *Who Killed Lynne Harper?*. Montreal, Canada: Optimum Publishing Company, 1979.

Vallee, Brian. *Life With Billy*. Toronto: McClelland-Bantam, 1986, 210 pp.
> This is the story of Jane Stafford who killed her husband Billy, after years of abuse.

Vallee, Brian. *Life After Billy*. Toronto: McClelland-Bantam, 1995, 335 pp.
> Ten years after Jane Hurshman (Stanford) killed her husband Bill Stanford she died from a gun shot. The author raises questions about why she died and by whom.

Vallée, Brian. *Edwin Alonzo Boyd: The Story of the Notorious Boyd Gang*. Toronto: Doubleday Canada, 1997.

Vallée, Brian. *The Torso Murder: The Untold Story of Evelyn Dick*. Toronto: Key Porter, 2001.

Vine, Cathy and Paul Challen. *Gardens of Shame: The Tragedy of Martin Kruze and the Sexual Abuse at Maple Leaf Gardens*. Vancouver: Greystone Books, 2002.
> Martin Kruze was born in 1962 and died in 1997. This book tells the story of his life as the target of sexual abuse in Maple Leaf Gardens and his struggle to expose child sexual abuse.

Williams, David R. *With Malice Aforethought: Six Spectacular Canadian Trials*. Victoria, B.C. : Sono Nis Press, 1993.
> The trials discussed are: Patrick James Whelan (assassination of Thomas D'Arcy McGee in 1868); Louis Riel; Ernest Chenoweth (the 8 year old tried for murder in 1900), Wilbert Coffin, Steven Truscott, and Peter Demeter.

Williams, Stephen. *Invisible Darkness: The Horrifying Case of Paul Bernardo and Karla Homolka*. Canada: Little, Brown and Company (Canada) Limited, 1996.
> Williams recounts the grisly kidnapping, torture, rape, and killing of Kristen French, Leslie Mahaffy and Tammy Holmolka. Three years after the book was published Williams was charged with disobeying a court order prohibiting anyone other than a limited number of people from viewing the video tapes made of the crimes. See Burnside for cases.

Williams, Stephen. *Karla: A Pact with the Devil*. Toronto: Cantos International, 2003.

Wilson, Garrett. *Deny, Deny, Deny*. Victoria, B.C.: Trafford Publishing, 2000 (updated from 1986 version).
> Case of Colin Thatcher charged with the murder of his ex-wife.

Worthington, Peter, and Kyle Brown. *Scapegoat: How the Army Betrayed Kyle Brown*. Toronto: Seal Books, 1997.

Glossary

30 day bail review An automatic review of a detained accused's custodial status where a trial on a summary conviction matter has not commenced within 30 days of the making of the detention order.

90 day bail review An automatic review of a detained accused's custodial status where a trial on an indictable matter has not commenced within 90 days of the making of the detention order.

a person who is capable of acting judicially A judicial or "quasi-judicial" authority able to act in a neutral and impartial manner.

absolute discharge A sentence which involves no penalty or conviction.

absolute jurisdiction A court has absolute jurisdiction over an offence if no other court is allowed to try the offence. In such cases, the accused has no election as to where to be tried.

admission A statement by a person which is adverse to the person's interests.

adversarial system A description of the legal system in Canada, in which both parties to a dispute (such as the Crown and the accused) are represented by counsel who argue opposing positions, and the dispute is settled by a judge who is an impartial arbiter.

affidavit A written statement, sworn on oath. For example, in the context of the wiretap provisions, this is the sworn document supporting the application for authorization to intercept private communications, setting out the grounds for the application and the other required information.

agreed statement of facts A statement filed, for example, in summary appeals from decisions in summary conviction matters, listing the facts as agreed to by both parties to the litigation.

alternative measures Formerly referred to as diversion; now codified in section 717 of the *Criminal Code*–ways of dealing with suspects outside of the court process.

amicus curiae Someone appointed to assist the court by making representations in matters of law or fact that might not otherwise be addressed.

appeal as of right A situation in which a party has an absolute right to appeal, and so does not require the permission of any court.

appeal by way of stated case A form of appeal in respect of summary conviction matters, restricted to a question of law alone, where the facts are not in dispute.

appearance notice A document given to a person accused of an offence before being released by an arresting police officer, requiring the person to attend court at a specified time and place.

application A request to a court to be allowed to do something. For example, in the context of the wiretap provisions, this is the request to a judge for authorization to intercept private communications.

arraignment/arraigned Arraignment is the process of calling the accused in court by name, reading the charge(s), and asking whether the accused pleads guilty or not guilty.

array The pool of people subpoenaed to court, from which a jury is selected.

arrest with warrant An arrest of a suspect pursuant to judicial authority.

arrest without warrant An arrest of a suspect without the authority of a judicially approved warrant, permissible in certain circumstances.

Glossary

autrefois acquit A special plea that the accused has already been acquitted of the charge before the court.

autrefois convict A special plea that the accused has already been convicted of the charge before the court.

bail review An appeal of the decision on judicial interim release.

basket clause A clause in a **judicial authorization** to intercept private communications which permits authorities to intercept the communications of unknown (and therefore unnamed) persons if doing so would further the investigation.

body-pack A device hidden on one's person which records or transmits conversations, thereby allowing their interception.

breach of probation The offence of failing to comply with the terms of a probation order.

bring the administration of justice into disrepute The test under section 24(2) of the *Charter of Rights* for whether improperly obtained evidence should be excluded. Such evidence must be excluded if its admission could bring the administration of justice into disrepute. See *Grant* framework.

burden of proof (see **onus of proof.**)

calling no evidence A process whereby the Crown proceeds to trial, but calls no evidence to support the charges, resulting in the acquittal of the accused.

case to meet The principle against self-incrimination means that an accused is not required to respond to allegations until the Crown has established a case to meet.

causal connection A doctrine which would require evidence to have been obtained as a direct result of a breach of *Charter* rights before it could be excluded. This is not the law in Canada.

certiorari An application to a higher court to quash the decision of a lower court.

challenge An objection to a particular member of the array sitting on the jury.

challenge for cause A challenge to a prospective juror, on a specified ground in section 638 of the *Criminal Code*. There is no limit to the number of challenges for cause which may be exercised.

challenge to the array A challenge directed at removing the whole array (panel) of jurors, on the basis of "partiality, fraud or wilful misconduct on the part of the sheriff or other officer" who put the array together.

change of venue Changing the location of the trial, from the usual location (which is normally where the offence was committed) to some other location.

character Whether the accused was the *type* of person who would commit the type of offence alleged.

circumstances Informal disclosure of the circumstances of an offence by the Crown to the Defence.

circumstantial evidence Evidence introduced to prove one fact, from which another ultimate fact in issue can be inferred.

codification The process of reducing the common law to a written, statutory form.

collateral issues Issues not related to those properly before the court.

common law The body of judge-made law, developed on a case-by-case basis, through the interpretation and extension of decisions in previously decided cases.

compellable Refers to whether a witness can be required to testify.

competent Refers to whether a witness is legally permitted to testify.

Glossary

conditional discharge Like an **absolute discharge**, but with the requirement that the accused complete a period of probation, which may involve certain conditions.

conditional sentence A sentence of imprisonment served in the community. See section 742.1 for the limits to such a sentence.

confession An informal **admission** made to a person in authority.

confession rule A confession made to a person in authority is admissible as evidence against the accused only if the Crown can show, beyond a reasonable doubt, that it was voluntary. The court will look at four factors: 1) threats or promises, 2) oppression, 3) operating mind of the suspect, and 4) other police trickery (*Oickle* para. 33).

confidential informant A person who provides information to the police about criminal activity. Such information may be cited in a police officer's **Information to Obtain**, without identifying the source specifically.

conscriptive evidence Evidence that an accused is forced to create or forced to participate in discovering.

consent committal A situation in which an accused consents to being committed to stand trial in superior court without the necessity of a preliminary inquiry, or without completing the preliminary inquiry.

consent surveillance The interception of private communications where one of the parties to the communication has consented to its interception.

corroboration "Confirmation from some other source that the suspect witness is telling the truth in some part of his story which goes to show that the accused committed the offence with which he is charged" (*Vetrovec* 16).

count A charge of a specific act of criminal misconduct. An information or indictment may contain multiple counts.

credibility/credit Addresses whether a witness is telling the truth; how much weight is to be attached to a witness's evidence.

crime control (Also see **due process**.) A possible purpose of the criminal justice system, emphasising the protection of the public.

criminal record The record of the accused's previous convictions for criminal offences.

dangerous offender A person so declared by a court following conviction for a "serious personal injury offence," who is consequently sentenced to an indeterminate period of incarceration.

default Failure to pay a fine.

deferral A judicial order permitting the authorities to delay informing a person that they have been the target of the judicially authorized interception of private communications.

demonstrative evidence Evidence which stands on its own, such as video from a bank surveillance camera.

deposit A sum of cash posted by the accused to ensure attendance in court.

derivative evidence Evidence found as a result of information contained in a (typically inadmissible) confession, or obtained in some other illegal way.

derivative-use immunity The right under the *Charter* not to have evidence derived from evidence given under oath at one proceeding used against oneself at another proceeding. (See **evidence immunity**.)

derived confessions rule Addresses the question of when a second statement should be excluded following an inadmissible first statement.

Glossary

detention Involuntary restraint of liberty, which may be sufficiently established by a reasonable perception of suspension of freedom.

dial number recorder A device which records the telephone numbers dialled in out-going calls from a particular telephone.

direct evidence Evidence used to prove a fact which is in issue.

direct indictment A procedure whereby the Crown bypasses the preliminary inquiry and compels the accused to go directly to trial in the superior court.

directed verdict At common law, where the trial judge was of the view that there was no evidence upon which to convict an accused, the jury was ordered to return a verdict of "not guilty"; this rule has been abrogated in Canada.

discharge of juror Dismissal of a juror during the course of the trial.

disclosure Crown disclosure is the process whereby the defence is apprised, before trial, of the evidence in the possession of the Crown. There is also limited defence disclosure in Canada which requires the accused to disclose certain information to the Crown.

discretion Where the authority making the decision is not absolutely bound to act in a certain way in certain circumstances, but rather may choose to do so or not.

diversion A process by which minor criminal matters are removed from the formal criminal justice system, to be dealt with informally and without resulting in a criminal record. Now codified as **alternative measures**.

doctrine of vagueness A constitutional doctrine of statutory interpretation that laws must not be so vague as to not give citizens fair notice of what the law is, or to not impose any limits on law enforcement discretion.

due process (Also see **crime control**.) A possible purpose of the criminal justice system, emphasising the protection of the rights of individuals.

endorsed warrant The provision of further authority on a warrant. For example, an arrest warrant may authorize the arresting officer to release the accused after arrest, or a warrant may be endorsed by a justice so that it can be enforced in other jurisdictions.

entry warrants Warrants that allow police officers to enter a private dwelling to make an arrest–also called "Feeney warrants."

estreatment An order that a judgement be made in favour of the Crown against a surety up to the amount the surety pledged as security for the accused's release, made as a result of the accused's breach of the terms of judicial interim release.

evidence immunity The right under section 13 of the *Charter* not to have evidence given under oath at one proceeding used against oneself at another proceeding. (See **derivative use immunity**.)

evidential burden (See **secondary burden**.)
ex parte Where a hearing is conducted in the absence of one or more of the interested parties.

exclusionary rule A rule excluding evidence which would otherwise be admissible.

exclusive jurisdiction (See **absolute jurisdiction**.)

exigent circumstances A possible exception to a general exclusionary rule for illegally obtained evidence, which may apply when it can be shown that the situation was so urgent that it was impossible to comply with the proper procedures, such as obtaining a search warrant.

factum/facta A document filed with the court by lawyers that contains their arguments. It may include reference to secondary sources, original research, or both.

Glossary

factual guilt (Also see **legal guilt**.) Whether the accused actually committed the criminal act alleged; whether he or she "did it." Our criminal justice system is not concerned with this sort of guilt (except in an incidental way).

failing to appear (or "failure to appear," "FTA") The offence of not attending court when required to do so pursuant to an appearance notice, summons, promise to appear, or recognizance.

fear of prejudice Part of the test of the admissibility of a **confession**; whether the accused believed or was led to believe that something negative would occur if no confession was made.

Feeney warrants (See **entry warrants**.)

fine A monetary penalty imposed as sentence for a criminal offence.

forfeiture An order that cash posted as part of bail be forfeited to the Crown, as a result of the accused's breach of the terms of judicial interim release.

formal admission An **admission** made in legal proceedings which relieves the other side of the burden of proving a certain fact or facts.

fresh evidence on appeal The consideration by the appellate court of evidence which was not presented to the trial court.

frivolous Without merit.

fruit of the poisonous tree (see **exclusionary rule**.) A doctrine under which all evidence obtained pursuant to an illegal act is automatically excluded.

fund of knowledge The social facts in a judge's possession, which may be considered in deciding a case. (See **mental context**.)

good faith Without malice, or without intent to breach rights or laws; with a belief that the actions taken were lawful. This is a possible exception to a general **exclusionary rule** for illegally obtained evidence.

hearsay An out of court statement, made by someone other than the witness, led to prove the truth of what the statement asserts.

hope of advantage Part of the test of the admissibility of a **confession**; whether the accused believed or was led to believe that something positive would occur if a confession was made.

hot pursuit A situation in which a police officer is actively chasing a suspect.

hybrid (or dual) offence An offence which may be prosecuted either by way of summary conviction or by indictment, in the discretion of the Crown.

illustrative evidence Evidence used by a witness to illustrate something.

imprisonment A period of time in custody imposed as a sentence for a criminal offence.

incarceration (See **imprisonment**.)

indeterminate sentence An indefinite period of incarceration, imposed on a **dangerous offender**.

indictable offence An offence which is prosecuted by indictment, under Parts XIX and XX of the *Criminal Code*. Usually a more serious offence. Only the federal government can create indictable offences.

indictment A written accusation of crime against a person or several persons, preferred by the Attorney General or by a prosecutor as agent of the Attorney General.

Glossary

inevitable discovery A possible exception to a general **exclusionary rule** for illegally obtained evidence, which may apply when it can be shown that the evidence would have been lawfully obtained in any event, whether the illegal steps had been taken by the authorities or not.

informal admission An admission not made in the course of a judicial proceeding.

informant The person who swears an **information** or the **Information to Obtain**.

information A written complaint, sworn under oath, alleging that the accused has committed a specific criminal offence.

Information to Obtain The document prepared and sworn by the person seeking a search warrant, setting out the grounds for believing that the evidence sought of the offence alleged will be found in the location targeted.

informational rights Those constitutional rights to be advised of the right to counsel, access to duty counsel, and so on.

intercept Listening to, recording, or obtaining the content of a private communication by use of a specified device.

interlocutory application An application during the course of a trial, on an issue other than the ultimate one. Generally, interlocutory decisions may not be appealed before the conclusion of the trial.

jail-house informant A person who, while in custody with the accused, claims that the accused made an admission of guilt about the accusations against the accused.

joint submission An agreed upon position by the Crown and Defence, for example, as to what is an appropriate sentence.

judicial admission (See **formal admission**.)

judicial authorization An authorization of a judge, for example, permitting the interception of private communications pursuant to Part VI of the *Criminal Code*.

judicial decision A decision made after considering and weighing the evidence presented, and determining whether it meets the required legal standard.

judicial interim release (or "Bail") The release of an accused who has been taken into custody, pending the determination of charges.

judicial notice The court accepting certain facts without formal proof.

jurisdiction The legal authority of a court to try a matter.

jurisdiction over the offence This concerns whether the court has the legal authority to try the offence charged (see **absolute jurisdiction**).

jurisdiction over the person Whether the court has the legal authority to try the person charged. For example, if no valid process has been issued to compel an accused to attend court, the court will not normally have the jurisdiction to try that person.

justice A justice of the peace or a provincial court judge.

knock on A police investigational tactic whereby an officer will approach a dwelling house and knock on the door, in the hope that when an occupant opens it, sufficient evidence will be immediately apparent to support an arrest or an application for a search warrant. Now considered an illegal search.

leave to appeal A situation in which a party may only appeal with the permission of the court.

legal burden (see **onus of proof**.)

Glossary

legal guilt (Also see **factual guilt**.) Whether the accused is legally responsible for his or her actions, in a criminal sense; whether the court finds him or her "guilty." This is the type of guilt with which our criminal justice system is concerned.

long-term offender The court may find that an offender is such under section 753.1 of the *Criminal Code*.

mandamus An application to a higher court to compel a lower court to do something.

mental context The state of mind of the judge considering a case. (Also see **fund of knowledge**.)

miscarriage of justice Justice was not done; something fundamental went wrong during the process.

mistrial Where a jury cannot reach a unanimous verdict, or where something else goes wrong so as to fatally flaw the process of the jury trial, the judge may declare a mistrial. This has the effect of terminating the trial, which must then start again with the selection of a new jury.

no deposit Not requiring cash deposit as a condition of judicial interim release or release by an officer in charge.

no evidence motion An application by the Defence at the close of the Crown's case to have the charges dismissed, on the basis that the Crown failed to lead any evidence on some essential element of the offence.

non-conscriptive evidence Evidence "which existed independently of the *Charter* breach in a form useable by the state" (*Stillman*, para. 75).

nullity Something is a nullity if it is of no legal force or effect, due to some defect such as a lack of jurisdiction.

oath A religiously or spiritually based promise by a witness to tell the truth.

officer in charge A supervising police officer; see section 493 of the Criminal Code for a definition.

onus of proof Refers to which party is required to prove something.

operating mind (Also see **confession rule**.) Whether the statement or decision of an accused was truly voluntary, in that the accused was capable of deciding to make the statement or choice. This involves an assessment of the accused's mental and physical condition.

opinion evidence Evidence in which a witness (usually an expert) goes beyond testifying as to the existence of certain facts, and offers an opinion as to what certain facts mean.

obtained in a manner Evidence must be obtained in a manner that violates an accused's rights before section 24(2) of the *Charter* can be invoked.

packet The sealed envelope containing the information the judge considered when authorizing electronic surveillance

panel (See **array**.)

pardon The Crown, as represented by the Governor General, may conditionally or absolutely pardon a convicted person, absolving them of their crime. Also refers to a **special plea** that the accused has already been pardoned in respect of the charge before the court.

participant surveillance (See **consent surveillance**.)

particulars Formal particulars, ordered by the Court, require the Crown to specify certain aspects of its allegations, which must then be proved as if they had been included in the original information or indictment. (See also **circumstances**.)

peremptory challenge A challenge to a prospective juror for which no reason need be given. Each side has a limited number of peremptory challenges.

Glossary

perimeter search A search of the outside of a dwelling house and its surroundings, sometimes used in an attempt to secure sufficient information to support an application for a warrant to search the house itself.

perpetuated evidence In case a witness dies, becomes ill or insane, is unavailable, or refuses to testify at the trial of the accused, the evidence taken at the preliminary inquiry can be read into evidence at the trial of the accused, pursuant to section 715 of the *Criminal Code.*

person in authority A person who has some influence over the criminal proceedings, or a person the accused reasonably believes has some degree of power over the accused.

placed in charge/placed in the charge of the jury The formal commencement of a jury trial, marked by the clerk's address to the jury placing the case in their hands.

plain view A doctrine whereby a police officer lawfully on premises may seize without warrant any evidence of a crime which is discovered in plain view (*i.e.,* without searching).

pre-inquiry A hearing before a provincial court judge to determine whether a private information should be allowed to proceed.

preferred The act of presenting an indictment to a court of superior jurisdiction by a prosecutor, usually as agent of the Attorney General.

preliminary inquiry In cases where the accused elects to be tried in superior court, this is the hearing conducted in provincial court to ensure that there exists sufficient evidence to justify putting the accused on trial.

presumption of fact A situation in which a certain fact will be presumed to be true in the absence of evidence or proof to the contrary. Thus, the Crown (for example) is relieved of the normal burden of proving a certain fact, and the onus instead shifts to the defence to disprove it. (See **reverse onus.**)

pretrial conference A hearing held before the commencement of a jury trial, and often in the case of lengthy non-jury matters, to "promote a fair and expeditious trial."

prima facie "On the face of it."

prima facie **case** (See *prima facie.*) At a preliminary inquiry, the Crown is required to show a prima facie case—that there is a case against the accused on the face of it—before the accused will be committed to stand trial in superior court.

prior authorization (Also see **search warrant.**) Permission granted by a judicial authority in advance to authorize some action such as a search or seizure. This is generally the minimum requirement for a valid search.

privacy A right protected under section 8 of the *Charter,* against unreasonable search or seizure.

private communication A communication made in circumstances where either party to it has a **reasonable expectation of privacy.**

privilege The right of a person or the state, and the corresponding duty of a witness, to withhold from the court evidence which would be relevant and admissible in the absence of that privilege.

probation order An order that the accused, for a specified period of time, "keep the peace and be of good behaviour," report to court or to a probation officer as required. The order may contain various other conditions and restrictions on the accused.

probation report or pre-sentence report A report ordered by the judge prior to sentencing, prepared by a probation officer, detailing aspects of the accused's background, etc.

probative value A fact has probative value if it is capable of proving, or has the tendency to prove something.

proceeds of crime The profits of criminal activity, which may be confiscated under anti-profiteering legislation.

Glossary

production order   An order from a justice or judge to compel a person   compelling a person (other than one under investigation) to produce documents or to prepare documents from existing data.

prohibition order   Orders used at sentencing to prohibit a convicted person from doing something (for example, owning firearms or having care or control of birds or animals).

promise to appear   A document signed by an accused, before an officer in charge, requiring an accused to attend court at a specified time and place.

prosecutor   The agent of the federal or provincial attorney general on whose behalf the charges are brought, or in the case of a private prosecution, the person who swore the information.

provincial court   A court staffed by judges appointed by the province, which tries all summary conviction offences and some indictable matters, and conducts preliminary inquiries in respect of the remaining indictable matters.

public interest   A consideration of broad policy issues. A discharge may be granted if it is not contrary to the public interest, which includes a consideration of general deterrence.

public interest ground   (See **secondary ground**.)

public interest immunity   A common law rule protecting against the disclosure of certain information where it would be contrary to the public interest to divulge it.

question of fact   A question to be decided on appeal, which involves only the determination of a factual issue or matter of evidence.

question of law   A question to be decided on appeal, which involves only the determination of a legal issue.

question of mixed fact and law   A question to be decided on appeal, which involves the determination of an issue which includes both legal and factual aspects.

real evidence   (Especially physical evidence.) Evidence other than the *viva voce* testimony of a witness, which is observed by the court.

reasonable expectation of privacy   The standard applied in determining whether there exists a **privacy** right to be protected.

reasonable and probable grounds   (See **reasonable grounds**.)

reasonable grounds   Prerequisite for an arrest or a search warrant; there must be objectively justifiable grounds (a "credibly-based probability") to believe the accused has committed an offence, or that evidence will be found in the location to be searched.

reciprocal disclosure   A concept involving the Defence informing the Crown of certain aspects of the its case before trial. Now more commonly referred to as **defence disclosure**.

recognizance   A document acknowledging that the accused owes Her Majesty a sum of money, which will be forfeited to the Crown if the person fails to attend court.

release pending appeal   Bail granted following conviction and sentence, pending the determination of an appeal.

relevance   The extent to which the evidence tends to prove, disprove, or explain some issue properly before the court.

renewal   A judicial order extending the period of time for which a **judicial authorization** to intercept private communications is in force.

restitution   A sum of money ordered paid to the victim of a crime, by way of compensation for loss of or damage to property.

Glossary

reverse onus Generally, a situation where the onus of proof shifts from requiring one party to prove something to requiring the opposing party to disprove it. For example, at a show cause hearing for certain offences, the accused is required to justify judicial interim release, rather than the Crown being required to justify detention, as is normally the case. (See also **presumption of fact**.)

right of appeal Whether a party to litigation may appeal a decision to a higher court.

Rule in *Hodge's case* A common law rule which required that the jury be warned that before they convicted an accused on purely **circumstantial evidence**, they must be satisfied not only that the circumstances are consistent with the accused having committed the offence, but also that the circumstances are inconsistent with any rational conclusion other than that the accused committed the offence.

rules (See **standards**.) An approach to the structuring of the law which specifies in a very precise way how judges must decide or act in certain situations. A rules-oriented approach leaves little room for the exercise of judicial discretion.

rules of statutory construction Common law or statutory rules governing the procedures used in determining what a statute means.

search incidental to arrest The common law power of a peace officer to search a person who is taken into custody for weapons (for the protection of the officer and the prisoner) or evidence.

search warrant Permission given by a judicial authority for the conduct of a search or seizure.

secondary burden May require a party to raise a particular fact or issue by evidence or by implication from the evidence.

secondary sources Writings of learned authors, or other such reference material.

show cause hearing The hearing at which the question of judicial interim release is decided. Normally, the Crown must show cause why the accused should be detained, or some restrictive form of bail should be imposed.

similar fact evidence Evidence not directly related to the charges before the court, but which tends to show that the accused has committed similar acts at other times, from which it may be inferred that he committed the acts charged.

social authority Social science research that is used to interpret or develop the law.

social facts The application of social science methodology to concrete situations in a case.

social framework Social science research which provides the social or psychological context for what people might do in certain circumstances.

solemn admission (See **formal admission**.)

solemn affirmation A solemn promise by a witness to tell the truth (See **oath**.)

solicitor-client privilege Communications with a lawyer acting in a professional capacity, made in confidence, are privileged (with a few exceptions), and neither party can be compelled to disclose them.

special pleas Pleas allowed by section 607 of the *Criminal Code* (see ***autrefois acquit, autrefois convict***, and **pardon**), other than the usual pleas of "guilty" or "not guilty."

stand by A direction by the judge that a particular juror who wishes to be excused on the ground of hardship wait, to see whether a complete jury can be selected without that person.

standards (Also see **rules**.) An approach to the structuring of the law which specifies only in a general way how judges must decide or act in certain situations. A standards-oriented approach leaves room for flexibility and the exercise of judicial discretion in handling a case.

Glossary

standard of proof Addresses the level of proof required; beyond a reasonable doubt in criminal trials.

statutory instruments Various forms of legislation, including statutes, regulations, orders-in-council, etc.

stay of proceedings A direction that proceedings be suspended.

subordinate legislation Legislation which is not passed by Parliament or the Legislature, but rather is passed by individuals or groups (such as the Governor General in Council) under powers delegated in statutes. Regulations are the most common form of subordinate legislation.

subpoena duces tecum A *subpoena* compelling a witness to attend court to testify, and to bring certain relevant documents.

summary conviction offence An offence which is prosecuted by summary conviction, under part XVII of the *Criminal Code*, usually a less serious offence. Summary conviction offences may be created both by the federal and provincial governments.

summons A document issued by the court, requiring an accused to attend court at a specified time and place.

superior court A court staffed by judges appointed by the federal government, which tries more serious indictable matters.

surety A third party who pledges to forfeit cash or property if the accused fails to attend court as required.

suspended sentence A disposition whereby the passing of sentence is suspended for a specified period of time, and the accused is placed on probation. If the accused breaches the terms of probation, the court may then pass sentence for the offence.

talesmen The name given to prospective jurors summarily collected by sheriffs off the street in situations in which the original array was not sufficiently large to allow a complete jury to be selected from it.

target The subject of a **judicial authorization** to intercept private communications.

telewarrant A type of search warrant which may be obtained over the telephone, in circumstances where it is impractical to obtain a warrant in person.

territorial jurisdiction This concerns whether the offence alleged was committed within the geographical limits of the courts authority.

time to pay A period of time granted within which the accused is to pay the fine imposed.

trial *de novo* A form of appeal which amounts to a new trial before the appeal court.

trier of fact The party in a trial who judges or decides questions of fact. In a jury trial, the jury determines what the facts are, while the judge is the "trier of the law." In a judge alone trial, the judge is the trier of both fact and law.

triers Jurors, or potential jurors, who are appointed to determine the validity of challenges to jurors.

tunnel vision a single-minded narrow focus on a theory of a crime or a suspect such that the investigator's or the prosecutor's evaluation of the evidence is unreasonable affected.

ultimate burden (see **onus of proof**.)

undertaking to appear (UTA) A signed promise by the accused to attend court as required.

unreasonable search or seizure A search or seizure which does not meet the Constitutional standards for a valid search or seizure.

Glossary

unsolemn admission (See **informal admission**.)

venue The location in which the accused is to be tried.

victim fine surcharge A sum of money which the accused is ordered to pay, in addition to any other penalty, which goes into a fund to assist victims of crime.

victim impact statement (VIS) A written statement filed by the victim of an accused's crime, detailing the effect of the crime on the victim, which may be considered by the judge on sentence.

video surveillance The use of video cameras or video tape recorders to obtain evidence of a person's activities or communications.

viva voce evidence Oral evidence given by a witness in court under oath or affirmation.

voir dire A hearing ("trial within a trial") usually conducted by the judge in the absence of the jury (if any) to determine legal questions, such as the admissibility of evidence.

waiver Waiver of charges involves an accused having charges transferred from the court in which they were brought to a court in another territorial jurisdiction, for the purpose of pleading guilty.

withdrawal of charges An action by the Crown which negates the laying of the charges; it is as if the charges had never been laid.

Subject Index

30-day bail review. *See* Judicial interim release

90-day bail review. *See* Judicial interim release

Abuse of process, 30, 85, 106-107. *See also Charter of Rights*
Admissions, 217-218
    admission of as exception to hearsay rule, 218
    conduct or gesture as, 218
    defined, 217
    formal, 217
    informal, 218
    judicial, 217
    solemn, 217
    unsolemn, 820
    See also Confessions; Hearsay
Alternative Measures, 127
amicus curiae, 10-11
Anti-Terrorism Investigative Hearings, 236
Appeals, 138-145
    appeal as of right, 139
    fresh evidence on, 143
    leave to appeal, 139
    preliminary inquiry, from, 94
    procedure by indictment, 140-142
        powers of the court of appeal, 140
        release pending appeal, 140
        sentence, appeal from, 141
    procedure on summary conviction, 142-143
        agreed statement of facts, 142
        appeal by way of stated case, 142
        further appeals, 143
        grounds, 142
        summary appeal, 142
        trial *de novo*, 142
    question of fact, 139
    question of law, 139
    question of mixed fact and law, 139
    rights of appeal, 138-139
Appearance notice, 44, 48, 54, 71. *See also* Judicial interm release
Arbitrary detention. *See* Detention, arbitrary
Arraignment, 24, 78-79, 92
Arrest
    anticipated breach of judicial interim release, 50
    civil liability, 54
    criminal liability, 54
    duty not to, 47
    endorsed warrant, 72
    entry of dwelling to, 48-49
    hot pursuit, 48
    prevent commission of an offence, to, 46
    racial profiling, 52
    reasonable grounds for, 47-48
    release by peace officer, 54-55, 72
    release by officer in charge, 55-56, 71
    warrant, 43, 48, 55, 71

without warrant, 43, 48
    by peace officer, 45-49
    by private citizen, 45-46
See also Detention
Autrefois acquit. See Pleas
Autrefois convict. See Pleas

Bail. *See* Judicial interim release
Bill of Rights. See Evidence, exclusion of
Burden of proof, 247-248. *See also* Onus of proof

Certiorari, 33
Change of venue. *See* Jurisdiction, territorial limitations
Charter of Rights
    section 1, 15, 53-54
    section 7: life, liberty and security of person
        derivative use immunity, 234-235
        disclosure by Crown, 85-88
        examination of complainant, 275
        evidence immunity, 234
        perpetuated evidence, 94-95
        right to silence, 227-231
        right to silence, witness's, 232-237
        screens, use of by child witness, 261-262
        videotaped evidence, 262-263
        See also Self-incrimination, privilege against
    section 8: unreasonable search or seizure, 171-174, 176-177, 179, 182-183, 185-187, 204
        consent surveillance; *see Charter of Rights,* section , participant surveillance
        participant surveillance , 206-207. *See also* Interception of private communications
        privacy rights, 171
        standards, 172-173
        See also Search by warrant; Search without warrant
    section 9: arbitrary detention, 52, 57
    section 10(b): right to counsel, 53-54, 224-227
        change in jeopardy, 226-227
        duty on police officers, 225
        informational rights, 225-226
        waiver, 226, 227
    section 11(b): right to trial within a reasonable time, 33-37
        delay between conviction and sentencing, 36
        delay due to appeal, 37
        factors, 35
        post-charge delay, 34-36
        pre-charge delay, 33-34
        waiver of, 226, 227
    section 11(d): presumption of innocence, 247-248
        perpetuated evidence, 94-95
        screens, use of by child witness, 261-262
        videotaped evidence, 262-263
    section 11(e): right to reasonable bail, 140
    section 11(f): right to a jury trial, 101
        citizenship requirements, 102
    section 13: privilege against self-incrimination, 232-234
        "any other proceedings," 232-233
        derivative use immunity, 234-235
        evidence immunity, 234
        purpose of, 233-234

re-trial, use of previous evidence at, 232-233
re-trial, use of previous evidence to cross-examine at, 233
transaction immunity, 270
use immunity, 270
witness, right of not to answer questions, 233-235
See also Self-incrimination, privilege against
section 24, 16, 159-165, 182-184
burden, 162
causal connection, 162
confessions, 218
factors, 164-165
Grant framework, 164-165
onus, 162
preliminary inquiry, at, 93
rationale, 161-162
test, 163-165
what remedies are available, 160-161
who qualifies for a remedy, 159-160
Classification of offences, 23-25, 26. *See also* Hybrid offences, Indictable offences, Summary conviction offences
Collateral issues, 250
Confidential informant, 176
Compelling the appearance of the accused, 43-56, 72. *See also* Appearance notice, Arrest, Judicial interim release, Promise to appear, Summons
Complainants' privacy, 274-275
Confessions, 218-231
burden on Crown, 221
defined, 218
derivative evidence, 223-224
common law, 223-224
derived confession rule, 223
historical background, 219
informational rights. *See Charter of Rights,* section 10(b)
person in authority, 218, 221-222
test of, 221-222
rationale for rule, 219-220
right to counsel. *See Charter of Rights,* section 10(b)
right to silence. *See Charter of Rights,* section 7
undercover police officer, 221
voir dire, 219, 221-222
voluntariness, 219-220
operating mind, 220-221
See also Admissions, *Charter of Rights,* section 7; section 10(b), Jailhouse Informants
"Court of competent jurisdiction," 93
Corroboration. *See* Evidence
Credibility
previous testimony, use of to impeach, 279. *See also* Witnesses; Evidence
Crown discretion, 75-76, 85-88. *See* Discretion, Crown

Dangerous offenders. *See* Sentencing
Delay. *See Charter of Rights,* section 11(b)
Deposit. *See* Recognizance, Judicial interim release
Derivative use immunity. *See Charter of Rights,* section 7; section 13
Detention, 50-54
arbitrary, 52
defined, 50-51
investigative detention, 51-52, 185
See also Charter of Rights, sections and 1; Judicial interim release

Subject Index

Direct indictment. *See* Indictments
Directed verdicts. *See* Trial by jury
Disclosure
    Crown, 85-88
        circumstances, 88
        particulars, 88
        disclosure court, 89
    defence, 89-91
    See also Charter of Rights, section 7
Discretion
    courts
        witness, to exempt from testifying, 235
    Crown. *See also* Election, Crown; Pre-charge screening
    search warrant, re:, 172, 175
    Sheriff, 102, 103-104
Diversion (see Alternative Measures)
DNA, 177-178
Dual offences. *See* hybrid offences

Election
    Crown, 24, 30. *See also* hybrid offences
    exercise, 29
    mode of trial, 26-28, 69, 79, 92, 101
        overriding, 28, 101
        re-election, 28, 79
Endorsed warrant. *See* Arrest
Evidence
    character, 280-282. *See also* evidence, similar fact
    corroboration, 257-258, 260
    credibility, 282-284
        prior consistent statements, 283
        prior inconsistent statements, 284
        See also Witnesses
    defined, 248
    eyewitness testimony, 251-252, 253
    exclusion of, 157-165
        American position, 157-158
        Bill of Rights, 158
        common law, 157
        See also Charter of Rights, section 24; Privilege
    hearsay, rule against, 276-280
        contemporary position on, 276-280
        defined, 276
        exceptions, 218, 276-280
        purpose of statement, depending on, 276
        rationale, 276
    intercepted private communications. *See* Interception of private communications
    illegally obtained, 157-166. *See also Charter of Rights*
    law reform, 251
    rape shield provisions, 258-260
    relevance, 249
        probative value, 249
    rule in *Hodge's Case,* 253-254
    rules of, 249-250
        purpose, 249-250
        source of, 250
    similar fact, 284. *See also* evidence, character

types of, 251-254
    circumstantial, 253-254
    commissioned evidence, 252
    demonstrative, 252
    direct, 253
    Eye-witness testimony, 251-252, 254
    illustrative, 252
    real evidence, 252
    testimonial evidence, 251-252
    viva voce evidence, 251-252
    See also Burden of proof; *Charter of Rights*; Onus of proof; Opinion evidence; Witnesses
Evidence immunity. *See Charter of Rights,* section 7; section 13
Exclusion of public. *See* Preliminary inquiry; Witnesses, children
Ex parte, 71

Failure to appear (FTA), 44, 45, 56, 58. *See also* Judicial interm release
Fingerprints, 44, 45
Focus Hearing, 92

Hearsay. *See* Evidence
Hot pursuit. *See* Arrest
Hybrid offences , 23, 24, 26, 30
    time limitations
    See also Election, Crown

Immunity. *See* Privilege
Indictable offences, 24, 26, 29-30, 45, 70
    time limitations, 29-31
Indictment, 30, 69-74
    amending, 74
    contents, 72-73
    counts, 72-73
    direct indictments, 76-77
        by Attorney General, 76-77
        by Court, 76
    historical example, 71
    information distinguished, 72
    objection to, 74
    preferring of, 72
    sample, 73
    sufficiency of, 73
    See also Information
Information, 30, 69-74
    amending, 74
    contents, 72-73
    counts, 72-73
    hearing by justice, 71-72
    indictment distinguished, 72
    laying of, 30, 43, 45, 69
    objection to, 74
    reasonable grounds for belief, 70-71
    sample, 70
    sufficiency of, 73
    timing of, 71
    See also Indictment

Subject Index

Interception of private communications, 197-210
    authorizations for, 202, 203-205
        affidavit, 203
        basket clause, 204
        application for, 203-204
        contents of, 203-204
        emergency, 205
        listed offences, 199
        requirements for, 203-204
        renewals of, 207\5
        target, 204
    common law, 197-198
    consent to, 206-207
        authorization of to prevent bodily harm, 207
        authorization with consent, 206
    dial number recorders, 179
    emergency, without authorization, 209
    evidence, admission as, 208
        notice requirements, 208
        See also Charter of Rights, section 8; section 24
    extent of, 198, 203
        report to Parliament, 203
    intercept, definition of, 199
    lawyers, conversations of, 205
    legislative history, 198
    legislative scheme, 198-210
    notification to target, 203
        deferral of, 203
    offences relating to, 201-202
        damages, 203
        exceptions, 203
    Ouimet Report, 198
    participant surveillance. *See* Interception of private communications, consent to
    private communication, 200-201
        cellular telephone calls as, 199, 201
        definition of, 200-201
        e-mail as, 201
    reasonable expectation of privacy, 200-201, 207
    review of authorizations, 209-210
        section 8 *Charter* application, 210
        Garofoli hearing, 209-210
        packet, access to, 209-210
    video surveillance, 207
        warrants for, 207
Judgements, Reasons for, 118-119
Judicial interim release, 56-62
    appeal, pending, 60
    conditions, 57-58
    delay, 57
    deposit, 58
    detention, 56, 58, 59-60
    estreatment, 61-62
    forfeiture, 61-62
    grounds, 58-59
    onus, 57-60
    presumption of innocence, 60
    provincial court judge, 57

public interest ground, 58-59, 60
recognizance, 58
reverse onus, 56, 59-60
review, 60-61
>30-day bail review, 61
>90-day bail review, 61
revocation, 60
show cause hearing, 56-60
superior court judge, 57
sureties, 58
undertaking to appear, 57
variation, 60
See also Appearance notice; Failure to appear; Promise to appear; Recognizance; Summons
Judicial notice, 293-299
assumptions, tacit, 294-295
facts, of, 295-299
>adjudicative facts, 297-298
>fund of knowledge, 298
>legislative facts, 298-299
>mental context, 298
law, of, 295
opinion evidence, 293-294
rationale, 296-297
secondary sources, 293-294, 299-300
See also Social science research
Jurisdiction, 26-28, 28-32
absolute, 26-28
over the offence, 26-28, 29
over the person, 29
territorial limitations, 31-32
>change of venue, 31-32
>waiver, 31
time limitations, 29-30
Jury trials. *See* Trial by jury
Justice. *See* Provincial Court
Law Reform Commission of Canada, 24-25, 46, 57, 69, 72, 75, 76, 77, 89, 111, 115, 158-159, 184, 185, 199, 200, 251, 297-298, 298

Mandamus, 33
Ministerial Review Applications, 144-145
pardon, 132
references to the courts, 144
See also Pleas; Appeals
Monahan and Walker. *See* Social science research

No evidence motion, 118

Normative approach, 174, 175
Nullity, 28

Officer in charge. *See* Arrest, release by officer in charge
Onus of proof. 247-248. *See also* Burden of proof
Opinion evidence, 301-307
expert, 293, 302-307
>hearsay, based on, 302-307
>requirements for, 306-307
general rule, 301

non-expert, 301-302
oath helping, 307

Panel. *See* Trial by jury, array
Pardon. *See* Pleas; Mercy
Participant surveillance. *See Charter of Rights,* section 8; Interception of private communications
Perpetuated evidence, 94. *See also* Preliminary inquiry
Pleas, 79-80
    plea bargaining, 79-80
Pre-charge screening, 75-76. *See also* Discretion, Crown
Preliminary inquiry, 92-96
    abolition of, 95-96
    absconding accused, 94
    appeal from, 94
    Charter remedies at, 93
    committal to stand trial, 92-94
    consent committal, 93
    exclusion of public, 94
    focus hearing, 92
    procedure on, 93
    test, 93
Presumption of fact, 246. *See also,* Onus of proof
Presumption of innocence. *See* Judicial interim release; *Charter of Rights,* s. 11(d)
Pre-trial conference, 88-89
Pre-trial publicity. *See* Trial by jury, jury selection; Jurisdiction, territorial jurisdiction, change of venue
Prima facie , 47, 93
Private prosecutions, 70, 74-75
    pre-inquiries, 74-75
Privilege, 867 269-274
    complainants' privacy, 274-275
        complainants' records, privacy of, 274-275
        examination of complainant by defence experts, 275
    "Crown privilege," 272-274
    extending, 272
    police informer's privilege, 274
    public interest immunity, 272-274
    solicitor-client, 181, 269-271
    spousal communications, 271-272
    Wigmore's criteria, 272
    See also Self-incrimination, privilege against
Procedure at trial, 118
    charge by judge, 118
    opening statements, 118
    order of addresses, 118
    See also Trial by jury
Proceeds of crime, 137-138
Proceeds from publications about crime, 138
Prohibition orders. *See* Sentencing
Promise to appear, 55, 71. *See also* Judicial interm release
Prosecutor, 74-76
    control over discretion, 77-78
    misconduct, 77-78
    Public Prosecution Service of Canada, 12-13
    See also Private prosecutions
Provincial Court, 23, 57, 92
    absolute jurisdiction of, 26-28

Subject Index

Publication bans, 116-118
    test, 117
    See also Preliminary Inquiry

Reasonable and probable grounds. *See* Reasonable grounds
Reasonable grounds, 47, 172, 174-175, 204
Recognizance, 55, 71
    deposit, 55
    surety, 55
    See also Judicial interim release
Relevance. *See* Evidence
Reverse onus, 246. *See also* Judicial interim release; Onus of proof; *Charter of Rights*, s. 11(d)
Right to counsel. *See Charter of Rights,* section 10(b)

Search by warrant, 175-184
    confidential informant, 176
    contents of, 176
    dial number recorder, 179
    execution of warrant, 180
    informant, 176
    information to obtain, 176
    investigative detention, during, 185
    lawyer's office, 181
    miscellaneous search warrant provisions, 177-180
    reasonable and probable grounds, 175-176
    reasonable expectation of privacy, 179
    section 487 warrants, 175-177
    telewarrant, 180
    tracking device, 18079
    video surveillance, 176, 178-179, 207
    See also Charter of Rights, section 8; Interception of private communications
Search without warrant, 184–189
    common law, 185-189
    consent, by, 188-189
    incident to arrest, 185-187
    knock-ons, 183
    legislation, 184
    perimeter search, 182-183
    plain view, 187
    roadside spot checks, 184
    to obtain information for a search warrant, 182-183
    See also Charter of Rights, section 8; Interception of private communications
Secondary sources, 293, 299, 306-307
Self-incrimination, privilege against, 232-237
    common law, 232
    Canada Evidence Act, section 5, 232
    witness, right to silence, 233-234
    See also Charter of Rights, section 13
Sentencing, 127-137
    appeal, 140, 141
    criminal record, 130
    breach of probation, 132
    compensation, 136
    conditional sentence, 133
    dangerous offenders, 140

    discharge, 131-132
        absolute, 131
        conditional, 131
        criteria, 131
        diversion, effect of previous, 131-132
        probation order, 132-133
        public interest, 131
    dispute as to facts, 128-129
    fines, 133
        default, 133
        victim fine surcharge, 136
    hearing, 128-129
    incarceration, 134-135
    indeterminate sentence, 140
    intermittent sentence, 135
    joint submission, 129
    pre-sentence report, 129
    principles of, 127-128
    probation report, 129-130
    procedure, 128-129
    proceeds of crime, 137-138
    prohibition orders, 137
    restitution, 136
    suspended sentence, 132-133
        conditions, 132-133
    time in custody, 131
    victim fine surcharge, 136
    victim impact statements, 130
Sex Offender Registration, 137
Show cause hearing. *See,* Judicial interim release
Similar fact evidence. *See* Evidence
Social science research. 4-5, 32, 107, 111-112, 113, 114, 163
    Monahan and Walker, 4-5, 299
    social authority, 4-5, 298-299
    social facts, 5, 297-298
    social framework, 5, 299
Solicitor-client privilege. *See,* Privilege, Solicitor-client
Special pleas. *See* pleas
Standard of proof. *See* Burden of proof
Statutory construction, rules of, 251
Stay of proceedings
    judicial, 33, 107
    See also Termination of proceedings
Summons, 43, 45, 54, 72. *See also* Judicial interim release
Summary conviction offences, 23, 24, 26, 29-30, 70
    time limitations, 29-30
Superior Court, 23, 57, 92, 204
    absolute jurisdiction of, 26-28
Surety. *See* Recognizance, Judicial interim release

Termination of proceedings, 76
    calling no evidence, 76
    stay of proceedings, 76
    withdrawal of charges, 76
    See also Stay of proceedings
Time limitations, 29-30

Subject Index

Trial by jury, 101-116
    abolishing, 115
    array
        challenging, 103-104
        exempting jurors, 102
        selecting, 102-103
    directed verdicts, 110
    discharging a juror, 109-110
    hung jury, 110
    jury selection, 104-109
        challenge, 105-109
        challenge for cause, 105-109
        defects in, 110
        excusing juror, 104
        peremptory challenge, 105
        pre-trial publicity, 106-107
        procedure, 104-106
        questioning jurors, 105-109
        role of jury, 113-115
        stand aside, 104
        systemic racism, 107-109
        stand by, 104
        talesmen, 109
        triers, 104, 105
    jury comprehension, 113, 114
    jury nullification, 113-115
    jury unanimity, 110
    loss of right to, 101-102
    mistrial, 110
    placing the accused in the charge of the jury, 109
    qualification of jurors, 102-103
    right to, 101-102
    secrecy of deliberations, 111, 113
    when available, 101
    See also Procedure at trial
Trial within a reasonable time. *See Charter of Rights,* section 11(b)

Undertaking to appear. *See* Judicial interim release

Venue. *See* Jurisdiction, territorial limitations
Victim fine surcharge. *See* Sentencing
Victim impact statements. *See* Sentencing
Voir Dire, 115-116, 163
    issue on, 116
    procedure, 116
    purpose, 115-116
    See also Charter of Rights, section

Witnesses, 254-263
    affirmation, solemn, 254
    compellability, 255
    competence, 256-258
    corroboration, 257-258
    disabilities, with, 260-263
    oath, 254-255
    privacy of complainant, 274-275
    refreshing memory, 283

restrictions on cross-examination, 262
safety of, 260-263
screens, use of, 261-262
shielding rape victims, 258-260
spouse, 256-257
support person, 261
videotape, use of, 262-263
See also Credibility

Index of Cases

A.(L.L.) v. B.(A.) (1995) 274, **290**

A.M. (2008) 174, 185, **193**

Abbey (1982) 306-307, **310**

Ah Wooey (1902) 255, **267**

Application under s. 83.28 (2004) 117, **123**, 236, **242**

Araujo (2000) 205, **214**

Arcuri (2001) 92, **99**

Arp (1998) 189, **193**

Asante-Mensahm (2003) 45, 54, **65**

Askov (1990) 34, 35, 36, **40**, 79

B. (C.R.) (1990) 284, **290**

B.(K.G.) (1993) 255, 268, 278, 279, 280, 284, **290**

B.G. (1999) – see *G.(B.)*

Baron et al. (1993) 172, **193**

Barrow (1987) 104, **123**

Basi (2009) 274, **290**

Belnavis (1997) 173, 186, **193**

Benji (2006) 10, **20**

Big M Drug Mart Ltd. (1985) 159, **168**

Biller (2005a) 129, **151**

Biller (2005b) 129, **151**

Biniaris (2000) 139, **151**

Bird (1996) 27, **40**

Biron (1975) 46, **65**

Bjelland (2009) 7, **20**, **99**, 160, **168**

Black (1989) 226-27, **242**

Blackmore (2009) 13, **20**

Bogiatzis et al. (2003) 204, **214**

Borden (1994) 188, **193**

Born with a Tooth (1993) 103, **123**

Boucher (1955) 85, **99**

Braich (2002) 119, **123**

Brais (2009) 207, **214**

Branch, British Columbia Securities Commission v. 236, **242**

Branco (1994) 60, **66**, 140, **151**

Briere (2004) 177, **193**

Brooks (2000) 228, **242**

Brown (1991) 129, **151**

Brown (2002) 270, **290**

Brown (2003) 52, **66**

Broyles (1991) 229-230, **242**

Brydges (1990) 225-226, **242**

Budai (1999) 112, **123**

Buhay (2003) 46, **66**, 139, **151**, 160, **168**, 173, **193**

Burke (1996) 141, **151**

Burlingham (1995) 163, 165, **168**, 224, 225, **242**

Butler (1984) *103*, **123**

C. (M.H.) (1991) 85, **99**, **290**

C.C.F. (1997) – see *F. (C.C.)*

C.R.B. (1990) – see *B. (C.R.)*

Index of Cases

Calder (1996) 223, **242**
Campbell (1999) 108, **123**, 271, **290**
Canadian Broadcasting (1996) 116, **124**
Canadian Newspapers Co. v. Canada (Attorney General) (1988) 116, **124**
Carey v. Ontario (1986) 273, **291**
Carter (1987) 33-34, **40**
Caslake (1998) 186, **193**
Cerniuk (1947) 196, **310**
Chatterjee (2009) 137, **151**
Chiu (1984) 132, **151**
Chromiak (1979) 50, **66**
Church of Scientology et al. (no. 1) 102, **124**
Cichanski (1976) 58, **66**
Cinous (2002) 247, **267**
Clarkson (1986) 227, **242**
Clayton (2007) 185, **193**
Cloutier v. Langlois (1990) 185-187, **193**
Cody and Langille (2004) 179, **194**
Collins (1983) 276, 277, **291**
Collins (1987) 161, 162, 164, **168**, 174, **194**
Collins (1989) 31, **40**
Connors (1998) 44, **66**
Cook (1998) 14, **20**
Cooper (1977) 253, **267**
Corbett (1988) 282-283, **291**
Coté (1978) 73, **84**
Couture (2007) 278, **291**
Craig (2009) **151**
D.(A.) (2003) 118, **124**
D.O.L. (1993) – *see L.(D.O.)*
D.R. (1996) – see *R. (D.)*
Dagenais v. Canadian Broadcasting Corp. (1994) 117, **124**
Darrach (2000) 260, **267**
Daubert (1993) 305, **310**
Davidson (2005) 74, **84**
Davie (1980) 201, **214**
Debot (1989) 176, **194**
Dell (2005) 46, **66**
Dennison (1990) 129, **151**
Dersch (1993) 209, **214**
Descoteaux et al. v. Mierzwinski and A.G. Quebec et al. (1982) 181, **194**
Dinardo, (2008) 119, **124**, 283, **291**
Dixon (1998) 87, **99**
Doiron (2004) 205, **214**
Doucet (2003) 205, **214**
Downey (1992) 247, **267**
Drew (1978) 131, **151**
Drybones (1970) 158, **168**
Duarte (1990) 204, 206, **214**
Dubois (1985) 218, 232-233, **242**
Dudley (2009) 30, **40**, 44, **66**
Duguay (1989) 163-164, **168**
Duguay (2003) 142, **151**
E.T. (1993) – *see I. (L.R.) and T. (E.)*

Index of Cases

Eccles v. Bourque (1975) 48, **66**

Edwards (1996) 160, **168,** 172-174, **194**, 200

Edwards Books and Art Ltd. (1986) 53-54, **66**

Egger (1993) 85, **99**

Elendiuk (1986) 132, **151**

Elias (2001) 135, **151**

Ellard, (2009) 283, **291**

English (1993) 32, **40**, **124**

Evans and Evans (1996) 176, 183, **194**

Evans (1994) 164, **168**

F. (C.C.) (1997) 263, **267**

F.(W.J.) (1999) 280, **291**

Fallofield (1973) 131, 132, **151**

Farinacci (1994) 60, **66**, 140, **151**

Feeney (1997) 47, 48, **66**, **168**, **194**

Feeney (1999a) 49, **66**

Feeney (1999b) 49, **66**

Feeney (2001) 48, 49, **66**

Ferguson (1996) 256, **267**

Ferguson (2008*)* 135, **151**

Fice (2005) 134, **151**

Find (2001) 106, 109, **124**, 296, **310**

Fitzpatrick (1994) 237, **243**

Flintoff (1998) 187, **194**

Fliss (2002) 206, **214**

François (1994) 303, **310**

French (1977) 307, **310**

G. (S.G.) (1997) 282, **291**

G.(B.) (1999) 223, **243**

Gardiner (1982) 129, **151**

Garofoli (1990) 204, 209, 210, **214**

Garrington et al. (1972) 58, **66**

Gayme and the Queen, Re (1991) – see *Seaboyer; Gayme*

Genest (1989) 176-177, 182, **194**

Gladue (1999) 128, **152**

Godin (2009) 34-35, **41**

Godoy (1999) 49, **66**

Golden (2001) 187, **194**

Goldhart (1996) 162, **169**

Gough (1999) 256, **267**

Graat (1982) 247, **267**, 302, **310**

Grandinetti (2005) 222, **243**

Grant (2009) 50-51, 52, **66**, 161, 162, 164-165, **169**, 175, **194**, 208, 224, 227, 235, 237

Greffe (1990) 188, **194**

Griffin (2009) 253-254, **267**, 277, **291**

Grinshpun (2004) 75, **84**

Gruenke (1991) 269, 272, **291**

Guess (2000) 112, **124**

H.S.B. (2008) **291**

Ha (2009) 179, **194**

Hall (2002) 59, **66**

Handy (2002) 284-285, **291**

Hanemaayer (2008) 252, **267**

Hanna (1993) 278, **291**

Index of Cases

Hape (2007) 14, **20**
Harrison (2009) 184, **194**
Hauser (1979) 12, **20**
Hawkins (1996) 257, **267**, 278, **291**
Hebert (1990) 229, **243**
Henry (2005) 233, 243, **260**
Hibbert (2002) 252, **267**
Hodge's Case (1838) 253-254, **267**
Hodgson (1998) 221-222, **243**
Hogan (1975) 158, **169**
Horvath (1979) 220, **243**
Hubbert (1975) 106, **124**
Hufsky (1988) 52, **66**, 184
Hunter (2001) 142, **152**
Hunter et al. v. Southam Inc. (1984) 171-172, 179, 182, **194**, 206
Hynes (2001) 93, **100**
I. (L.R.) and T. (E.) (1993) 223, **243**
Ibrahim (1914) 219, **243**
J.L.J. (2000) 305, 306, **311**
Jamieson (2004) 208, **214**
Jewitt (1985) 85, **100**
Jones (1999) see *Smith v. Jones*
K.G.B. (1993) – see *B.(K.G.)*
Kalanj (1989) 34, **41**
Kalavar (1991) 255, **267**
Kang-Brown (2008) 185, **195**
Keegstra (1991) 106, **124**
Kehler (2004) 2580, **268**
Kenny (1992) 106-107, 113, **124**
Kerr (1982) 132, **152**
Khadr (2008) 14, **20**
Khan (1990) 277-278, **291**
Khela (2009) 258, **268**
Khelawon (2006) 276, 280, **291**
Kirkham (1998) 104, **124**
Koh (1998) 108, **124**
Kokesch (1990) 182, 183, **195**
Krieger (2002) 78, **84**
Krieger (2006) 114, **124**
Kripps Pharmacy Ltd. and Kripps; sub nom. Wetmore (1981) 12, **20**
Krymowski (2005) 296, **311**
Kuldip (1990) 232, **243**
L.(D.O.) (1993) 262, **268**
L.(W.K.) (1991) 34, **41**
L.E.D. (1989) – see *D. (L.E.)*
L.R.I. (1993) – see *I. (L.R.) and T. (E.)*
Laba (1994) 248, **268**
Lagiorgia (1988) 161, **169**
Landry (1986) 48, **67**
Laporte (1972) 176, **195**
Latimer (1997) 104, **124**
Latimer (2001) 114, **125**
Lavallee (1990) 302-03, 306, **311**
Lavallee et al. (2002) 181, **195**

Index of Cases

Laws (1998) 102, **125**
Leduc (2003) 161, **169**
Lee (1989) 102, **125**
Leipert (1997) 274, **291**
Lerke (1986) 46, **67**
Levogiannis (1993) 261, **268**
Liew (1999) 231, **243**
Lifchus (1997) 247, **268**
Lloyd and Lloyd (1981) 271, **291**
Lucas (2009) 205, **214**
Ly (1993) 129, **152**
Lyons (1987) 136, **152**
M.(R.M.) (1998) 173, **195**
M.H.C. (1991) – see *C. (M.H.)*
MacDougall (1998) 35, 36, **41**
MacFarlane, U.S.A. v. (1989) 55, **67**
Macooh (1993) 48, **67**
Makin v. Attorney-General for New South Wales (1894) 284, **291**
Makow (1974) 106, **125**
Malott (1998) 303, **311**
Mann (2004) 51, **67**, 164, **169**
Mannion (1986) 225, **243**
Marquard (1993) 256, **268**
Marquard (1994) 307, **311**
McClure (2001) 270 274, **292**
McDonnell (1997) 141, **152**
McIntosh (1997) 304, **311**
McIntyre (1994) 229, **243**
McIsaac (2005) 200, **214**
McNeil, (2009) 11, **20**, 85, 86, 87, **100**, 292
Mellenthin (1992) 184, **195**
Mentuck (2001) 117, **125**
Miazga v. Kvello Estate (2009) 78, **84**
Milgaard (1971) 284, **292**
Miller (1987) 186, **195**
Mills (1986) 93, **100**
Mills (1999) 11, 275, **292**
Mohan (1994) 304, 305, 306, **311**
Monney (1999) 188, **195**
Moore (1992) 93, **100**
Morales (1992) 59, 60, **67**, 140, **152**
Morgan (1997) 118, **125**
Morgentaler (1988) 114, **125**
Morin (1988) 304, **311**
Morin (1992) 34-35, 36, **41**
Morin (1993) 61, **67**
Morrison (1987) 187, **195**
Murray (2000) 91, **100**
Naglik (1993) 110, **125**
Napope (1992) 200, **214**
Nelles (1989) 78, **84**
Ng (2003) 28, **41**
Nguyen (2004) 179, **195**
Nikolovski (1996) 252, **268**

Index of Cases

Noël (2002) 7, **20**
O'Connor et al., Re Regina and (1993) 9-10, **20**, 87, **100**
O'Connor (No. 2) (1994) 274, **292**
Oakes (1986) 15, **20**
Oickle (2000) 219-220, **243**, 300, **311**
Olscamp (1994) 275, **292**
Ouellette (2009) 137, **152**
Palmer and Palmer (1979) 143, **152**
Pan (2001) 111, **125**, 247, **268**
Pangman (2000) 205, **215**
Parks (1993) 107, 108, **125**
Paterson et al. (1987) 204, **215**
Patrick (2009) 172-174, **195**
Pearson (1992) 59-60, **67**
Pearson (1994) 110, **125**
Perron (1990) 270, **292**
Phillips (1995) 107, **125**
Pires (2005) 209-210, **215**
Plant (1993) 182-183, **195**
Potvin (1989) 94-95, **100**
Potvin (1993) 37, **41**
Prosko (1922) 219, **244**
Prosper (1994) 225-226, **244**
Proulx (2000) 134, **152**
Proulx (2001) 78, **84**
R. (D.) (1996) 279, **292**
R. v. R.E.M. (2008) 119, **125**, 283, **292**
R.J.S. (1995) – see *S. (R.J.)*
R.W. (1992) – see *W.(R.)*
Raynier (1992) 271, **292**
Read (1993) 270, **292**
Regan (2002) 75, 77, **84**
Richard (1995) 176, **195**
Riley (2008) 207, **215**
Risby (1976) 276, **292**
Robb (2005) 75, **84**
Roberge (1983) 47, **67**
Roblin, (1994) – see *Rowbotham; Roblin*
Rodgers (2006) 178, **195**
Rodney (1984) 200, **215**
Rose (1998) 118, **125**
Rowbotham; Roblin (1994) 110, **125**
Rudyk (1975) 130, **152**
S.A.B. (2003) 178, **195**
S.(R.D.) (1997) 294-295, **311**
S. (R.J.) (1995) 233-235, **244**
S.G.G. (1997) – see *G. (S.G.)*
S.J.L. (2009) 92, **100**
S.P.I. (2005) 92, **100**
Salituro (1991) 249, 257, **268**, 271
Samson (1983) 200, **215**
Sander, Canada (Attorney-General) v. (1994) 272, **292**
Sarson (1996) 139, **152**
Schnell v. British Columbia Electric Railway Co. (1990) 296, **311**

Index of Cases

Seaboyer and the Queen; Re Gayme and the Queen, Re (1991) 235, **244**, 249, 258-259, 261, **268**
Semayne's Case (1604) 182, **195**
Seo (1986) 53, **67**
Sheppard (2002) 118, **125**
Shepherd (2009) 184, **195**
Sheppard, U.S.A. v. (1976) 93, **100**
Sherratt (1991) 102, 103, 104, 106, 107, **125**
Shoker (2006) 133, **152**
Simpson (1993) 51, **67**
Singh (2007) 217, 222, 223, **244**
Six Accused Persons (2008) 207, **215**
Skalbania (1997) 139, 141, **152**
Skogman (1984) 92, 93, 94, **100**
Smith (1992) 276, **292**
Smith (2009) 258, **268**
Smith v. Jones (1999) 270, **292**
Smythe (1971) 25, **41**
Southam (1997) 139, **152**
Southam Inc. v. Coulter (1990) 72, **84**
Spence (2005) 109, **125**, 298, **311**
Spencer (2007) 220, **244**
St. Jean (1976) 272, **292**
St. Lawrence (1949) 223-224, **244**
Steinberg (1967) 197, **215**
Stinchcombe (1992) 85, 86, 87, 89, **100**
Stillman (1997) 164, **169,** 175, **196**, 224
Stirling (2008) 283, **292**
Storrey (1990) 47, **67**
Strachan (1988) 162, **169**
Streu (1989) 218, **244**
Suberu (2009) 51, 53, **67**
Symes, M.N.R. v. (1994) 4, **20**
T. (E.) (1993) – see *I. (L.R.) and T. (E.)*
Taillefer (2003) 79, **84**, 87, **100**
Tallman (1980) 131, **152**
Tam (1993) 200, **215**
Tan (1974) 131, **152**
Taylor (1997) 133, **153**
Tele-Mobile Co. (2008) 181, **196**
Terceira (1998) 305, **311**
Terry (1996) 14, **20**
Teskey (2007) 36, **41**
Tessling (2004) 172-174, **196**, 200, **215**
Thatcher (1987) 88, **100**, 110, **125**
Therens (1985) 15, **20**, 50, **67**, 160, 161, 162, **169**
Thomsen (1988) 52, 53, **67**, 184
Threinen (1976) 32, **41**
Tremblay (1987) 165, **169**
Trochym (2007) 306, **311**
Trotta (2007) 88, **100**, 305, **311**
Truscott (1967) 255, **268**
Turcotte (2005) 227-228, **244**, 253, **268**
Turpin (1989) 27, **41**, 101, **126**
Urbanovick (1985) 130, **153**

Index of Cases

Underwood (1998) 283, **292**
Vancouver Sun, Re (2004) 117, **126**, 236, **244**
Vetrovec (1982) 257-258, **268**
Voss (1989) 218, **244**
W.(D.) (1991) 283, **292**
W.(R.) (1992) 141, **153**
Walker (2008) 118, **126**
W.J.F. (1999) – see *F.(W.J.)*
W.K.L. (1991) – see *L(W.K.)*
Ward (1979) 220, **244**
Warsing (1998) 143, **153**
Weir (1998), 201, 202, **215**
Wells (1998) 222, **244**
White (1999) 160, 161, **169**, 236, **245**
Whittle (1994) 220, **245**
Wholesale Travel Group Inc. (1991) 159, **169**
Whyte (1988) 247, **268**
Wighton (2003) 237, **245**
Wilder (2000) 237, **245**
Wildman (1984) 272, **292**
Wiles (2005) 137, **153**
Williams (1998) 108, **126,** 293, 295, **311**
Wilson (1996) 108, **126**
Wise (1992) 179, **196**
Wise (1996) 180, **196**
Wong (1990) 176, 178, **196**, 206, **215**
Wray (1970) 157, 159, **169**, 223-224, **245**
X (1983) 9, **20**
Yakeleya (1985) 218, **245**
Yebes (1987) 141, **153**
Zelensky et al. (1978) 136, **153**
Zundel (1987) 106, **126**, 296-297, 299, **311**
Zundel (1990) **311**
Zylstra (1995) 272, **292**